CONFLICT AND COOPERATION ON THE SCHELDT RIVER BASIN

ENVIRONMENT & POLICY

VOLUME 17

The titles published in this series are listed at the end of this volume.

Conflict and Cooperation on the Scheldt River Basin

A Case Study of Decision Making on International Scheldt Issues between 1967 and 1997

by

Sander V. Meijerink

Delft University of Technology,
Delft, The Netherlands

KLUWER ACADEMIC PUBLISHERS
DORDRECHT / BOSTON / LONDON

A C.I.P. Catalogue record for this book is available from the Library of Congress.

ISBN 0-7923-5650-0

Published by Kluwer Academic Publishers,
P.O. Box 17, 3300 AA Dordrecht, The Netherlands.

Sold and distributed in North, Central and South America
by Kluwer Academic Publishers,
101 Philip Drive, Norwell, MA 02061, U.S.A.

In all other countries, sold and distributed
by Kluwer Academic Publishers,
P.O. Box 322, 3300 AH Dordrecht, The Netherlands.

Printed on acid-free paper

Printed in the Netherlands.

"Cooperation is in a dialectical relationship with discord, and they must be understood together. Thus to understand cooperation, one must also understand the frequent absence of, or failure of, cooperation."

Robert O. Keohane, International Studies Quarterly

To Chantal

TABLE OF CONTENTS

Acknowledgements

Working on one research project for four years is an ambivalent undertaking. On the one hand as a Ph.D. student you sometimes realize the luxury of being able to concentrate on one research topic, and to plan your own work. On the other hand, because of the solitary work, the mostly slow progress, and the permanent uncertainties involved you sometimes wonder why you have ever taken up this academic challenge. I have not yet found the definite answer, but intellectual interest and ambition certainly have been important drives. Although writing a Ph.D. thesis in the nature of things is a solitary task, it is impossible to complete a thesis without the help of many others contributing to the results in as many different ways.

First of all, I would like to thank my supervisor, prof.mr. J. Wessel, professor of water law and administration at the Faculty of Civil Engineering, who has always put trust in me and my work. His catching enthusiasm for everything related to water and rivers formed an important stimulus for writing the book at hand, and his extensive international network enabled me to orientate myself internationally. During my studies I was given the opportunity to visit France, Germany, and Hungary twice, Canada, and our southern neighbours countless times. Even after his retirement prof. Wessel was prepared to continue our frequent and intensive interactions on my Scheldt research.

Furthermore, I am much indebted to the Directorate-General for Public Works, and Water Management of the Dutch Ministry of Transport, Public Works, and Water Management (Rijkswaterstaat). This organization enabled me to participate in a long term research project on the integrated management of the river Scheldt basin, and generously supplied the normally poor Ph.D. payment. The intensive contacts with Rijkswaterstaat made it possible to address the practise of international water management, and preserved me from becoming an armchair scholar. An organization, however, always is represented by persons. During the past years I have contacted many civil servants of the Ministry who all have one thing in common: their fascination for the river Scheldt. I have not forgotten the persons from the very beginning: Eelke Turkstra of the Regional Directorate Zeeland, and Roel Zijlmans at that time employed at the General Directorate, who introduced me to the technical and political problems in the Scheldt basin. In a later stage of the research, they were succeeded by Frans de Bruijckere and Leo Santbergen of the Regional Directorate Zeeland, and Carel de Villeneuve of the General Directorate. I would like to thank Leo for being helpful with the establishment of cross-border contacts, and Carel for the careful reading of my manuscripts and the many interesting conversations giving me a glimpse behind the scenes of international diplomacy. Henk Smit and Albert Holland of the Research Institute for Coastal and Marine Management arranged a temporary working place for me in Middelburg, and with that largely facilitated the data collection. Finally, I would like to thank Wim van Leussen, at that time employed at the same institute, who has always shown interest in me and my research.

Apart from the representatives of the Dutch Ministry of Transport, Public Works, and Water Management, I am grateful to mister J.-M. Wauthier of the Ministry of the Walloon Region, and mister M. Grandmougin of the Water Agency Artois-Picardie (France). They were willing to welcome me to their organizations, and with that enabled me to study the water management in the upstream parts of the Scheldt basin.

The research was carried out at the Department of Water, Environmental, and Sanitary Engineering of the Faculty of Civil Engineering, and the Centre for Research on River Basin Administration, Analysis and Management (RBA Centre). Although I am not xiv

a civil engineer by education, and not even an engineer, the colleagues at the department and the RBA Centre made me feel home in Delft. I would like to mention three persons in particular: Pieter Huisman, Herbert Berger and prof. dr. M. Donze. Pieter made an important contribution to my research by pointing out to the work of Hofstede on cultural differences between nation states. Herbert, who was involved in the project during the first years, introduced me to the peculiarities of working in an academic environment, and gave a short course on managing relationships with professors. I will never forget the statements made by prof. Donze who, among other things, taught me that a Ph.D. thesis consists of porridge and the best bits ('pap en krenten'), and that a Ph.D. student should write one page of his thesis a day at least.

Prof.dr.ir. W. Thissen of the Faculty of Systems Engineering, Policy Analysis, and Management, and prof. dr. J. Soeters of the Royal Military Academy (KMA) were my main academic contacts out of the Faculty of Civil Engineering. I am much indebted to professor Thissen for faithfully attending the meetings of my supervising team, and for always putting his finger on the problems. Several discussions in Breda with professor Soeters on the 'cultural factor' and the methodological aspects of social science were really inspiring, and have contributed much to the final results. The other members of the steering group I did not mention above did all one way or another make a contribution to this research. I am grateful to prof. dr. J. de Jong of the Faculty of Civil Engineering, prof.dr. P. Glasbergen of the University of Utrecht, and prof.dr. E. Somers of the University of Ghent (Belgium) for their willingness to become a member of the defence commission.

Besides the experts involved in the supervision, I am grateful to the forty respondents who were willing to spend their valuable time to inform me on the latest developments in the international arenas. In the references their names are included.

My direct colleagues in the Scheldt project were Jeroen Maartense, Paul Verlaan and Jos Timmermans. With Jeroen I spent many hours discussing possibilities of interdisciplinary research and formulating joint research proposals. Although our efforts did not bear much fruit, they belonged to the most enjoyable parts of my Ph.D. study. Paul, always good-humoured, a great sense of perspective, and the expert in administrative ranks. Unfortunately, most research topics he addressed were too difficult for me. Jos joined the project in a later stage. Our discussions on decision making and policy analysis helped me a lot to better understand my own research.

In addition to the many professional contacts, my family contributed to the completion of my studies. Jeroen, thank you for stimulating me to cross the river IJssel, and to move to the western part of the Netherlands. For more than one reason you were right! Although I have never regretted the choice to start a Ph.D. study in Delft, born in the 'Achterhoek' I can hardly suppress a feeling of relief now I am able to leave the ant hill called 'Randstad'. My parents have always stimulated me to continue my studies simply because they observed that I really enjoy learning. Finally, I mention Chantal, to whom I dedicate this book, for her invaluable practical and moral support. I am fortunate to have a partner who knows what it means to be a Ph.D. student.

Sander Meijerink

Zwolle, December 1998

Part I:
Introduction, Theory, and Case Study Design

CHAPTER 1

INTRODUCTION

1.1 Background

1.1.1 International river issues

More than 200 major river basins are shared by two or more countries.[1] These basins account for about 60 percent of the earth's land area[2], and constitute a significant portion of the world's fresh water resources. They are a great asset to the nations sharing them, and at the same time a potential source of conflict.

Global economic development and population growth cause an increase in the number and severity of problems emerging in these international river basins.[3] Notorious examples of international river issues can be found in the Middle East. Shortage of clean fresh water in this region causes conflicts on water use in the basins of the Euphrates, Tigris, and Nile.[4] Because of the urgency of the situation in these basins some do expect the next war in the middle east to be about water.[5] Others do predict a global water crisis caused by a mondial shortage of clean fresh water.[6]

Although fortunately the situation is less critical in Western Europe, examples of international river issues can also be found concerning the four rivers flowing through the Netherlands to the North Sea: the rivers Rhine, Meuse, Scheldt, and Ems. Examples of actual international issues in these river basins are flood alleviation in Rhine and Meuse since the high waters of 1993 and 1995, water allocation during periods of low flow in the river Meuse, water and sediment pollution in Rhine, Scheldt and Meuse, and the management of the navigation channel in the estuary of the Scheldt basin. Finally, because of the increasing awareness of environmental degradation in these rivers, ecological river rehabilitation has become an important international issue. Because of their ecological values, and the many user functions assigned to them, knowledge about the management of international river basins is of vital importance for the quality of the environment and economic welfare in the Netherlands.

---

1. LeMarquand (1977, p. 1), The World Bank (1993, p. 38).

2. The World Bank (1993, p. 38).

3. Frey (1993, p. 56).

4. Biswas (1994), Newson (1992, pp. 185-186), Wallensteen and Swain (1997).

5. See, for example, Knoppers and van Hulst (1995, p. 43).

6. Saeijs and van Berkel (1997, pp. 3-20).

1.1.2 Decision making on international river issues

International river issues do have several important characteristics, which have implications for decision making on these issues.[7]

First, because cause-effect relations in a river basin do easily exceed national state boundaries, the issues emerge at geographical scales that do not coincide with national scales. The basin state having a problem, for example because it suffers from transboundary water pollution, depends on one or more other basin states as regards the solution of this problem. Consequently, effective problem solving mostly requires cooperation between the basin states.[8] In an international context neither of the parties involved in decision making is able to rely on hierarchical steering as regards the solution of their problems. Because of the absence of central authority, international cooperation has to develop voluntarily.[9] In some cases, the European Union is an exception to this general rule. Presently, however, possibilities to enforce cooperation in European river basins are still limited.

Secondly, in an international context decision making has a so-called multi-level character. Decision making on the international level, i.e. between the basin states, is linked to *intra*national decision making processes on the strategies to be used in the international negotiations or on the implementation of international policies.

Thirdly, the international character of decision making implies that the actors involved are embedded in different institutional environments. Each basin state, for example, has a different organization of water management, different water management policies, and a different (decision making) culture, which may complicate the decision making process.

Fourthly, river issues are natural resources issues, and unlike policy fields, such as social welfare or education, water management is a technical policy sector. Therefore, decision making on these issues often is dominated by technical experts, and technical expertise plays a relative important role in decision making.[10]

Fifthly, most international river issues are characterized by clear upstream-downstream relations. Activities in the upstream parts of a river basin may affect the functions or activities in the downstream part of the basin and vice versa.[11] For example, in a transboundary river basin the downstream basin states depend on upstream basin states as regards the supply of a reasonable amount of water of a reasonable quality. Consequently, for these issues the upstream basin states do have a relative power advantage, and are able to exert influence on downstream basin states. The upstream basin states, however, may depend on downstream basin states as regards the management of a navigation channel in the river, or possibilities of fish migration. For these issues the downstream basin states have a relative power advantage, and are able to exert influence on the upstream basin states. In many cases several issues emerge simultaneously, and basin states are mutually dependent.

---

7. In Chapter 2 the characteristics of decision making on international river issues will be discussed more extensively.

8. See for example Wessel (1996, pp. 17 and 23).

9. Van Dam (1992, p. 15).

10. Newson (1992, p. 188).

11. Carroll (1986, p. 213), LeMarquand (1977, p. 10).

1.1.3 International policies and steering concepts

The international community is gradually recognizing the urgency of the problems emerging in international river basins, and the amount of legal and other agreements on the management of international rivers has been increasing continuously during the past years.[12] To mention several examples, the International Law Commission of the United Nations (UN-ILC) has worked for many years on the Convention on the Non-Navigational Uses of International Watercourses. This UN-Convention which contains provisions for transboundary water management, was adopted in May 1997.[13] In 1992 the member states of the United Nations Economic Commission for Europe (UN-ECE) concluded the Convention for the Protection and Use of Transboundary Watercourses and International Lakes. This convention, which provides a framework for international cooperation on the management of shared water resources, came into force in 1996.[14] While this thesis is written, the EU is preparing a Framework Directive for Water Resources Management aiming at the integration of several existing EU water directives.[15]

An important steering concept in many international regulations is the so-called river basin (management) approach, which puts the emphasis on the basin scale as the best scale for dealing with (international) river issues.[16] An international river or drainage basin may be defined as " [...] *a geographical area extending over two or more states, determined by the watershed limits of the system of waters, including surface and underground waters, flowing into a common terminus"*.[17] Because of the hydrological relations within a river basin between land and water, upstream and downstream parts of a river basin, water quality and water quantity, and groundwater and surface water, it is generally recommended to manage river basins as a functional unity.[18] Also in the Netherlands the river basin approach is becoming an influential steering concept. The third policy document on water management[19], the report 'Space for Water' (*Ruimte voor Water*)[20], and the governmental proposal for the fourth policy document on water management do all emphasize the need for a river basin approach to solve the problems emerging in (inter)national river basins.[21] A main topic of discussion among water

---

12. See for example Nollkaemper (1993 a and b).

13. Draft Articles on the Law of the Non-Navigational Uses of International Watercourses, Draft Report of the International Law Commission, U.N., GAOR, 43rd Sess., at 1, U.N. Doc. A/CN/.4/L.463/Add.4 (1991).

14. United Nations Economic Commission for Europe, Convention on the Protection and Use of Transboundary Watercourses and International Lakes, Geneva, E/ECE/1267, 1992.

15. Commission proposal for a council directive establishing a framework for community action in the field of water policy. 26.02.1997.

16. Caponera (1985; pp. 566-567; 1992, pp. 185-186), Lammers (1984), Wessel (1991). Downs and Gregory (1991) give an interesting overview of several similar terms that are used for river basin management.

17. This is the definition presented in Article II of the Helsinki rules of 1966. Since than numerous other definitions have been made.

18. For an extensive description of the various functional relations within a river basin, see Newson (1992).

19. V&W (1989).

20. Berends *et al.* (1995, p. 17).

21. V&W (1997).

management experts is the design of institutional arrangements for international river basin management. Should the basin states aim at the development of supranational basin authorities with legal and financial means to develop, implement, and monitor river basin policies, or should they aim at cooperative policy development within the existing institutional settings?[22] Among other things, this research aims to contribute to this intellectual debate.

Another influential steering concept in international river basin management is the concept of sustainable development. Perhaps most well known is the definition of the Brundtland report: "*Sustainable development is defined as a development that meets the needs of the present without compromising the ability of future generations to meet their own needs.*"[23] Even more than the concept of a river basin approach the concept sustainable development has influenced policy makers. Almost any policy document or statement concerning environmental or water issues aims to contribute to sustainable development.

In spite of several (inter)national regulations and policies, and the steering concepts treated above, there still are severe problems in many international river basins. One of these basins, the basin of the river Scheldt, serves as a case study in this research. In the next section this case study will be introduced shortly.

1.1.4 International Scheldt issues

The Scheldt is a relatively small international river, which rises in France, and flows through Belgium and the Netherlands to the North Sea (See Appendix 8). Since the Belgian part of the Scheldt basin extends over the Walloon, Flemish, and Brussels Capital regions, and as Belgian water management competencies are distributed among the federal state and these three regions, there actually are six main parties involved in the management of this river. The basin of the Scheldt has a surface area of 21.863 km^2 [24], and the average annual discharge of the river is 120 m^3/s.[25] The Scheldt estuary is that part of the basin where tidal influence is present, and consists of a salt, brackish, and fresh water zone (See Appendix 9).

Major international river issues in the Scheldt basin that were placed on the international agenda were issues related to the maritime access to the Belgian port of Antwerp, and issues related to water and sediment pollution. Water quantity issues have hardly been addressed internationally.

The port of Antwerp is one of the largest ports in the world. The part of the Scheldt estuary connecting this port to the North Sea is called the Western Scheldt, and is situated on Dutch territory. There is a long history of Belgian-Dutch conflict and cooperation concerning the improvement and maintenance of the navigation channel in the Western Scheldt. In the past, conflicts on the management of this navigation channel were strongly related to the competition between Dutch and Belgian ports, and the issue of the sovereignty over the Western Scheldt. More recently, controversies concern the environmental impacts of dredging works and the dumping of sediments in the Western Scheldt. It is now generally recognized that the improvement and maintenance of the

---

22. See for example Newson (1992, Ch. 7) and Wessel (1989, p. 166; 1992a, p. 24).

23. World Commission on Environment and Development (WCED) (1987).

24. ICBS/CIPE (1997).

25. ISG (1994a).

navigation channel in the Western Scheldt has negative consequences for the ecology of the Scheldt estuary. Two reasons for this are the geomorphological disturbances, and the dissemination of pollutants in the estuary, which are caused by the dredging works and the dumping of dredged material. Continuous improvements of the maritime access to Antwerp, however, are of the utmost importance to the economy of the Flemish region. Next to the management of the navigation channel in the Western Scheldt, in 1967 the construction of two channels improving the maritime access to the port of Antwerp were placed on the international agenda: the Baalhoek canal connecting the port of Antwerp to the Western Scheldt near the village Baalhoek, and the construction of a by-pass near Bath (See Appendices 10 and 11). Although it has a low priority, the former project still is on the international agenda. The latest important event in the history of Belgian-Dutch relations concerning the management of the navigation channel in the Western Scheldt is the conclusion of the Convention on the deepening of the navigation channel in the Western Scheldt on 17 January 1995[26], which entails the implementation of a deepening programme significantly improving navigation possibilities to the port of Antwerp.

A second main cluster of international Scheldt issues are the issues related to water and sediment pollution. Because of the high population density and industrialization in the basin, the low discharge of the river, and the relatively late development of water quality policies in the upstream basin states and regions, the river belongs to the most polluted rivers in Europe. Fortunately, the water quality of the river Scheldt has been improved slightly recently. Because pollutants adhere to fine sediment fractions, severe water pollution also caused contamination of the water beds. River bed contamination, in turn, complicates infrastructure and maintenance dredging works in the estuary, because environmental regulations have made the storage of heavily polluted dredged material costly. Unlike navigation issues, which have characterized the relationship between the former northern and southern Netherlands (the present Netherlands and Belgium) for centuries, issues of water and sediment pollution only reached the international agenda in the second half of the 20th century. In 1961 the first international agreement addressing pollution in the Scheldt estuary was signed.[27] Since 1985, when the Belgian government applied for a permit for dumping dredged material in the Western Scheldt, the Dutch government has been formulating conditions to this permit obliging the Belgian (Flemish) government to improve the water quality of the Scheldt, and to extract large amounts of polluted sediments from the Lower Sea Scheldt.[28] In 1994, the Scheldt basin states and regions were able to reach an agreement on a multilateral Convention on the protection of the Scheldt against pollution[29], and the basin states installed a provisional International Commission for the Protection of the Scheldt (ICPS). The main task of this commission is the preparation of the first Scheldt Action Programme (SAP) with the first basin-wide water quality policies.

Ecological issues are strongly related to the issues discussed above. Geomorphological

---

26. Verdrag tussen het Vlaams Gewest en het Koninkrijk der Nederlanden inzake de verruiming van de vaarweg in de Westerschelde. Trb. (Treaty Series of the Netherlands) 1995 Nr. 51.

27. Verdrag betreffende de verbetering van het kanaal van Terneuzen naar Gent en de regeling van enige daarmede verband houndende aangelegenheden. Trb. 1960 Nr. 105.

28. This is the part of the estuary between Rupelmonde and the Flemish (Belgian)-Dutch border.

29. Verdrag inzake de bescherming van de Schelde, Trb. 1994 Nr. 150. The Flemish region signed this convention on 17 January 1995.

disturbances caused by the continuous dredging works and dumping of sediments in the Scheldt estuary, and the severe pollution of the river Scheldt do both harm the ecology of the river. Plans for restoration of the Scheldt ecosystem are discussed in the ICPS, and the Convention on the deepening of the navigation channel in the Western Scheldt of 1995 contains provisions for a compensation of the nature losses implementation of the deepening programme is expected to cause. In the Netherlands discussions concerning the development of a plan for nature restoration focused on the policy alternative 'ontpolderen', which implied a partial landward movement of the seawalls. Because public and administrative support in the Province of Zeeland for this policy alternative was lacking, an alternative plan has been developed.

Water quantity issues concern both situations of a high river discharge, and the structural low river discharge of the river Scheldt. Unlike the other estuaries in the Rhine-Scheldt Delta, the Scheldt estuary is not closed. Consequently, the combination of a storm surge and an extremely high fresh water discharge of the river Scheldt may cause problems in the upper estuary near the city of Antwerp. According to experts sea level rise and the implementation of the deepening programme will increase flood risks along the estuary of the Western Scheldt. Discussions on the safety along the Scheldt estuary and flood protection are strongly related to plans for 'ontpolderen' along the estuary, because the partial landward movement of the sea walls increases the storage capacity of the estuary, and consequently may lower flood risks. More recently, also the structural low discharge of the river Scheldt is getting more attention. According to experts, the many artificial diversions of Scheldt water to the North Sea via canals diminishes the fresh water flow into the estuary, which in turn harms the unique tidal fresh water zone of the estuary. Although awareness of water quantity issues seems to be growing, in 1997 these issues have not yet been addressed internationally.

1.2   Research objective and problem statement

The objective of the study is twofold:
- To acquire knowledge of decision making on international river issues.
- To contribute to the solution of upstream-downstream problems in international river basins.

The research objective leads to the formulation of a threefold problem statement:
- How does decision making on international river issues develop?
- How can this development be explained?
- Which strategies can be used to contribute to the solution of upstream-downstream problems in international river basins?

The first component of the problem statement addresses the course of the decision making process. The second component refers to the explanation of the course of the decision making process. The third and final component relates to the prescriptive part of this research. In Chapter 3 the problem statement will be elaborated, and seven research questions will be formulated.

1.3   Research characteristics

In this section the main research characteristics will be discussed shortly.

The research at hand is:
- A single case study covering a long time span
- An empirical rather than a normative study
- An international study

First, the research comprises a single case study of decision making on international Scheldt issues. It focuses on the strategic interactions between the Scheldt basin states and regions during the past 30 years. Some periods are characterized by conflict, whilst others are characterized more by cooperation. The case study covers the notorious Belgian-Dutch negotiations on the water conventions dealing with Scheldt and Meuse water management and infrastructure issues, but addresses the recent interactions in the International Commission for the Protection of the Scheldt against pollution (ICPS), and the implementation of the deepening programme of the navigation channel in the Western Scheldt as well.

A second characteristic of the research is its empirical rather than its normative character. The research aims to contribute to the knowledge of decision making on international river issues, and to indicate possibilities to influence this decision making. It does, however, not address normative questions, such as: what is good decision making, or what is sustainable river basin management? The main reasons for this choice are that: (1) different actors do have different opinions on these issues, and (2) the author does not believe in objective, scientific answers to these questions, because the answers depend on values, norms and interests, and consequently are largely subjective and political.[30] On the other hand, the author is aware that all research inevitably comprises subjective elements. In the end, the choice for empirical rather than normative research also is a subjective one. Furthermore, the selection of theories and methodologies often is largely subjective. Finally, also some concepts, such as pollution, have a subjective connotation.

A final research characteristic is its international character. The research entails the collection and analysis of written and oral data stemming from three different countries, and makes it necessary to reflect on the structural and cultural differences between them.

1.4   Relevance

1.4.1 Theoretical relevance
The research at hand is part of the large scale research on the management of

---

30. As an example, two subjective elements of the concept of sustainability will be elaborated. The concept refers to a balance between economy and ecology, and between the welfare of the present generation compared to future generations. These trade-offs depend, among other things, on (1) the *value* one assigns to environmental protection or economic welfare, and (2) whether one *believes* that technical solutions will eventually solve all environmental problems, or whether one is a so-called 'eco-pessimist', who is *of the opinion* that economic growth by definition leads to environmental degradation, mainly because it leads to ever increasing consumption. The words printed in italics emphasize the highly normative, and consequently subjective character of the operationalization of this complicated concept.

international river basins, which has been carried out all over the world during the last years.[31] Partly because of the increasing practical relevance of international river basin management, scientists with different disciplinary backgrounds applied themselves to the study of international river issues. These disciplines use different concepts to describe and analyze international river issues, and address different aspects of these issues. Traditionally, natural scientists, engineers, economists, and lawyers studied (international) water management. More recently, the number of social scientists studying water management issues is increasing. Together, these disciplines, and there undoubtedly are many more, do contribute to the development of a multidisciplinary knowledge base of the management of international river basins. The thesis research presented in this book belongs to the category of social research, and focuses on the process dimension of decision making on international river issues. Neither legal provisions for decision making, nor the content of international river policies are at the core of this study. Its focus is on the interactions between the basin states, and their strategic behaviour.

In addition to the contribution to a multidisciplinary knowledge base of international river basin management, the research fits in the research on policy networks and complex decision making, which presently is an important field of study for scholars in public administration. Research on policy networks and complex decision making, however, did hardly address decision making on *international issues*. The international character of decision making on many river issues makes it necessary to address the influence of institutional differences between the basin states on the decision making process, and problems of intercultural management. This research may contribute to the integration of these aspects in the theory of policy networks and complex decision making. Because upstream-downstream power asymmetries are characteristic to most international *river issues*, the study contributes specifically to the knowledge of decision making on issues characterized by power asymmetry.

1.4.2 Practical relevance
In Section 1.1.1 the increasing number and severity of problems emerging in international river basins were already mentioned. The solution of these problems does not only require technical, ecological, and legal knowledge, but knowledge about decision making processes as well. A better understanding of these processes enables water managers to develop more effective strategies of problem solving. In this book, strategies are discussed that may be used by basin states perceiving upstream-downstream problems, by the EU, and by independent parties involved in decision making on international river issues, such as the chairman of an international river commission or an independent policy analyst.

1.5   Outline of thesis

In Figure 1.1 the contents of this book are schematized.

---

31. For an overview of this research see, for example, the Proceedings of the ninth World Water Congress in Montréal (Canada) of 1997. Collection environnement de l' Université de Montréal (1997).

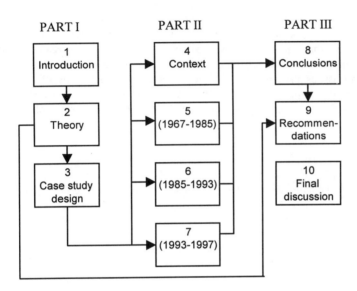

Figure 1.1 *Outline of thesis*

The thesis consists of three parts. Part I contains an introduction to the field of study, and the theoretical and methodological discussions. After this introductory chapter presenting the objective and problem statement of the research, in Chapter 2 the theory of policy networks will be treated, and the characteristics of decision making on international river issues touched upon in Section 1.1.2 will be discussed in more detail. Based on these theoretical considerations a conceptual framework is developed for the analysis of decision making on international river issues. In Chapter 3 the problem statement is elaborated, and seven descriptive, explanatory, and prescriptive research questions are formulated. In addition some philosophical and methodological problems of research on decision making are discussed.

Part II contains the case study of decision making on international Scheldt issues. Chapter 4 starts with a description and analysis of the context of decision making on these issues. After a short description of the Scheldt basin and its main functions, the main international Scheldt issues are introduced. The following dimensions of the context of decision making on these issues are distinguished: the underlying hydrological structure of the issues at stake, the international context, the *intra*national developments in France, Belgium, and the Netherlands, the (decision making) cultures in each of these basin states, and the history of conflict and cooperation on international Scheldt issues. The Chapters 5,6, and 7 contain chronological descriptions and analyses of decision making on international Scheldt issues. These chapters successively address the periods from 1967 to 1985, 1985 to 1993, and 1993 to 1997. In the first part of each chapter the descriptive research question is answered, and the strategic interactions between the Scheldt basin states are described. In the second part the explanatory research questions are answered, and decision making is analyzed using the conceptual framework presented in Chapter 2.

Part III contains the conclusions and recommendations of the research, and a final discussion. In Chapter 8 an overall answer is given to the descriptive and explanatory research questions formulated in Chapter 2, and the conclusions of the thesis research are presented. Chapter 9 presents an answer to the prescriptive research question, and discusses strategies that may be used to solve upstream-downstream problems in international river basins. This chapter draws on the theory of decision making on international river issues presented in Chapter 2, and the conclusions of the case study research presented in Chapter 8. Chapter 10 concludes this book with a discussion of the research process and results, and some recommendations for further research.

CHAPTER 2

DECISION MAKING ON INTERNATIONAL RIVER ISSUES

2.1 Introduction

The objective of this chapter is to present a conceptual framework for the analysis and explanation of decision making on international river issues. In Chapter 3 this framework will be used to elaborate the problem statement introduced in Chapter 1, and to formulate several research questions. Subsequently, these research questions will be answered in part II of this book, which contains the case study of decision making on international Scheldt issues.

In Section 2.2 the main theoretical concepts of the network perspective on decision making are treated. In the network perspective, decision making is conceived of as a series of games or rounds. The discussion focuses on the perceptions and strategies of the individual actors, the strategic interactions between them, and the learning processes that may develop during these interactions. In addition, the relationship between games and networks, and between network characteristics and the structural and cultural characteristics of the context are discussed. Finally, the problems related to the evaluation of decision making are discussed shortly.

Section 2.3 focuses on the use of the theory presented in Section 2.2 for the analysis and explanation of decision making on international river issues. First, the characteristics of decision making on *international issues* are addressed. Subsequently, the characteristics of decision making on international *river issues* are discussed.

In Section 2.4 a conceptual framework is presented for the analysis and explanation of decision making on international river issues. This framework draws on the theoretical concepts of the network perspective on decision making treated in Section 2.2, and the discussion of the characteristics of decision making on international river issues in Section 2.3.

2.2 Main theoretical concepts of the network perspective on decision making

In this section the main theoretical concepts of the network perspective on decision making will be treated, and the relations between these concepts will be clarified. In Section 2.2.1 the network or pluricentric perspective on decision making is compared with the classical or unicentric perspective on decision making, and with that the theoretical perspective used in this research is shortly introduced. In Section 2.2.2 networks are defined as multi-actor structures of interdependence. Interdependencies, which are shaped by the distribution of resources among actors, are the fundament of interactor relations, because they are the reason why actors start interacting with each other. Section 2.2.3 discusses the types of strategies that actors may use to reduce or manage their dependence on other actors. In Section 2.2.4 several dimensions of interaction are treated, and it is argued that the strategic interactions between actors may be conceived of as series of decision making games or rounds. Section 2.2.5 relates the strategies of the players of a decision making game to their perceptions of the issues at stake and the games that are played. In Section 2.2.6 the relationships between games and networks, and between network characteristics and structural and cultural characteristics of the context are addressed. Section 2.2.7 explains that decision making may be conceived of as a learning process as well, distinguishes between strategic or interactive learning and cognitive learning, and addresses the strategic use of knowledge in a decision making process. Finally, in Section 2.2.8 the problems related to the evaluation of decision making when using the network perspective on decision making are shortly addressed.

2.2.1 Positioning of the network perspective on decision making

The network or pluricentric perspective on decision making often is compared with the classical or unicentric perspective on decision making.[1]

In the classical or unicentric perspective on decision making, there is one central decision maker, and decision making is generally modelled with a stages model. A typical stages model comprises the following stages: agenda setting, policy preparation, policy formulation, policy implementation, policy evaluation, feedback, and policy termination.[2] A basic assumption in this perspective is that for each problem an optimal solution can be found.[3] Therefore, this perspective is also called the rational perspective on decision making. "*It is as if government could run the social system like an engineer, on the basis of its understanding of the natural world.*"[4]

In the network or pluricentric perspective on decision making policies are not the result of an intellectual process at a central level, but the result of a strategic interaction process between multiple interdependent actors.[5] These interactions do not develop according to predefined stages, but may be modelled better as series of games that are played between the actors.

Because each perspective highlights different aspects of decision making, the network perspective does not replace the classical perspective on decision making. Nevertheless, there are three main reasons why the network perspective is chosen in this research. First, empirical research on decision making processes has indicated the descriptive inaccuracy of the stages heuristic. "*Although proponents of the stages model often acknowledge deviations from the sequential stages in practice, a great deal of recent empirical study suggests that*

---

1. See, for example, Frances *et al.* (1991, pp. 1-19), Kickert *et al.* (1997, pp. 7-10), Teisman (1992, pp. 26-30), and in 't Veld (1995, pp. 16-24). With the exception of in 't Veld, these authors introduce a third perspective, which is called the multicentric or market perspective. This perspective emphasizes the autonomy of local actors, and the 'invisible hand', or market as a mechanism to coordinate choices made by local actors.

2. Jenkins-Smith and Sabatier (1993, pp. 1-4).

3. Teisman (1992, p. 31).

4. Glasbergen (1995, p. 8).

5. Teisman (1992, p. 35). Levacic (1991, p. 47) speaks about "[..]a pluralist system in which a wide range of political interests influence decision making."

deviations may be quite frequent [...]."[6] Secondly, the network perspective on decision making does not exclude the possibility of hierarchical steering. Rather it emphasizes that most hierarchical relations are characterized by interdependence as well, because the actor imposing a policy programme depends on the actors that have to implement that program. The third, and probably most important reason is that in an international context central authority mostly is absent (See also Section 2.3.1).

The network perspective on decision making can be used as an analytical model and as a prescriptive or steering model.[7] The aim of the analytical use of the network perspective is to analyze and explain a decision making process. The use of the network perspective as a prescriptive model aims at the formulation of recommendations, for example concerning the strategies to be used by individual actors or the organisation of a decision making process. The primary focus in this thesis is on the analytical use of the network perspective, i.e. on its analytical and explanatory power. Only in Chapter 9 the network perspective will be used as a prescriptive model, and recommendations will be formulated concerning the strategies that may be used by individual actors in decision making on international river issues.

2.2.2 Multi-actor structures of interdependence

In the literature on policy networks several definitions of *policy networks* can be found. Some definitions, in order of increasing length, are: *"multi-actor structures of interdependence."*[8], *"[...] patterns of interactions between interdependent actors which take shape around policy issues or policy programmes."*[9] or *"(Changing) patterns of social relationships between interdependent actors which take shape around policy issues and/or policy programmes, and that are formed, reproduced and changed by series of games in which actors try to influence policy processes as much as possible by strategic behaviour."*[10] Two central concepts in most definitions of policy networks are *actors* and *interdependence* or mutual dependence.

Actors are the basic units in a decision making process. They are defined as units that by a certain unity of action act as an influencing party.[11] Actors can be states, organizations, departments of organizations or individuals. In international river basin management, many different actors can be distinguished. Important actors may be the national governments and parliaments of the basin states, delegations of the national governments, ministers, and the ministries involved. In addition to these national actors, regional and local government agencies, and environmental or other NGOs may be involved in the decision making process.

The concept of interdependence refers to the situation that actors at least partly depend on each other for solving their problems or achieving their objectives. Interdependence is caused by the distribution of the (control of) resources among the actors.[12] *Resources* are all

---

6. Jenkins-Smith and Sabatier (1993a, p. 3).

7. Hanf and O' Toole (1992, p. 163), Kickert (1991, p. 13), Tatenhove and Leroy (1995, p. 131).

8. Bressers *et al.* (1995, pp. 5-6).

9. De Bruijn, Kickert, Koppenjan (1993, p. 19).

10. Klijn and Teisman (1992, p. 36).

11. Teisman (1992, p. 50).

12. Godfroij (1981, p. 74). Other terms that are used to indicate the distribution of the (control of) resources among multiple actors are the dispersion or disaggregation of problem-solving capacity (Glasbergen, 1995, p. 10), (Hanf and O' Toole, 1992, p. 166).

things that may help actors solving their problems or achieving their objectives. Two broad classes of resources are money and authority.[13] Generally, actors do not possess all resources they need for the solution of their problems or the achievement of their objectives, and therefore they are at least partly dependent on other actors. As an example, a water management agency in a downstream part of an international river basin may perceive problems of water and sediment pollution, and may know that these problems are partly caused by the transboundary load of pollutants in the river. The agency is aware of its dependence on actors in the upstream basin states, because these actors do possess the indispensable resources to reduce the transboundary load of pollutants, such as the authority to formulate emission or water quality objectives, to levy taxes on waste discharges, or to issue environmental permits. For other issues, such as navigation, however, the actors in the upstream basin state may perceive a dependence on the downstream water management agency.

2.2.3 Strategies and positions

Because in the pluricentric perspective decision making is conceived of as a strategic interaction process, *strategy* is a key variable in this perspective on decision making. The literature contains as many definitions of strategy as definitions of networks. Some definitions are: *"patterns of behaviour aiming at the realization of individual or common objectives"* (*"Gedragspatronen gericht op het realiseren van al of niet gemeenschappelijke doeleinden")*[14], *"a pattern in action over time"*[15], or *"empirically observed regularities of behaviour."*[16] Unlike in the rational perspective on decision making the concept strategy does not by definition refer to planned behaviour. *"An organization can have a pattern (or realized strategy) without knowing it, let alone making it explicit. [...] A realized strategy can emerge in response to an evolving situation, or it can be brought about deliberately, through a process of formulation followed by implementation."*[17] The types of strategies that actors may use strongly depends on their *positions*. Actors may have three different positions, which are created by the distribution of resources among the network: the *interaction position*, the *incentive position*, and the *intervention position*.[18] Table 2.1 produces information on the types of strategies that may be used by actors in each of these positions. Actors that have an interaction position are only able to exert influence on other actors in an interaction process. Table 2.2 produces information on four broad categories of strategies that may be used by actors that have an interaction position. Below, these strategies will be discussed.

---

13. Benson (1975, p. 229).

14. Godfroij (1992, p. 368).

15. Mintzberg (1987, p. 67).

16. Crozier and Friedberg (1980, p. 25).

17. Mintzberg (1987, p. 68).

18. Teisman (1992, p. 55).

Table 2.1 *Positions and strategies, Modified after Teisman (1992)* Positions, characteristics of positions, and types of strategies that may be used from positions

Position	Characteristics of position	Type of strategies that may be used from position
Interaction position	Actor is able to solve his problems or achieve his objectives in interaction with other parties	Offensive, reactive Interactive, aimed at maintaining autonomy
Incentive position	Actor is able to influence the behaviour of other actors by creating (dis)incentives	Support behaviour by providing resources Block behaviour by taking away resources
Intervention position	Actor is able to influence patterns of interaction between other actors by intervention	Influence decision making procedures (rules of interaction) and structures Mediate

Table 2.2 *Categories of strategies that can be used in interaction position (Modified after Teisman, 1992)*

	Interactive	Maintenance of autonomy
Offensive	Actor takes initiative to start interactions with actors possessing indispensable resources	Actor tries to acquire new resources, or to find substitutes
Reactive	Actor reacts on proposals made by other actors, and tries to solve his problems or to achieve his objectives in strategic interactions with the other actors	Actor does not want to start interactions with other actors, because he does not expect that these interactions will contribute to the solution of his problems or the achievement of his objectives

Actors generally strive toward maintenance of autonomy, and therefore want to avoid or reduce their dependence on other actors.[19] If an actor is able to control all necessary resources, he can develop projects that aim at the achievement of his objectives independently, and interactions with others are no longer necessary. *Strategies aimed at maintaining autonomy* may be *offensive* or *reactive*.[20] Actors may try to increase their control of resources, or to develop alternative resources that may substitute resources controlled by others. As an example, a downstream basin state that would like to produce drinking water out of a polluted international river may decide to extract water from a less polluted river.

---

19. Klijn and Teisman (1992, p. 42).

20. Teisman (1992, p. 88).

Furthermore, actors may reduce their dependence by lowering their level of ambition.[21] Obviously, if an actor has ambitious objectives, he is more dependent on other actors than in case his objectives are more modest. If an actor refuses to interact when other actors propose to start interactions on an issue, this is called a reactive strategy aimed at maintaining autonomy.[22] Actors may use this strategy if they do not perceive a dependence on the actors, and do not expect that joint action may contribute to the achievement of their objectives.[23]

If an actor perceives and accepts a dependence on other actors, he may activate the network, and start interactions with actors possessing indispensable resources.[24] The selection of the relevant actors, i.e. the actors possessing indispensable resources, and the establishment of relations with them is an important strategy of network management.[25] For a careful selection the actor has to make an assessment of the resources he needs for the achievement of his objective, and the distribution of resources among the network. The counterpart of the strategy of network activation is the strategy of ending interactions with actors, because these actors do no longer possess indispensable resources. Part of the craftsmanship of network management is to find a balance between the costs of maintaining numerous relationships, and their contribution to the achievement of an objective. An actor can become involved in a process of strategic interaction in two different ways.[26] He can use an *offensive interactive* strategy, and start interactions himself, or a *reactive interactive* strategy, and react on proposals of others. In the latter case the actor uses the opportunity of interaction to achieve his objectives.[27]

Actors that have an *incentive position* are able to influence the behaviour of other actors by giving incentives. They can do so in two different ways. First, they may support behaviour by providing resources. Secondly, they may hinder or block behaviour by taking away resources. In other words, they may use "carrots" or "sticks" to influence the behaviour of other actors. Incentives mostly are directed to individual actors, and not to the whole network.

Actors that have an *intervention position* are able to intervene in the interactions between other actors. First, they may influence decision making procedures (or rules of interaction), and the structure of the network. As an example, a central government may decide that regional and local actors have to develop an action program to reduce non-point source pollution. The central government may decide on the terms of the plans to be developed, and on the actors that have to be involved in the planning process. Secondly, it may facilitate the decision making process, for example by providing information. Finally, if the regional planning process would get in an impasse, the central government might try to mediate.

---

21. Ibid.

22. Ibid.

23. To be more precisely: If actors do not expect that the benefits of the interactions will outweigh the costs of these interactions.

24. Klijn and Teisman (1992, p. 42).

25. Termeer (1993a, p. 106).

26. Teisman (1992, p. 88).

27. Ibid.

2.2.4 Strategic interaction, games, and decision making rounds

Strategic interaction
Interdependence is the fundament of interactor relations, because it is the reason why actors start interacting which each other. Relevant dimensions of *interaction* can be communication, negotiation, decision making and exchange.[28] The first dimension, *communication*, is a fundamental dimension of interaction, and comprises the exchange of information, demands, announcements, claims and opinions.[29] The second dimension, *negotiation*, refers to mutual influencing actors, and is defined as a type of communication in which the actors try to find exchange formulas that are attractive to both parties.[30] The third dimension of interaction, *decision making*, is a type of communication in which the actors agree (implicitly or explicitly) on an exchange formula. This definition of decision making is a most restricted one. Unless indicated otherwise, in this thesis *decision making* refers to all dimensions of interaction. The final dimension, *exchange*, concerns the implementation of the exchange agreements that are made between the actors. Types of exchange are the exchange of goods, information, personnel, money, or a restriction of behaviour the actors have reached agreement on.[31] In interaction processes, each actor tries to influence the structure, process and outcome of decision making in such a way that it contributes most to the solution of the problem he perceives, or the achievement of his objective. Therefore, interactions mostly are called *strategic* interactions.

Decision making games or rounds
For a better understanding of the strategic interactions between actors, the analogy with a *game* that is played by the actors can be useful. The game played between the actors is the whole of interactions, incentives and interventions that develop in a *policy arena*.[32] A policy arena, which sometimes is called the *activated network*, is the part of the network where the interactions take place.[33] The actors that have an interaction position are playing the game, whilst the actors that have an incentive or intervention position have a more peripheral role, which, however, does not imply that this role is less important. The games played between the actors in a policy arena may be characterized by *game types*. In the literature several game types are distinguished.[34] The probably most predominant game types are *conflict* and *cooperation*. According to Frey, a conflict exists *"if one actor attempts to exert power over another to overcome that actor's perceived blockade of the first actor's goal, and faces significant resistance."*[35] A game may be characterized by cooperation if the actors are *"working together for common benefits"*.[36]

---

28. Godfroij (1981, p. 86).

29. Ibid.

30. Ibid.

31. Godfroij (1981, p. 87).

32. Teisman (1992, p. 63).

33. Ibid, p. 62.

34. Godfroij (1981, p. 88).

35. Frey (1993, p. 57).

36. Ibid. For an excellent discussion of different definitions of conflict and cooperation, see Godfroij (1981, pp. 160-166).

O' Toole mentions three inducements to cooperation.[37] First, actors may start cooperation if they have the same objectives (or face the same problems). Secondly, they may start cooperation if they perceive possibilities of exchange, which could make both parties better off with an exchange agreement than without one. Such a situation is often indicated a win-win situation. Thirdly, authority may be an inducement to cooperation. Whilst in the first two cases cooperation develops voluntarily, in the last case cooperation is imposed. For a discussion of other game types, see for example Godfroij.[38]

In real life interaction processes game types may be present in succession. In a negotiation process actors may solve conflicts and establish cooperation, whilst in a situation of cooperation between actors conflicts may emerge. Game types may be present simultaneously as well. Some parts of an interactor relationship may be characterized by cooperation, whilst other parts may be characterized better by conflict.

In Section 2.2.1, the classical perspective was compared with the pluricentric perspective on decision making. Whilst in the classical perspective decision making is generally modelled with a stages model, in the pluricentric perspective decision making is modelled with a rounds model of decision making. Because there is not one central decision maker, but there are multiple actors taking decisions that influence the course and outcome of a decision making process, in the network perspective decision making is conceived of as "*a complex of series of decisions, taken by different actors.*" ("*een kluwen van reeksen van beslissingen, genomen door verschillende actoren.*")[39] The outcome of decision making depends on the relations between these decisions. Decisions that are taken by the individual actors in a decision making process may relate to each other in three different ways.[40] First, an actor may take into account the decisions that have been made by other actors. Secondly, he may anticipate decisions that will be taken by other actors. Thirdly, actors may take joint decisions.

In the series of decisions, some decisions are more important than others, because they are used as a point of reference by the actors involved. Therefore, these decisions are called *crucial decisions*.[41] In the period between two crucial decisions the actors involved play a decision making game. In the network perspective, decision making is conceived of as a series of *decision making games* or *decision making rounds*. Figure 2.1 illustrates the rounds model of decision making.

---

37. O 'Toole (1988, pp. 417-441). Compare to Young who distinguishes between three paths to regime formation: spontaneous, in which regimes emerge from the converging expectations of many individual actions; negotiated, in which regimes are formed by explicit agreements; and imposed, in which regimes are initially forced upon actors by external imposition (Young, 1984, pp. 98-101).

38. Godfroij (1981, pp. 88-93).

39. Teisman (1992, p. 33), In 't Veld (1995, p. 19).

40. In 't Veld (1995, p. 21).

41. Ibid., p. 20.

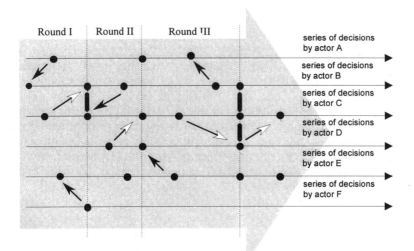

Figure 2.1 *The rounds model of decision making (Modified after In 't Veld, 1995, p. 21)*
In the rounds model decision making is conceived of as a complex of series of decisions
taken by different actors. These decisions may relate to each other in three ways. First,
an actor may take into account decisions taken by other actors (Black arrows). Secondly,
he may anticipate decisions that will be taken by others (Open arrows). Thirdly, actors may
take joint decisions (Vertical connections between two or more series of decisions). The
period between two crucial decisions is called a decision making round.

2.2.5 Perceptions, issues, and issue-areas

The main reason for the development of policies are problems. A *problem* may be defined
as a discrepancy between an actual and a desired situation. Because in the pluricentric
perspective on decision making there is not one central decision maker, there is not one
single problem as well. Different actors involved in a decision making process do mostly
have different perceptions of the actual situation and different preferences concerning the
desired situation. Consequently, their problem definitions mostly are diverging. Therefore,
it is better to speak of a *problem perception*, which refers to a subjective observation, and
is defined as a discrepancy between a *perceived* actual situation and a desired situation.
Unlike the classical perspective on decision making, the pluricentric perspective takes into
account the different problem perceptions, and the action rationality of the individual actors.

The strategies and objectives selected by an actor involved in a decision making process are not based on its *problem* perception only. They should be related to the broader concept *perception*, which mainly relates to three aspects of a game situation[42]:

- the interdependencies between the actors arising from the division of the resources needed to fulfil their ambitions;
- the ambitions, and 'stakes in the game' of the actors;
- the policy game and the importance of the policy problem being addressed in the game, compared with the policy problems being addressed in other games.

Because the actors involved in a decision making process may have different problem perceptions, and different perceptions of the games that are played, the concept *issue* will be introduced. According to Frey an issue exists whenever an actor perceives that some of its goals are being blocked (frustrated, denied) by another actor.[43] He adds that this blockade may be intentional or unintentional.[44] For example, a downstream water management agency may perceive problems of water and sediment pollution. For the solution of these problems, it depends partly on the upstream actors. Therefore, one may speak of an international pollution *issue*. A more detailed observation, however, may learn that there are several separate issues. The actors may, for example, discuss the impacts of the transboundary load of pollutants on the water beds, on the ecosystem, or may discuss the implementation of international water quality policies. These separate issues may be converted by the actors themselves into an *issue-area*, a *recognized* cluster of concerns involving interdependence not only among the actors but among the issues themselves.[45] In our example, the actors may convert the separate issues mentioned above into an issue-area water and sediment pollution.

2.2.6 Games, networks, and context

Games and networks
The relationship between networks and the games that are played within these networks is rather complicated. On the one hand structural and cultural network characteristics, such as the distribution of resources among the actors, and formal or informal rules of behaviour, influence the games that are played. At the same time, network characteristics may be reproduced or changed by the games that are played. These last processes are called *institutionalization* processes. Because of institutionalization processes, in Section 2.2.2 networks were defined as "*(changing)* patterns of social relationships between interdependent actors which take shape around policy issues and/or policy programmes, *and that are formed, reproduced and changed by series of games* [...]".[46] In a game the players may reach an

---

42. Klijn and Teisman (1992, p. 39; 1997, p. 100).

43. Frey (1993, p. 57).

44. Ibid. Other authors, such as Caldwell and Haas, add an element in their definitions of an issue. According to them the situation described above may be called an issue only, if it leads to political action, i.e. if it is placed on the agenda of negotiators. Caldwell defines an environmental issue as "[...]*a result of an emerging change in the environment , caused or influenced by human activity, with consequences arousing social concern and creating problems leading to political action.*" (Caldwell, 1990, p. 11). Haas defines issues as "*separate items that appear on the agenda of negotiators.*" (Haas, 1980, p. 364).

45. Haas (1980, p. 365). In the literature the term 'problem-area' can be found as well (Sabatier, 1993, pp. 20-21).

46. For a discussion of this 'duality of structure', see Klijn and Teisman (1997, p. 103).

agreement on policy objectives and/or on policy programs, which restricts their behaviour in the succeeding games. Furthermore, they may reach an agreement on the development of institutional arrangements, which are defined as organizational arrangements structuring and regulating interactions.[47] Institutional arrangements, for example, are an agreement on the exchange of information, or a project organization to carry out joint research. Finally, also informal rules of behaviour may change during the games that are played. Because the players know that institutional arrangements may influence the course and outcome of future decision making processes, these arrangements are the result of a *strategic* interaction process.[48] Each actor will try to influence the institutional arrangements in such a way that the outcomes of the games to be played contribute most to the achievement of his objectives.

Some authors have developed stages models of institutionalization processes. According to these models interorganizational relations have, just like organizations, life stages. In a model developed by Mijs institutionalization processes are conceived of as two analytically distinct subprocesses that continuously interact.[49] One of them (sometimes smooth and gradual, otherwise jerky) involves the extending, intensifying, structuring and organizing of the interactions between the actors. The other involves the emergence, simultaneous to and by way of these increasingly dense interactions, of an interorganizational culture: a pattern of shared expectations, goals, norms and values pertaining to the ways in which organizations cooperate or ought to cooperate.[50] Mijs modelled the development of inter-organizations. In these 'second-order' organizations, the 'first-order' member organizations have joined together to constitute a foundation or association with the intention of collaboration within a strictly defined area.[51] In the first stage there are various types of interactions among the members of the organizations in question, but no appreciable degree of cooperation among the organizations in the network as a whole. In this way a subnetwork is created. In the second stage institutionalized but not yet formalized coordination occurs among the members of the subnetwork. In the third stage, the integrated subnetwork tries to mobilize the entire organizational network by way of conferences or an *ad hoc* coalition of some kind. In the fourth stage a critical majority of the organizations in the larger network starts to realize that common problems, goals and ambitions can only be addressed by the establishment of a more elaborate and permanent form of interorganizational cooperation.[52]

Soeters developed a model of the development of Euregional networks.[53] In the 'zero-stage' of network development one or more parties recognize the existence of common interests and/or mutual dependencies. In the first stage of network development, the 'expressive stage', the expression of mutual respect, trust and common interests takes place. In this stage often a declaration of intent is made. In this joint declaration the necessity of

---

47. Teisman (1992).

48. Godfroij (1981, p. 96).

49. Mijs (1992, p. 159).

50. Ibid.

51. Ibid., pp. 155-169.

52. Ibid.

53. Soeters (1993, pp. 644-646), Driessen (1995, pp. 32-33), and Driessen and Vermeulen (1995, pp. 160-161).

joint action is expressed, and a general policy objective is defined.[54] According to Soeters, after the 'expression stage' network development mainly has a cognitive character. In the 'cognitive stage' exchange of information and new ideas take place. The aim of this stage is to map the different perceptions of a problem and to come to a common problem definition. In this stage research projects or projects aiming at the exchange of information are developed. In the third stage, the 'productive stage', compromises on the directions of solutions and policy measures are made. In addition, agreements on financing are made. Finally, in the 'formalizing stage', a plan or covenant is produced, and decisions on how to elaborate and implement the policies agreed upon are made. After the formalizing stage new dependencies between the parties are formed. According to Soeters networks continue to develop as long as the parties recognize their mutual dependencies. If they do no longer recognize such dependencies, the network may easily fall back into a preceding stage.[55] Empirical research has shown that the transition from the cognitive to the productive stage is most difficult.[56]

Trust as a network characteristic
Trust is an important aspect of interactor relationships. Bradach and Eccles define trust as "[..] *a type of expectation that alleviates the fear that one's exchange partner will act opportunistically.*"[57] If actors have reached an exchange agreement, they have to trust that their exchange partners will observe this agreement. Because trust is a kind of expectation of the strategic behaviour of others, it is related to the concept of reputation. A *reputation* is a conviction of others that an actor will use a specific strategy.[58] If an actor continuously uses interactive strategies, others will expect this actor to continue his interactive strategies, whilst a permanent use of strategies aimed at maintaining autonomy may give an actor the reputation of being unwilling to cooperate. Reputations are a type of information that actors may use in decision making on their strategies. Lengthy conflicts or deadlocks are characterized by rigid mutual strategies, which may give the actors involved negative reputations. These reputations shape relations characterized by distrust.

Context and network
The structural and cultural characteristics of a network are strongly related to the structural and cultural characteristics of the context of this network, which may be called the society[59], macro-environment[60], or macro-context.[61] This context comprises the structural and cultural characteristics of the broader society, and of other (related) networks. Structural characteristics, for example, concern the economic situation, and the state and administrative organization in a country. Cultural characteristics concern, for example, the environmental

54. Whilst Soeters sees the 'expressive stage' as a separate stage, Driessens treats the expressions as the results of the 'zero stage'.

55. Soeters (1993, p. 646).

56. Ibid, p. 648.

57. Bradach and Eccles (1991, p. 282).

58. Axelrod (1990, p. 126).

59. Godfroij (1981, p. 74).

60. Mijs (1992, p. 161).

61. Glasbergen (1990, p. 46).

awareness of a society. These structural and cultural characteristics influence the distribution of resources, interests, and values among the network. Furthermore, organizations in the macro-environment may have an incentive or intervention position, and therefore be able to control the crucial resources of the actors in the policy arena, to grant subsidies, or to impose rules of interaction.[62] Apart from the social context, also the natural context may influence network characteristics.[63] In section 2.3.2 it will be argued that for that reason the underlying hydrological structure of a river issue may be an important explanatory variable for decision making on that issue.

2.2.7 Learning

Strategic learning
In the network perspective decision making is conceived of as a *strategic or interactive learning* process as well. During the interactions the actors learn about interdependencies, (problem) perceptions of other actors, score possibilities, and the effectiveness of their strategies. Consequently, their perceptions are changed. Based on the new information, the actors may adapt their objectives and/or strategies. Existing strategies shown to have failed may be excluded from consideration, whilst successful strategies may be built upon.[64] Objectives may be changed if they appeared to be much too ambitious, or in case an actor has learned that these objectives do not contribute to the achievement of a more central objective. Mintzberg describes interaction processes as processes of strategic learning in which actors respond to the behaviour of other actors.[65] During the interactions actors discover new resources, new relations and new objectives, which they often find more interesting than the *a priori* defined ones. Crozier and Friedberg define learning as "[...] *the discovery, creation, and acquisition by the actors concerned of new relational models, new modes of reasoning, and similar collective capacities*".[66] Teisman distinguishes between strategic learning and cognitive learning.[67] Strategic learning is based on the experiences that the actors have gained during the interaction process, whilst cognitive learning is based on an *a priori* rationality according to which alternatives of behaviour are formulated and evaluated.[68]

Cognitive learning
Although the emphasis on strategic or interactive learning certainly is an interesting feature of the network perspective on decision making, *cognitive learning* may take place in decision making as well. Cognitive learning concerns the increase of knowledge on the intellectual relationships in an issue-area, i.e. on cause-effect relationships, policy alternatives, impacts of policy alternatives, or on the intellectual relationships between issues. In the network perspective, however, this type of knowledge has a different function than in the hierarchical

---

62. Mijs (1992, pp. 155-169).

63. Godfroij (1981, p. 74).

64. Mitchell (1995, p. 245).

65. Mintzberg (1987, p. 69).

66. Crozier and Friedberg (1980, p. 221).

67. Teisman (1992, p. 82).

68. Ibid.

perspective on decision making. In the hierarchical perspective this knowledge is used to improve the "policy theory" of a central policy maker, which enables him to develop more effective strategies for the achievement of his objectives. Research improves the rationality of the policies that are developed at a central level. In the network perspective on decision making, research that produces new knowledge on the intellectual relationships in an issue-area influences the perceptions of the actors involved in decision making on that issue, and consequently their objectives and/or strategies.[69]

If actors recognize new intellectual relationships between issues, they may link these issues into packages, called issue-areas (See Section 2.2.5).[70] This type of linkage is called *substantive issue-linkage*[71], and should be distinguished from tactical issue-linkage, which will be discussed in Section 2.3.2.

The influence of new knowledge on the intellectual relations in an issue-area on the perceptions and strategies of the actors involved in decision making in that issue-area may be illustrated with the influence of research on the German strategy in the Rhine river negotiations.[72] In the Rhine basin, the Netherlands are situated downstream of Germany. Because the Dutch faced problems of water and sediment pollution, for example because of the severely polluted water beds in the port of Rotterdam, they aimed at a reduction of the transboundary load of pollutants in the river Rhine. Therefore, the Dutch used offensive interactive strategies, and aimed at a further basin-wide effort to reduce the discharge of pollutants into the Rhine basin than agreed in the Chemical convention of 1976. In first instance, Germany did not see the necessity of the Dutch plans to orient the sanitation efforts to the sediment quality for cost reasons. Later, however, research on the relationship between Rhine river pollution and the pollution of the North Sea indicated that pollutants discharged into the Rhine and other rivers are transported by currents in the North Sea, and also have a negative impact on the water quality of the North Sea along the German coast. In other words, research had clarified the intellectual relationship between Rhine and North Sea pollution issues. By that, it indicated that in the Rhine basin Germany is not only an upstream basin state, but has a downstream position as well. Because the Germans perceived an interest in solving the pollution problems along the German coast, they perceived an interest in a further clean-up of the river Rhine as well. Consequently, they modified their objectives and strategies in the Rhine river negotiations.

This example illustrates that research indeed may have an important impact on a decision making process. Empirical research, however, has shown that it generally takes a period of a decade or more until research results do have an impact on decision making.[73] "*A focus on short-term decision making will underestimate the influence of policy analysis because such research is used primarily to alter the perceptions and conceptual apparatus of policy makers over time.*"[74] This influence of decision making over time often is indicated the

---

69. For the impact of research on policy making, see also Wisserhof (1994).

70. Haas (1980, p. 361).

71. Ibid.

72. The author is grateful to P. Huisman for this interesting example of cognitive learning in decision making. See Huisman (1998) and Dieperink (1997, p. 220).

73. Sabatier (1993, p. 16).

74. Ibid.

'enlightenment function' of scientific research.[75] A corollary of this view is that it is the *cumulative* effect of findings from different studies and from ordinary knowledge that has the greatest influence on policy.[76]

Strategic use of knowledge and joint learning

Actors know that knowledge on the intellectual relationships in an issue-area may influence the perceptions of the other actors involved in decision making in that issue-area, and consequently their objectives and/or strategies. Knowledge is an important resource in a decision making process, and therefore is often used strategically. Learning occurs in the context of a *political* process, which is not a disinterested search for "truth".[77] Sabatier and Jenkins-Smith address the strategic use of policy analysis or technical information in a decision making process in their "advocacy coalition framework".[78] In the following, the strategic use of technical information will be illustrated with the example of transboundary water and sediment pollution. A downstream water management agency and some other actors situated downstream in a river basin (the advocacy coalition A) perceive pollution problems. Therefore, they start research on the seriousness and the causes of the problems they perceive. They may identify several causes, and conclude that the transboundary load of pollutants belongs to the main causes. Because of the research results the downstream parties may propose to develop an action program to reduce the discharge of certain substances into the river. Upstream actors however, who feel themselves aggrieved by the proposed policies and have the resources to do something (the advocacy coalition B) have a number of options. They may[79]:

1. Challenge the validity of the data concerning the seriousness of the problem;
2. Challenge the causal assumptions concerning:
   a. The validity of technical aspects, such as the links between emissions, water quality, and effects on the water beds and the ecosystem;
   b. The efficacy of institutional arrangements that will provide for the necessary changes in behaviour;
3. Attempt to mobilize political opposition to the proposal (to develop an action program) by pointing to costs to themselves and others, that is, by creating or enlarging their coalition.

According to Jenkins-Smith and Sabatier, the original group normally responds to these challenges, thus initiating a political and analytical debate.[80] It may sponsor research to defend evidence for the seriousness of the problem; defend the importance of the causes indicated before, and defend the efficacy of the proposed policies the solve its problems. In the debate, the coalitions may reach an agreement on policies to be implemented. If they do not, and the first coalition still perceives a serious problem, and stresses the importance of the indicated causes, the coalitions may either continue their research efforts, or develop an

---

75. Weiss (1977), cited in Sabatier (1997, p. 16).

76. Sabatier (1993, p. 16).

77. Jenkins-Smith and Sabatier (1993b, p. 45).

78. Ibid.

79. Ibid.

80. Ibid., p. 47.

action program with a strong research component and weak coercion.[81]

Jenkins-Smith and Sabatier also formulate conditions under which a productive analytical debate between members of different coalitions is likely to occur. According to them learning across coalitions is a function of at least three variables[82]:

- the level of conflict;
- the analytical tractability of the issue;
- the presence of a professionalised forum.

If an issue-area is characterized by a high level of conflict learning across coalitions is less likely. First, the higher the level of conflict, the greater the incentives are to commit resources to provide and use analysis to defend ones interests. Secondly, as conflict rises the receptivity to analytical findings that threaten perceived interests declines. Obviously, when the objectives of the coalition are hardly conflicting, no analytical debate may develop at all. Therefore, it is hypothesized that learning across coalitions is most likely when there is an intermediate level of conflict.[83]

Another hypothesis of the advocacy coalition framework is that problems for which accepted quantitative data and theory exists are more conducive to learning than those in which data and theory are generally qualitative, quite subjective, or altogether lacking.[84] Therefore, it is also hypothesized that problems involving natural systems are more conducive to learning than those involving purely social or political systems.[85]

Jenkins-Smith and Sabatier also discuss several analytical fora where issues may be discussed.[86] First, issues may be discussed in so-called open fora, which are characterized by the lack of shared norms of scientific investigation and resolution of competing empirical claims, and its openness for the expression of a wide range of different points of view. An example of an open arena is the legislature, in which floor debates and committee hearings provide ample opportunity for all sides to be heard. Beside these open fora, one may distinguish professionalised fora, which admit participants on the basis of professional training and technical competence.[87] *"Ideally, such a forum would be made up of analysts committed to scientific norms who share common theoretical and empirical presuppositions and could resolve a wide range of analytical disputes."*[88] Examples of professionalised fora are conferences and journals of professional groups, advisory committees, and perhaps interagency technical committees.[89] In the advocacy coalition framework it is hypothesized that learning across coalitions is most likely when there exists a forum that is prestigious enough to force professionals from different coalitions to participate, and that is dominated by professional norms.[90]

---

81. Ibid., p. 46.

82. Ibid., p. 48.

83. Ibid., p. 51.

84. Ibid., p. 52.

85. Ibid.

86. Ibid., p. 53.

87. Ibid.

88. Ibid.

89. Ibid., p. 54.

90. Ibid.

2.2.8 Evaluation of decision making

The use of a pluricentric or network perspective does not only have consequences for the analysis of decision making, but for the evaluation of decision making as well. Although this study focuses on the description and analysis of decision making on international river issues, and does not aim at an evaluation of these processes, this topic is shortly addressed. The main reason for this is that a discussion on evaluation criteria clarifies the pluricentric perspective on decision making, and the action rationality of individual actors involved in decision making. The description and explanation of decision making inevitably involves subjective elements, such as the choice for a theoretical perspective on decision making. Nevertheless, the evaluation of decision making is much more subjective. Different persons may easily disagree on the evaluation criteria. The best and only way to deal with this subjectivity is to make explicit the criteria used in evaluation studies.

The choice of evaluation criteria relates to the perspectives on decision making discussed in Section 2.2.1. In the hierarchical perspective on decision making evaluation criteria are based on official policy objectives formulated by the central decision maker. Evaluation studies focus on goals-achievement, the effectiveness or the efficiency of formulated policies.[91] In the pluricentric perspective on decision making evaluation criteria are less obvious, and subject of extensive debate between policy scientists.[92] Because in this perspective there is not one central decision maker, and decision making is conceived of as a strategic interaction process between multiple actors, there is not one single obvious evaluation criterion. Each actor perceives different problems, has different objectives, and consequently has a different perception of the success of decision making. Some actors may perceive the outcome of decision making as very successful, whilst others perceive it as disastrous. Apart from the problem that each actor uses different criteria to evaluate decision making, there are several other drawbacks of classical evaluation studies.[93] First, official policy objectives are often formulated in very general terms. Secondly, in many cases the official objectives are different from the real objectives of an organization. Thirdly, official objectives sometimes are lacking. A fourth and probably most important drawback of classical goals-achievement studies is that goals are dynamic. In Section 2.3.6 it was argued that decision making may be conceived of as a strategic learning process. In the games that are played an actor may adapt his goals or discover new ones. Finally, in some cases actors try to create a win-win situation by linking decision making on an issue of small interest to them to decision making on an issue of much interest to them (See Section 2.3.2). Because of such linkages new objectives are introduced. Classical evaluation studies mostly do not take into account such 'project-related' or 'project-external' objectives.[94]

A first possibility to deal with the multiplicity of problem perceptions and objectives in decision making is to choose the perspective of one of the participants in a decision making process.[95] Evaluation studies may focus on the contribution decision making makes to the solution of the problems perceived by this actor, or the achievement of his objectives. Secondly, if actors have reached an agreement on a policy objective, one may evaluate the

---

91. See, for example, Bressers (1989, p. 171-172) and Honigh (1985).

92. See, for example, Kensen and Abma (1994), Pröpper (1996).

93. Herweijer (1985, p. 27).

94. Teisman (1992, p. 97).

95. Termeer (1993b, p. 298).

achievement of this common or negotiated objective. Thirdly, the problems involved in tracing the real objectives of actors, the dynamics of objectives, and the existence of project-external objectives have stimulated the development of alternative evaluation criteria, which are based on the process of decision making. A frequently used process criterion is the degree of consensus that is reached among the actors involved in a decision making process. *"[...] real consensus does not imply total agreement. Rather, it means that the parties are generally willing to collaborate because it is in their own interest to do so. Thus their readiness to cooperate does not exclude the possibility that interests will continue to diverge on crucial points".*[96] Fourthly, Teisman introduces institutional criteria for the evaluation of complex decision making.[97] According to this criterion decision making is more successful if the actors have established institutional arrangements to regulate their interactions.[98] Like the traditional evaluation criteria these newly developed criteria are disputed as well. The main criticism on these criteria is that they would neglect the contents of policies, and focus entirely on the decision making process.

Finally, each researcher has the freedom to define his or her own evaluation criteria, and may, for example, evaluate the sustainability of developed policies.

2.3   Characteristics of decision making on international river issues

In this section the characteristics of decision making on international river issues will be discussed. Section 2.3.1 starts with a discussion of the characteristics of *international* decision making. Although international decision making processes in many respects are very similar to *intra*national decision making processes, they may be distinguished from *intra*national decision making in three respects. First, international interactions mostly are characterized by the absence of central authority. Secondly, international decision processes may be conceived of as *multi-level games*, because on the international level the games that are played within the individual states are connected to the international games that are played between states. Thirdly, the actors playing the international decision making game are embedded in a different institutional context. Structural and cultural differences between the basin states may explain partly why the perceptions of representatives of different states are diverging. Section 2.3.2 continues with a discussion of two characteristics of decision making on international *river* issues. A first characteristic concerns the influence of the underlying hydrological structure of a river issue on the (perceived) distribution of interests and resources among the actors involved in decision making on that issue. The diverse upstream-downstream relations in a river basin may cause an asymmetric distribution of interests and resources. Secondly, river issues are natural resources issues. Therefore, they are rather conducive to learning, and environmental values and natural or environmental disasters may have a considerable impact on the course and outcome of decision making. The section concludes with a discussion of institutional arrangements for international river (basin) management.

---

96. Driessen and Vermeulen (1995, p. 172).

97. Teisman (1992, p. 99).

98. Ibid.

2.3.1 Decision making on international issues

Absence of central authority
In the theory on international relations, the absence of central authority is a generally used assumption. *"On the international level, analysts have never taken seriously the notion that external enforcement can resolve collaboration problems. A core assumption of dominant models of international relations, one that seems a sound description of reality, is that no central authority exists."*[99] The absence of central authority, sometimes indicated international anarchy[100], does not imply that international relations are always characterized by equality. Because of an asymmetric distribution of resources among states, some states may be able to exert more influence in an international decision making process than others. *"On the international level, decisions about the institutions to be adopted are not made through a formalized constitutional process. Instead, they result from bargaining among the major players in an issue-area, so that the interests of the most powerful are sure to be reflected in the types of institution chosen."*[101]

Because of the absence of central authority, neither of the states involved in a decision making process does have an intervention position, and is able to impose game rules or policies. The dominant decision making rule in decision making on international issues is unanimity, and international institutional arrangements (and policies) only develop voluntarily. Whether institutional arrangements further develop or erode depends on the interests the actors have in fostering or hindering their development. Therefore, the network perspective on decision making, which focuses on the action rationality of individual actors, and the need for mutual beneficial agreements, and unlike the hierarchical perspective does not rely on hierarchical steering, seems to apply well to cases of international decision making.

In Europe the EC in some cases is an exception to the general rule discussed above. The EC is an supranational organization that has an intervention and incentive position in several policy fields, such as in the field of environmental policies. Nevertheless, one should keep in mind that also EC policies are preceded by a strategic interaction process in which the EU member states have to reach an agreement.

Multi-level games
Decision making on international issues may be conceived of as a *multi-level game*. Representatives of the states are playing the international decision making game. This game is linked to various *intra*national decision making games. The linkages concern agenda setting, the formulation of objectives and strategies to be used in the international decision making game, and the implementation of international agreements.

The players of the international game mostly are representatives of national governments and the national bureaucracies.[102] These actors are able to place issues on the international agenda. Ideas to put an issue on the international agenda, however, mostly do not stem from the national governments or bureaucracies, but from the actors that are affected by an

---

99. Martin (1995, p. 77).

100. Sebenius (1992b, p. 323).

101. Martin (1995, p. 75).

102. Caldwell (1990, p. 13).

international issue, i.e. the actors perceiving   problems.[103] Environmental NGOs, for example, generally play an important role in the agenda setting of environmental issues. Parties that expect benefits from international cooperation on certain issues, have to convince the national government of the need to place these issues on the international agenda, and to start international interactions.

Once an issue is placed on the international agenda, apart from national bureaucracies, also regional and local councils, executives, and agencies (*lower level governments*), and NGOs may try to influence *intra*national decision making on the objectives and strategies to be used in the international decision making game. These actors do mostly perceive different problems, and therefore propose different strategies. Generally, states use one set of strategies aiming at the achievement of several national objectives. These objectives and strategies are the result of an *intra*national decision making game that is played between the national government, national bureaucracies, lower level governments, and NGOs.

Finally, the implementation of international policies is a national concern. The implementation may be conceived of as an *intra*national decision making game. The resources that are needed for the implementation of international policies mostly are distributed among national bureaucracies, lower level governments, and NGOs. If an actor possessing indispensable resources opposes the international agreement, and refuses to cooperate on the implementation of that agreement, a so-called *implementation gap* may emerge. In such cases provisions for monitoring and enforcement may be helpful.

Although the conceptualization of international decision making as a two-level game is analytically attractive, and may contribute to a better understanding of decision making on international issues, it is a rather simplified model of real life decision making. The main reason for this is that not only national governments and bureaucracies interact with each other, but that lower level governments and NGOs establish cross-border contacts as well. Haas states that: "*the numbers and types of participating actors are greater than ever. In addition to foreign and economic ministries, almost every agency of modern government has a stake in some aspect of international relations and maintains direct contacts- often by-passing the foreign ministry- with its opposite numbers.*"[104], and: "*More non-governmental groups of all kinds maintain continuous contact with their counterparts elsewhere, and seek to shape the foreign policies of their home countries. In short, the channels of international communication are more numerous, decentralized, and diverse than ever.*"[105]

Structural and cultural differences
A third characteristic of international decision making is that the actors involved in the decision making process are embedded in a different institutional context. The states involved may have rather different (decision making) cultures and structures. Although cultural differences may be present in *intra*national interaction processes as well, for example because there are different organization cultures, in an international context these differences mostly are more predominant. Culture can be defined as; "*the social heritage of a society that is transmitted to each generation; learned behaviour that is shared with others*"[106], or as: "*the*

---

103. Caldwell (1988, p. 17).

104. Haas (1980, p. 357).

105. Ibid.

106. Champion *et al.* (1984, p.31).

collective programming of the mind which distinguishes the members of one group or category of people from another."[107] The cultural and structural characteristics of a state or society cannot always be meaningful distinguished, and the causal relationships between the two often are not clear. On the one hand a decision making culture influences the organization of decision making, whilst on the other hand a decision making structure may influence the decision making culture. The problem of which one comes first, which has initially produced the other, is a chicken-and-egg debate without definite answer.[108] For example, a state may have a hierarchical culture accompanied with a hierarchical organization of decision making.

Culture may influence the perceptions of actors, their strategic behaviour, and consequently the course and outcome of a decision making process. Faure and Rubin, who studied the influence of the cultural factor in several cases of international negotiations, concluded that culture is not "[...] *the only explanation of outcome, nor necessarily the determining element of the process. Rather the diverse cases invite the observation that any reasonable explanation of what happens in international negotiation must include the cultural aspects of the negotiation relationship*"[109]

Structural and cultural differences between states may influence decision making in several ways. First, these differences may complicate the international interactions. Because of structural differences, such as a different division of tasks among levels of government or different planning systems, actors may face difficulties to recognize interdependencies, and to find out which actors they need for the solution of their problems or the achievement of their objectives. Secondly, cultural differences may cause misunderstanding and misinterpretations of the behaviour of others, and therefore lead to ineffective strategic behaviour. A rather obvious example concerns the use of a different language, but representatives from different states may have different negotiation styles as well. Thirdly, different decision making cultures may explain different preferences concerning the organization of international decision making processes. Finally, structural and cultural differences may influence international decision making indirectly, because they influence the *intra*national decision making games discussed in the previous section. As an example, differences between environmental awareness and economic prosperity influence decision making on the objectives and strategies to be used in decision making on international environmental issues.

2.3.2 Decision making on international river issues

The underlying hydrological structure of international water management issues
Before discussing types of international water management issues, and their underlying hydrological structure, several types of international waters will be introduced first. International waters may be distinguished into international surface waters and international groundwater basins (aquifers). International surface waters comprise international seas, estuaries, rivers, and lakes. Finally, international rivers may be distinguished into *successive*

---

107. Hofstede (1991, p. 5).

108. Faure and Rubin (1993, p. 223).

109. Ibid., p. 212.

or *transboundary rivers*, and *contiguous* or *boundary rivers*.[110] The former type of river crosses the border between two or more states, whilst the latter type forms the border between two or more states. Many rivers are both a boundary and a transboundary river, because some sections of these rivers form the border between states, whilst other sections cross national borders.[111] As regards the geographical situation in a (transboundary or successive) river basin, Frey distinguishes between three absolute positions: upstream, midstream and downstream basin states.[112] One should, however, bear in mind that there are only two relative geographical positions, namely upstream and downstream positions.

The underlying hydrological structure of an international water management issue influences the (perceived) distribution of interests and resources among the actors involved in decision making on that issue, and consequently their perceptions of (inter)dependence. The underlying hydrological structure of international water management issues may create either rather symmetric or asymmetric relations between the actors.[113]

Examples of symmetric issues are pollution issues in an international lake or boundary river, or the joint exploitation of an international groundwater basin. If two states share an international lake or boundary river, and both discharge waste water in the shared water body, they may both perceive an interest in pollution reduction, whilst they do both have problem solving capacity. Similarly, if they share an international groundwater basin, and both extract water for drinking water production, they may both perceive a problem of depletion, whilst they do both have problem solving capacity. In these cases, the basin states are mutually dependent for the solution of their problems. Such problems are mostly called collective action or commons problems.[114]

Many international water management issues, however, are characterized by asymmetry. Some examples concern water pollution, water distribution, and flow regulation in transboundary river basins. Upstream basin states may discharge large amounts of pollutants into the river, which causes problems related to water and sediment pollution in the downstream parts of the river basin. Furthermore, they may divert water from the river thus creating water shortages in the downstream parts during periods of low flow. Finally, they may carry out canalization projects thus increasing the flood risks in the downstream parts of the basin. In these cases, the actors situated upstream do possess the resources that are

---

110. LeMarquand (1977, p. 7). Others use other terms or classifications. Frey (1993, p. 55) distinguishes between *international* and *transnational* river systems. The former category comprise all rivers that form the border between two or more nations. The latter category comprises the rivers that flow across international borders, thus creating upstream-downstream riparian states. Prescott (1987) cited in Van de Bremen (1992) distinguishes, beside boundary waters and successive waters, tributaries of boundary waters.

111. LeMarquand (1977, p. 7) gives the example of the river Rhine. This river begins in Switzerland; a short distance downstream its sources it forms the boundary between Liechtenstein and Switzerland and then between Austria and Switzerland before it enters into Lake Constance. The lake is divided between those two countries and Germany. Below the lake the river roughly follows the German-Swiss border to Basel, in some places forming the border, in others cutting across Swiss territory. From Basel the river forms the French-German border to Karlsruhe where for the next 550 km the river becomes entirely German. Finally the river crosses the Dutch border and flows through the Netherlands for its final few miles before it reaches the North Sea.

112. Frey (1993, p. 61).

113. LeMarquand (1977, pp. 8-10).

114. Three influential models of this type of problems are Hardin's tragedy of the commons (Hardin, 1968), Olson's logic of collective action (Olson, 1965), and the Prisoners Dilemma game of formal game theory (Axelrod, 1990, p. 15).

necessary for the solution of the problems perceived by the actors situated downstream in the basin, but may not perceive an interest in the solution of these problems at all. For other river issues, such as navigation or fish migration, there may be an opposite power asymmetry.[115] An upstream basin state may depend on a downstream basin state as regards the management or improvement of a navigation channel. Furthermore, dams or barrages constructed in the downstream parts of a basin may hinder fish migration, and therefore fish production in the upstream parts of the basin as well. In these cases, the actors situated downstream do possess the resources that are necessary for the solution of the problems perceived by the actors situated upstream in the basin. A water management issue characterized by an asymmetric distribution of interests and resources among the actors involved in decision making on that issue, which is caused by upstream-downstream relationships, is called an *upstream-downstream issue*. International upstream-downstream issues may be present in transboundary or successive river basins.

As regards the influence of the underlying hydrological structure of international water management issues, four additional remarks have to be made. First, it should be noted that in one international river basin several issues may be present simultaneously, each of them characterized by a different (perceived) distribution of interests and resources among the basin states. In the Rhine basin, for example, the downstream basin state, the Netherlands, perceives a dependence on the upstream basin states as regards the issues of water and sediment pollution. As regards the issue of fish migration, however, the upstream basin states and regions perceive a dependence on the Netherlands as well. Secondly, two states may share more than one international water resource. In one of these basins the basin state A may be situated upstream of the basin state B, whilst in the other basin, basin state A may be situated downstream of basin state B. Thirdly, the distribution of resources between the basin states does not only depend on the underlying hydrological structure of the international water management issues at stake, but on the general distribution of resources between them. Beside international water management issues, there may be other international issues at stake, either characterized by power asymmetries or not. Furthermore, one of the basin states may have sufficient political, economic or military resources to impose its policy proposals. A fourth, and probably most important remark is that one should bear in mind that in the end the (inter) dependencies between the actors involved in decision making on an international water management issue are *perceived* (inter)dependencies. As an example, the government of an upstream basin state A discharging pollutants into a river may perceive an interest in the water quality in the part of the basin situated in the upstream basin state only, but may perceive an interest in the water quality of the entire basin and the sea into which the river flows as well. In the former case, it may perceive no dependence on the downstream basin state at all, whilst in the latter case it may perceive a dependence on the governments of the downstream basin states. This example illustrates that the perceptions of the actors are related to their (environmental) values. One may even argue that if the upstream basin state perceives a dependence on the downstream basin states, because it perceives an interest in a clean-up of the entire basin, it perceives a commons problem.

Decision making on international upstream-downstream issues: compensating potential losers
Many international river issues are upstream-downstream issues. To solve these issues, cooperation between the upstream and downstream actors mostly is necessary. In Section

---

115. Linnerooth (1990, pp. 629-660).

2.2.4 three inducements to cooperation were distinguished. First, actors may have the same objectives or face the same problems. Secondly, they may start cooperation if they perceive possibilities of exchange, which could make both parties better off with an exchange agreement than without one, i.e. when they are able to create a win-win situation. Thirdly, authority may be an inducement to cooperation. In decision making on upstream-downstream issues, the first inducement to cooperation is absent, because actors do not have similar objectives or perceive the same problems. Whilst some parties may win from an international agreement, others stand to lose from that agreement. Because in an international context central authority mostly is absent, and the dominant decision rule is unanimity, the last inducement to cooperation mostly is lacking as well. The actors perceiving upstream-downstream problems may try to influence the perceptions of the actors having problem solving capacity. These strategies, however, may not suffice or only be effective in the long run. In these cases the party perceiving an interest in the establishment of international cooperation faces the challenge to create a win-win situation. Two possibilities to compensate potential losers are to make tactical linkages or to compensate financially.

As argued in the preceding section, the (inter)dependencies between actors are not limited to a specific issue-area. In different issue-areas actors may perceive different interdependencies. Sometimes, a more comprehensive approach taking into account the issues of more than one issue-area can contribute to international agreement and the establishment of cooperation on upstream-downstream issues. The heterogeneity of preferences of actors in different issue-areas enable them to gain from trade.[116] As Sebenius argues: *"though many people instinctively seek "common ground" and believe that "differences divide us", it is often precisely the differences among negotiators that constitute the raw material for creating value."*[117] For the creation of potential value in cases of heterogeneity the negotiators have to go beyond the boundaries of a specific issue-area, and have to introduce issues to the agenda that are not connected by any intellectual coherence at all. This type of linkage is called *tactical issue-linkage*, and should be distinguished from substantive issue-linkage, which was discussed in Section 2.2.7. Whilst substantive linkages are based on the intellectual coherence between issues, tactical linkages are based on their strategic coherence. Unless indicated otherwise, in this thesis the concept issue-linkage means tactical issue-linkage. Whilst within the scope of individual issues heterogeneity can create conflict of interests that reduce available gains from cooperation, heterogeneity may create opportunities for gains from exchange across issue-areas, thus enhancing the scope and potential of cooperative arrangements.[118] By linking issue-areas the actors may prepare *package deals* or exchange agreements that are beneficial to all of them.

Martin argues that the chance that movement away from the status quo will involve issue-linkage depends on the distribution of preference intensities across issue-areas and an institutional characteristic, namely the decision rule. According to her, in cases where decision making is based on unanimity, and the actors have different preference intensities in different issue-areas the probability that movement away from the status quo will involve tactical issue-linkage is high (See Figure 2.2).[119] In these cases issue-linkage becomes one of

---

116. Martin (1995, p. 81).

117. Sebenius (1992a, p. 29).

118. Martin (1995, p. 81).

119. Ibid.

the key elements in understanding international cooperation.[120]

Decision-making rule

	Majoritarian	*Unanimity*
Homogeneous	very low	low
Heterogeneous	low	high

Figure 2.2 *Probability that movement away from status quo will involve issue-linkage* (Martin, 1995, p. 87)

As an example, an agreement with a neighbouring country on an international river scheme that country wants may be used to gain concessions for other bilateral river or non-river issues.[121]

Individual actors face temptations to renege on cross-issue deals, hoping to achieve concessions on those issues of most intense interest to themselves and then back down from commitments of more benefit to their bargaining partners.[122] Institutional arrangements, such as provisions for monitoring and enforcement, may reduce temptations to renege, thus solidifying issue linkages and encouraging heterogeneous states to cooperate with one another.[123] Beside these formal arrangements, there is another important social mechanism that may stimulate cooperative behaviour, namely the *shadow of the future*, which concerns the likelihood that actors will need each other in the future to solve their problems or achieve their objectives. The shadow of the future may explain why an actor is willing to build up a "reservoir of goodwill" to draw on when he needs support or a concession from his neighbour for a policy of greater national interest.[124] Axelrod argues that mutual cooperation can be stable if the future is sufficiently important relative to the present.[125] The actors can each use an implicit threat of retaliation against the other's defection - if the interaction will last long enough to make the threat effective. "*Prolonged interaction allows patterns of cooperation which are based on reciprocity to be worth trying and allows them to become established. Making sure that defection on the present move is not too tempting relative to the whole future course of the interaction is a good way to promote cooperation.*"[126] According to Axelrod, there are two basic ways to enlarge the shadow of the future[127]: (1) to make the interactions more durable, and (2) to make them more frequent.

---

120. Ibid. See also Marty (1997, p. 58).

121. LeMarquand (1977, p. 13).

122. Martin (1995, pp. 83-91).

123. Ibid.

124. LeMarquand (1977, p. 13)

125. Axelrod (1990, p. 108).

126. Ibid.

127. Ibid., p. 111.

Beside issue-linkage there is another possibility to compensate potential losers from an international agreement, namely to compensate them financially. Actors may be willing to cooperate on an issue, although they do not perceive a direct interest in that issue, if they are compensated financially. As regards pollution issues, however, such compensations are controversial, because the polluttee paying the polluter for stopping or reducing pollution clearly contravenes the polluter pays principle.[128]

River issues as natural resources issues
One of the hypotheses of the advocacy coalition framework is that problems involving natural systems are more conducive to learning than those involving purely social or political systems (See Section 2.2.7). Because river issues are natural resources issues, one may expect cognitive learning to play an important role in decision making. Because of the increased awareness of the diverse intellectual relationships in a river basin, water management experts have introduced the steering concept *river basin approach*, which emphasizes that river basins should be managed as a coherent unity (See also Section 1.1.3). These relationships may concern relations between:
- water quality and water quantity
- groundwater and surface water
- water and sediment quality
- water and water bed quality
- the quality of water and water beds and the ecology
- land-use and water
- upstream and downstream parts of a basin

Because of the increased knowledge and awareness of these relationships, the relationships between policy fields, such as land-use planning, water management, and environmental management are increasingly addressed as well.

Beside the relative importance of cognitive learning, there is another main characteristic of decision making on natural resources issues. The attitudes toward the environment or environmental values of the actors involved in decision making on these issues may be important explanatory variables for the course and outcome of the decision making process. In addition, principles of international environmental law, such as the polluter pays principle or the precautionary principle may influence the perceptions and strategies of the actors involved in decision making on international river issues. Because most river issues are environmental issues 'sustainable development' is an influential steering concept in international river basin management (See also Section 1.1.3).

The impact of natural and environmental disasters
Natural or environmental disasters may have a considerable impact on decision making on international river issues. Two important examples concern the impact of the fire in the Swiss chemical enterprise Sandoz on the development of institutional arrangements and policies to combat Rhine river pollution, and the impact of the river floods in the Meuse and Rhine of 1993 and 1995 on the development of institutional arrangements and policies aiming at flood protection and alleviation in the basins of these rivers.

A theoretical model that applies well to these cases is Kingdon's theory of policy

---

128. Marty (1997, p. 57).

windows.[129] This theory, which focuses on agenda setting processes, is based on in depth empirical research on various cases of decision making. On the basis of the research results Kingdon developed an alternative to the classical stages model of decision making (See Section 2.2.1). He distinguishes three relatively independent streams or processes in decision making: the problem stream, the policy stream, and the political stream. The problem stream concerns the recognition, expression, and formulation of problems that would have to be solved by public policy. In this process, pointing out problems, triggers, and feed back loops of policies play an important role.[130] The policy stream, which is dominated by scientists and professionals, relates to the development of policy alternatives to solve policy problems. In this process important selection criteria are the technical feasibility, the extent to which the solution matches prevailing values, and the extent to which policy makers expect problems with political decision making.[131] Finally, the political stream concerns political developments, such as the election of a new government, personnel changes or reorganizations of government agencies, and changes of the public opinion. According to Kingdon, only when a problem is linked to a policy alternative, and there is political support for that problem, i.e. when the three streams are linked, there is an opportunity to get political attention for that problem or policy alternative, and to place that problem or policy alternative on the political agenda. Such an opportunity is called a *policy window* or *window of opportunity*.

The theory of policy windows explains why it may take a long time until research results have an impact on decision making (See Section 2.2.7). Research that produces new (knowledge on) policy alternatives is part of the policy stream. Only if this stream is linked to the other streams, and a policy window occurs, the research results may actually influence decision making.

The impact of the fire in the Swiss chemical enterprise Sandoz on the development of institutional arrangements and policies to combat Rhine river pollution is a perfect illustration of Kingdon's theory of policy windows. Many parties had been aware of the problems related to Rhine river pollution since long. Furthermore, scientists and governmental experts knew very well which policy measures would have to be taken to solve these problems. The Rhine river pollution, however, had a relative low priority on the agendas of the basin states. The Sandoz-fire was a turning point in the history of the Rhine river cooperation, and a main trigger for decision making on the Rhine pollution issues.[132] The enormous negative impact of the accidental discharge of pollutants on the ecology of the river caused a wave of publicity, and increased awareness of the severity of the problems caused by the Rhine river pollution. The publicity and increased problem awareness lead up to attention at the highest political level for the pollution problems. The Sandoz-fire created a policy-window, which enabled professionals to present their policy proposals to combat Rhine river pollution, and to get political support for these proposals. Therefore, the Sandoz-fire is generally considered to be one of the most if not the most important explanatory variable for the rapid development of water quality policies in the Rhine basin states after 1986, and the development of the Rhine Action Program.

---

129. Kingdon (1984). For a discussion of Kingdon's theory of agenda setting, see also Edelenbos and Van Twist (1997, pp. 28-31) and Hendriks and Meijerink (1996).

130. Kingdon (1984, pp. 75-81).

131. Ibid., pp. 138-151.

132. Dieperink (1997, p. 328).

The second example concerns the impact of the floods in the Meuse and the Rhine basins in 1993 and 1995.[133] In these years the water levels in large parts of the basins rose dramatically, and extensive areas were flooded. Although no expert had been able to forecast the occurrence of such high water levels, they knew about the considerable flood risks in some parts of the two basins. Furthermore, they knew that for a reduction of the flood risks, dikes would have to be strengthened or partially moved landward, and that for flood alleviation international collaborative action would be necessary, for example to prevent or undo canalization projects, or to create flooding areas in the upstream parts of the basin. Unlike water quality issues, however, the Rhine and Meuse water quantity issues had a rather low priority on the political agendas, the implementation of policies to strengthen the dikes went laboriously, and international collaborative action for flood alleviation was absent. The two river floods were a main trigger for decision making on flood protection and alleviation. The increased awareness of the flood risks in the basins, and the dramatic consequences of high waters and river floods, lead to political attention for that issue. Because of the occurrence of a policy window, experts were able to present their policy proposals, and to get these proposals accepted. In the Netherlands, a Delta Act for the large rivers, which provides for a rapid implementation of programs to strengthen the dikes along some river sections in the Netherlands, was quickly adopted by the parliament.[134] Furthermore, the events induced the development of international institutional arrangements to discuss the water quantity issues in the Rhine and the Meuse.[135]

Institutional arrangements for international river (basin) management
In Section 2.2.6 it was argued that the actors playing the decision making game may develop institutional arrangements to regulate their interactions, i.e. to regulate their cooperation and to prevent conflicts. In this section several institutional arrangements that may be developed in decision making on international river issues are shortly discussed.
- *Arrangements for the exchange of information* The basin states may inform each other on the developments in the water system on their respective territories, and on the planned policies for these water systems. This exchange of information may lead to a process of mutual adjustment, which implies that the basin states take into account the preferences and (planned) activities of the other basin states. The exchange of information may either have an ad-hoc or a more structural character. In the former case one may think of symposia. In the latter case the basin states may establish a project-organization.
- *Arrangements for joint research* The basin states may develop a project-organization or commission to carry out joint research. Joint research may stimulate agreement on the cognitive aspects of the issues at stake.
- *Arrangements for joint planning* The basin states may develop a commission and procedures for joint planning in an international river (basin). These plans may contain objectives and the terms within which these objectives have to be achieved, and/or policies that have to be implemented.
- *Arrangements for joint or coordinated monitoring* The basin states may develop

---

133. Wessel (1994a and b), Waterloopkundig laboratorium (1996), V&W (1997, pp. 59-63).

134. Ministerie van Verkeer en Waterstaat, Directoraat-Generaal Rijkswaterstaat (1995).

135. Declaration of Arles. By the environment ministers of France, Germany, Belgium, Luxembourg and the Netherlands on tackling the problems caused by the high water level of the Rhine and the Meuse, Arles, 4 February 1995.

procedures and methods for the joint monitoring of a shared water system. Joint monitoring enables the basin states to evaluate and control the implementation of joint policies.

- *Arrangements for the implementation of joint policies* The basin states may develop a project-organization or commission for the implementation of joint policy-programs.
- *Arrangements for the evaluation of implemented policies* The basin states may develop procedures to be followed for the evaluation of planned policies.
- *Warning systems for cases of accidental pollution, and high or low water levels* The basin states may develop warning systems for cases of accidental pollution and/or periods with a high or low river flow.
- *Arrangements for conflict resolution/ dispute settlement* The basin states may develop procedures that will be followed if they are unable to settle a dispute themselves.
- *Arrangements for enforcement* The basin states may develop procedures or methods for the enforcement of joint policies.

These arrangements may have an informal or legally binding character. In many cases the basin states conclude international treaties or conventions, which provide for one or more of the institutional arrangements discussed above.

2.4 A conceptual framework for the analysis and explanation of decision making on international river issues

Figure 2.3 presents a conceptual framework for the analysis and explanation of decision making on international river issues. It draws on the theory of policy networks treated in Section 2.2, and the discussion of the characteristics of decision making on international river issues in Section 2.3. The framework is a simplified representation of real-life decision making, and highlights several important relationships that may be distinguished in a decision making process. In the following sections, these relationships will be discussed shortly.

Relationship 1: Perceptions and strategies
A main point of departure of the conceptualization is the action rationality of the individual participants in the decision making process. Therefore, the perceptions of these actors are at the core of the model. As explained in Section 2.2.5 these perceptions concern:

- the interdependencies between the actors arising from the division of the resources needed to fulfil their ambitions;
- the ambitions, and 'stakes in the game' of the actors;
- the policy game and the importance of the policy problem being addressed in the game, compared with the policy problems being addressed in other games.

The players of the decision making process may be the governments of the basin states, delegations of these governments, ministries, lower level governments or NGOs. In the international decision making game the national governments and ministries mostly play a dominant role. Because of the multi-level character of decision making on international river issues, however, lower level governments and NGOs may have a considerable influence on decision making on international river issues as well.

The perceptions of the actors involved in decision making on an international river issue influence their strategic behaviour. In Section 2.2.3 several types of strategies that may be used in a decision making process were discussed.

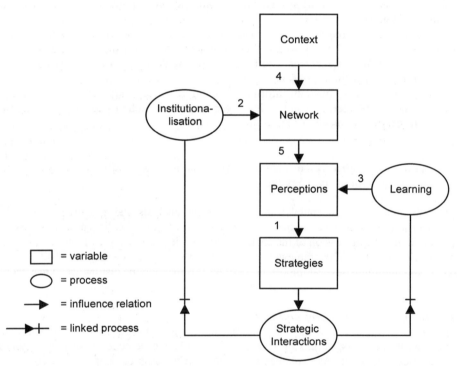

Figure 2.3 *Conceptual framework for the analysis and explanation of decision making on international river issues* The perceptions of the actors involved in decision making on an international river issue influence their strategic behaviour (Relationship 1). In the games that are played the structural and cultural network characteristics may change (Relationship 2). During the strategic interactions strategic and cognitive learning processes may develop as well (Relationship 3). Finally, the structural and cultural characteristics of the context influence the structural and cultural characteristics of the network (Relationship 4), which in turn influence the perceptions of the actors (Relationship 5).

Relationship 2: Strategic interaction and institutionalization
The interacting actors form the policy arena, and play the decision making game. Sometimes this game may be characterized more by conflict, and sometimes more by cooperation. Most interactor relationships do have elements of conflict and elements of cooperation simultaneously.

During the interactions institutionalization processes may develop. First, the actors may reach an agreement on the development of international institutional arrangements to regulate their interactions. In Section 2.3.2, some types of institutional arrangements in river basin management were discussed shortly. Secondly, the actors may reach an agreement on international policies. Policies comprise policy objectives and/or policy measures concerning an issue. The basin states of an international river may, for example, reach an agreement on water quality objectives for an international watercourse, or the joint implementation of a concrete policy program. Thirdly, in addition to these formal rules of behaviour, informal rules of behaviour may change. As an example, a continuous unwillingness of certain actors to cooperate may give these actors bad reputations, and cause a relationship characterized by distrust.

Relationship 3: Strategic interaction and learning

During the interactions learning processes may develop as well. In Section 2.2.7 two types of learning were distinguished: strategic or interactive and cognitive learning. Strategic learning takes place if during the interactions one or more actors gain information on the (inter) dependencies and problem perceptions of other actors involved in a decision making game, discover score possibilities, or possible win-win situations. Cognitive learning takes place if one or more actors gain knowledge on the intellectual relationships in an issue-area, i.e on cause-effect relationships, policy alternatives, impacts of policy alternatives, or on intellectual relationships between issues. Learning processes imply a change of the perceptions of actors, and therefore may lead to a modification of their objectives and/or strategies.

Relationships 4 and 5: Context, network, and perceptions

The fourth and fifth relationships concern the influence of structural and cultural characteristics of the context on the structural and cultural characteristics of the network, and of these network characteristics on the perceptions of the actors involved in decision making. The following interrelated dimensions of the context of decision making on international river issues can be distinguished:

- *the intranational context* The probably most important dimension of the context of decision making on international river issues is the *intra*national context. Structural characteristics of the basin states, such as their economic prosperity, the state organization, and the *intra*national institutional arrangements and policies to deal with the issue at stake influence their perceptions.
- *the international context* The basin states involved in decision making on an international river issue mostly are involved in decision making on other bilateral or multilateral (river) issues as well. The distribution of interests and resources in other issue-areas influences the perceptions of (inter) dependence of the basin states. Furthermore, the basin states may be involved in decision making on more generic policies that apply to the issue at stake. In the case of a European river basin, for example, the basin states may be involved in decision making on the EU-level.
- *the underlying hydrological structure of the issue at stake* In Section 2.3.2 it was argued that the underlying hydrological structure of a river issue, such as an upstream-downstream relationship, influences the (perceived) distribution of interests and resources among the actors, and consequently their perceptions of (inter)dependence.
- *the cultural context* Cultural characteristics, such as the environmental awareness in the basin states, or the decision making cultures, may influence the perceptions of the actors

involved in the decision making process
- *the historical context* The final contextual dimension distinguished in this research is the historical context of the decision making process. The history of conflict and cooperation between the actors may shape reputations, and consequently mutual expectations (See Section 2.2.6).

Whilst the first two dimensions of the context, i.e. the *intra*national and international context may be highly dynamic, the third and fourth ones, i.e. the underlying hydrological structure of the issues at stake and cultural context, mostly are very stable. The historical context of a decision making process is both stable and dynamic. On the one hand it is stable because no actor is able to change what happened in the past. On the other hand it is a rather dynamic dimension, because in the succeeding games new history of conflict and cooperation is produced continuously.

CHAPTER 3

CASE STUDY DESIGN

3.1 Introduction

This chapter contains the research design for the case study presented in Part II of the thesis. First, in Section 3.2 some difficulties of developing predictive theories of decision making are discussed shortly. Subsequently, in Section 3.3 the rationality of the use of a case study approach is discussed, and the selection and delineation of the case study are clarified. In Section 3.4 the problem statement formulated in Chapter 1 is elaborated, and using the theory treated in Chapter 2 seven descriptive, explanatory, and prescriptive research questions are formulated. Section 3.5 discusses the methods of data collection, and the main data sources consulted. Section 3.6 concludes this chapter with a guide for the reader of the case study reports presented in Part II.

3.2 Toward a predictive theory of decision making on international river issues?

An important issue in social science concerns the (im)possibility of the prediction of social phenomena, such as decision making. Prediction should be based on a predictive theory. Two types of predictive theories that may be distinguished are deterministic theories and probabilistic theories. Deterministic theories consist of hypotheses of the type "A causes B". Among social scientists there is a broad consensus that the development of deterministic theories of human behaviour is hardly possible. A main reason for this is that actors do have a so-called 'margin of liberty', which refers to the fact that "*human behaviour is always the expression and consequence of freedom, no matter how minimal that freedom may be. It reflects choices made by the actor in order to take advantage of available opportunities within the framework of constraints imposed upon him. For that reason, behaviour is never predictable, since it is not determined but, on the contrary, always contingent.*"[1] "*When observations are made of the actor in his actual situation, it appears that he does not base his behaviour on any sort of balance sheet of investments and returns, but rather on what opportunities he is able to discern in the situation and his capacity to comprehend them.*"[2] Because of the freedom of actors to choose between

---

1. Crozier and Friedberg (1980, p. 19).

2. Ibid., p. 21.

alternative courses of action, their behaviour is never determined. Consequently, prediction of human behaviour is fundamentally limited.

Because of the impossibility to develop deterministic theories, social research mostly aims at the development of *probabilistic* theories. Section 2.3.2 contained an interesting example of a probabilistic hypothesis. There, it was argued that in cases where actors do have different preference intensities in different issue-areas, and the decision rule is unanimity, the *probability* that movement away from the status quo will involve issue-linkage is high. This is not a deterministic hypothesis stating that if the conditions A and B are fulfilled, event C will occur, but a probabilistic hypothesis stating that if A and B are fulfilled the *probability* is high that event C will occur. Probabilistic hypotheses have the form: "A increases the chance on B". Most probabilistic theories do not apply generally, but to well defined situations only. These probabilistic theories also are indicated 'theories of the middle range.'

The (im)possibility of developing predictive theories of complex decision making is subject of debate among scientists in the field of public administration. Some of them do hardly believe in the possibility to develop predictive theories[3], whilst others are of the opinion that the development of such theories is possible, and is a *conditio sine qua non* for increasing scientific knowledge.[4]

Generally, one may contribute to the development of predictive theories in two different ways. First, one may formulate hypotheses on the basis of available theories, and subsequently try to test these hypotheses. This approach is called a deductive approach. The alternative is to start with empirical observations, and to formulate hypotheses on the basis of observed regularities, which is called an inductive approach. Most research, however, starts with some theoretical notions at least, and only few research starts without a quick reflection on the social phenomenon to be studied. Therefore, in most research a combination of these two approaches is used.

The research at hand has a rather inductive character. It concerns an in depth case study of strategic behaviour and interaction, and aims at the formulation of hypotheses concerning the course of a specific decision making process. First, the strategic behaviour of the actors involved in that decision making process is reconstructed. Subsequently, these strategies are related to their perceptions. Finally, it is tried to explain the perceptions and strategic interactions. Although the hypotheses formulated are largely case specific, some regularities found may apply to other cases as well, and therefore may contribute to the development of a theory of decision making on international river issues. The advantage of using a more inductive approach is that case specific circumstances can be taken into account, and that possibly relevant explanatory variables are not excluded beforehand. The disadvantage of such an approach, however, is that it is more difficult to relate the research results to one specific theory, and therefore to contribute to the further development of that specific theory.

---

3. Crozier and Friedberg (1980), Teisman (1992), Klijn and Koppenjan (1997).

4. Pröpper (1996).

3.3   A case study of decision making on international Scheldt issues

3.3.1 A case study strategy
A case study is an empirical inquiry that
*   investigates a contemporary phenomenon within its real-life context, especially when
*   the boundaries between phenomenon and context are not clearly evident.[5]

According to Yin the major rational for using this method is "[...] *when your investigation must cover both a particular phenomenon and the context within which the phenomenon is occurring, either because (a) the context is hypothesized to contain important explanatory variables about the phenomenon or (b) the boundaries between phenomenon and context are not clearly evident.*"[6] In this research the phenomenon studied is decision making on international river issues. The context of this phenomenon, among other things, concerns the structural and cultural characteristics of the societies of the basin states, and the (inter)national networks shaped around these issues. These variables influence the perceptions of the actors involved in the decision making process, and consequently their strategic behaviour. In addition, the boundary between context and phenomenon is not clearly evident. Structural and cultural network characteristics form the context of the strategic interactions between the actors involved in a decision making process. These characteristics, however, do also change because of the interactions between the actors, and therefore can hardly be distinguished from the studied phenomenon "decision making" (See Section 2.2.6). For these two reasons a case study approach seems to be suitable to answer the problem statement formulated in Chapter 1.

3.3.2 A single case study covering a long time span
The case study of decision making on international Scheldt issues is a so-called single case study. It covers a time span of about 30 years, and addresses the recent decision making on international Scheldt issues as well as decision making in the past. The period studied starts in 1967, when the Dutch government decides to link Scheldt infrastructure issues to Meuse water quality and quantity issues, and ends in 1997, when the provisionally installed International Commission for the Protection of the Scheldt has started with the preparation of the first Scheldt Action Programme.

In Chapter 4, on the basis of the intellectual coherence between the international issues, two major issue-areas in the Scheldt basin are distinguished: the issue-area of water and sediment pollution, and the issue-area of the maritime access to the port of Antwerp. The two issue-areas distinguished show a strategic coherence, which is caused by the many tactical linkages made between them. Decision making on the water quality of the Scheldt cannot be understood without an understanding of the decision making on infrastructure projects improving the maritime access to the port of Antwerp, and vice versa. For that reason, decision making in the two issue-areas is presented as one case study. In addition to the many tactical linkages between international Scheldt issues, several tactical linkages were made between international Scheldt issues and international Meuse issues, and even non-river issues. Therefore, the case study addresses decision making on these other issues as well.

The Scheldt case covers the Belgian-Dutch negotiations on the so-called water

---

5. Yin (1994, p. 13).

6. Ibid.

conventions, notorious for their length and deadlocks. It, however, also comprises the conclusion of several international conventions, the establishment of international river commissions for the Scheldt and the Meuse, and a gradually increasing willingness to cooperate internationally of all basin states and regions.

Most international Scheldt and Meuse issues discussed are typical upstream-downstream issues. Although possibilities to generalize the case study results by definition are limited, both because the study is a *single case study*, and because of the difficulties of developing predictive theories of decision making discussed in Section 3.2, some case study results may be relevant for decision making on upstream-downstream issues in other international river basins as well.

3.4   Elaboration of the problem statement and formulation of research questions

In this section the problem statement formulated in Chapter 1 is elaborated. Using the theory treated in Chapter 2 seven descriptive, explanatory, and prescriptive research questions are formulated. The threefold problem statement of the research was:

- How does decision making on international river issues develop?
- How can this development be explained?
- Which strategies can be used to contribute to the solution of upstream-downstream problems in international river basins?

In the Sections 3.4.1, 3.4.2, and 3.4.3 successively the descriptive, explanatory, and prescriptive research questions are formulated.

3.4.1 Descriptive research questions
The first research question resembles the first component of the problem statement, but is specified for the case of decision making on international Scheldt issues:

1.   *How did decision making on international Scheldt issues develop?*

Using the rounds model of decision making treated in Section 2.2.4 three procedural steps can be distinguished.

First, the decisions taken by the various actors involved in decision making on an issue have to be inventoried. The inventory of decisions is based on an analysis of written data sources and interviews with actors involved in the decision making process. The chronological ordering of the decisions taken by the various actors gives an overview of the course of the decision making process.

Secondly, the so-called crucial decisions marking the beginning and/or end of decision making rounds have to be traced. Crucial decisions were defined as decisions that have turned out to be a point of reference for the actors involved. Because actors do have different perceptions of the issues at stake and the games that are played, they may easily disagree on which decisions are the crucial decisions. Therefore, in this research it is tried to make objective the selection of crucial decisions by defining two important criteria. Decisions are crucial if they do have consequences (1) for the number and types of issues discussed, and/or (2) for the composition of the international arena, i.e. when actors enter or leave the arena.

Thirdly, the strategic interactions preceding crucial decisions have to be reconstructed. The crucial decision marking the end of a decision making round is preceded by a process of strategic interaction between the actors participating in the decision making round. These actors form the policy arena or activated network. The descriptions of the strategic interactions in each decision making round are so-called 'thick descriptions' including statements and opinions of the actors involved. The main reason for the use of such 'thick descriptions' is that in a study of social interactions 'feelings are facts', or as Crozier and Friedberg argue: "*Objective knowledge can be constructed only through analysis of "subjective" experiences*".[7] Feelings, perceptions, and opinions may explain why actors behaved the way they did. The description focuses primarily on the strategies used by the (representatives of the) governments of the Scheldt basin states and regions. In other words, the basin states and regions are the main units of analysis. In many cases, however, the strategic differences between the various government agencies and NGOs involved in the decision making process are addressed as well, mainly because knowledge on these differences contributes to a better understanding of international decision making.

Following the three procedural steps described above, three subquestions can be formulated:

1a.   *Which decisions were taken by the actors involved in decision making on international Scheldt issues?*
1b.   *Which decisions appeared to be crucial decisions in the decision making process?*
1c.   *How did the strategic interactions preceding these crucial decisions develop?*

The second research question addresses the context of decision making on international Scheldt issues. Because the context is assumed to contain important explanatory variables for the course of decision making on international Scheldt issues, like the decision making process itself, the context within which this process takes place has to be described carefully as well. Therefore, the second research question is:

2.   *What were the characteristics of the context of decision making on international Scheldt issues?*

In the conceptual framework presented in Section 2.4 five interrelated dimensions of the context of decision making on international river issues were distinguished, which makes it possible to formulate five subquestions:

2a.   *What was the underlying hydrological structure of the international Scheldt issues at stake?*
2b.   *In which other (related) international decision making processes were the Scheldt basin states and region involved?*
2c.   *What are the characteristics of the (decision making) cultures of the Scheldt basin states and regions?*

---

7. Crozier and Friedberg (1980, p. 264).

2d.   *Which intranational developments took place within each of the basin states in the case study period?*

2e.   *What is the history of conflict and cooperation on international Scheldt issues?*

3.4.2 Explanatory research questions

The explanatory research questions draw on the conceptual framework for the analysis of decision making on international river issues presented in Figure 2.3.

The third research question addresses the relation between the perceptions of the actors involved in decision making on international Scheldt issues and the strategies they have used (Relationship 1):

3.   *How did the perceptions of the actors involved in decision making on international Scheldt issues influence their strategies?*

As already mentioned in the previous section, the emphasis in this research is on the perceptions and strategies of the (representatives of the) governments of the Scheldt basin states and regions. Nevertheless, because lower level governments and NGOs may play an important role in decision making, their perceptions and strategies are addressed as well.

3a.   *How did the perceptions of the (representatives of the) governments of the Scheldt basin states and regions influence the strategies they have used in decision making on international Scheldt issues?*

3b.   *How did the perceptions of lower level governments and NGOs influence the strategies they have used in decision making on international Scheldt issues?*

The fourth research question addresses the institutionalization processes that may develop in decision making processes (Relationship 2):

4.   *How did the strategic interactions between the actors involved in decision making on international Scheldt issues influence structural and cultural network characteristics?*

On the basis of the three interrelated institutionalization processes distinguished in Section 2.4 three subquestions can be formulated:

4a.   *How did the strategic interactions influence the development of international policies?*

4b.   *How did the strategic interactions influence the development or erosion of international institutional arrangements?*

4c.   *How did the strategic interactions influence cultural characteristics of the international relationship?*

The fifth research question addresses the learning processes that may develop in the decision making process (Relationship 3):

5.   *Did learning processes develop?*

In Section 2.2.7 two types of learning were distinguished, which makes it possible to formulate two subquestions:

5a.   *Did strategic learning processes develop?*
5b.   *Did cognitive learning processes develop?*

The sixth and final explanatory research question addresses the relationship between the characteristics of the context of decision making on international Scheldt issues, and the perceptions and strategies of the actors involved in decision making on these issues (Relationships 4 and 5):

6.   *How did the context of decision making on international Scheldt issues influence the perceptions and strategies of the actors involved in decision making on these issues?*

On the basis of the five contextual dimensions distinguished in Section 2.4 five subquestions are formulated:

6a.   *How did the underlying hydrological structure of the international Scheldt issues at stake influence the perceptions and strategies of the actors involved in decision making on these issues?*
6b.   *How did other international decision making processes in which Scheldt basin states and region were involved influence the perceptions and strategies of the actors involved in decision making on international Scheldt issues?*
6c   *How did the (decision making) cultures of the Scheldt basin states and regions influence the perceptions and strategies of the actors involved in decision making on international Scheldt issues?*
6d.   *How did intranational developments within each of the Scheldt basin states and regions influence the perceptions and strategies of the actors involved in decision making on international Scheldt issues?*
6e.   *How did the history of conflict and cooperation on international Scheldt issues influence the perceptions and strategies of the actors involved in decision making on international Scheldt issues?*

3.4.3 Prescriptive research question
The final research question is similar to the third component of the problem statement and, unlike the other research questions, is formulated in general terms:

7.   *Which strategies can be used to contribute to the solution of upstream-downstream problems in international river basins?*

The answer to this research question will be based on the theory of decision making on international river issues presented in Chapter 2, and the conclusions of the case study of decision making on international Scheldt issues presented in Chapter 8.

3.5   Data collection

3.5.1 Methods of data collection

For the reconstruction and explanation of decision making on international Scheldt issues two methods of data collection were used. The main part of the case study research comprised the analysis of written data to reconstruct the strategic interaction between the Scheldt basin states and regions. The other part comprised a series of interviews. The interviews had three important functions in the research. First, they confirmed information, which was collected by the analysis of written data, thus enabling triangulation.[8] Secondly, interviews sometimes filled the knowledge gaps that remained after the analysis of written data. Finally, interviews enabled the researcher to collect types of information that can hardly be collected by the analysis of written data. The interviews made it possible to address personal experiences of those involved in decision making. This type of subjective information contributed much to an understanding of real life decision making. An important limitation of the use of interviews is that they can only be used to gather information on decision making during the last years. If the interviews address events and developments that date back earlier, the results are much less reliable than the results of the analysis of written data. Therefore, interviews were mainly used to collect information about the last decision making rounds distinguished. The description and analysis of the other decision making rounds were primarily based on the analysis of written material.

3.5.2 Data sources

The application of the methods discussed above made it necessary to consult two types of data:
1. written data;
2. oral data.

Written data may be distinguished into written data stemming from secondary sources and from primary sources.

Primary sources consulted were:
- files of international negotiations including reports of negotiations and letters;
- press releases;
- minutes of meetings of international commissions and projects;
- reports of Parliament;
- organization charts and maps.
- research reports

Secondary sources consulted were publications about decision making on international Scheldt, published in:
- journals;
- newspapers;
- study reports;
- newsletters.

---

8. Swanborn (1987, p. 332).

Although the author was unable to consult all files of the negotiations on the water conventions[9], the combination of all available sources enabled him to present a rather detailed reconstruction of decision making on international Scheldt issues during the past 30 years.

Oral data were collected by interviews with forty experts, most of them directly involved in decision making on international Scheldt issues. Interviews were held in all basin states and regions, and addressed decision making in the two main international issue-areas in the Scheldt basin: water and sediment pollution, and the maritime access to the port of Antwerp. The items to be addressed in the interview were formulated beforehand. The inductive character of the research, however, made it necessary to be flexible with the items to be addressed and the questions to be posed. Respondents were given the opportunity to 'ride their horse', as long as their information seemed to contribute to a better understanding of decision making on international Scheldt issues. Interview reports were only used or cited if the respondent allowed to do so. The references of this book include the list of respondents (List E).

3.6   Guide for the reader

Table 3.1 presents a guide for the reader of the case study reports presented in Part II of the thesis. Chapter 4 begins with a description and analysis of the context of decision making on international Scheldt issues, and therefore presents an answer to research question 2.

The description and analysis of the decision making process is divided over three chapters, each covering several decision making rounds. Chapter 5 begins with the Dutch decision to make a tactical linkage between Scheldt infrastructure and Meuse issues in 1967, describes the decision making rounds I to IV, and concludes with the declaration of intent of 1985. After a long impasse, in this declaration the Belgian and Dutch Ministers of Foreign Affairs decided to restart negotiations on Scheldt and Meuse issues. Chapter 6 begins with the declaration of intent of 1985, describes the decision making rounds V to IX, and concludes with the Dutch decision to unlink bilateral and multilateral Scheldt and Meuse issues in 1993. Finally, Chapter 7 begins with the bilateral and multilateral negotiations on Scheldt and Meuse issues, describes the decision making rounds Xa to XIIb, and ends in 1997, when the multilateral Scheldt and Meuse conventions are ratified, the Dutch parliament approves special legislation on the implementation of the deepening programme, and the Commission of wise persons issues its advice concerning the compensation of nature losses caused by the implementation of the deepening programme.

In the Sections 5.2, 6.2, and 7.2 the decision making rounds are described, and research question 1 is answered. In the Sections 5.3, 6.3, and 7.3 the four explanatory research questions are answered. In the first subsection of each section the influence of the perceptions of the actors involved in decision making on international Scheldt issues on their strategies are discussed, and research question 3 is answered. In the second subsection of the Sections 5.3, 6.3, and 7.3 the institutionalization processes are

---

9. The files of the negotiation commissions Biesheuvel-Davignon and Briesheuvel-Poppe, the bilateral and multilateral negotiation commissions installed in 1993, and the provisionally installed ICPS were confidential, and therefore could not be consulted.

discussed, and research question 4 is answered. In the third subsection of the Sections 5.3, 6.3, and 7.3 the learning processes are discussed, and research question 5 is answered. Finally, in the fourth subsection of the Section 5.3, 6.3, and 7.3 the influence of the context on the perceptions and strategies of the actors involved in decision making on international Scheldt issues is discussed, and research question 6 is answered.

Table 3.1 *Guide for the reader* Schematic presentation of the research questions that will be answered in the chapters, sections, and subsections of Part II of the thesis. Q = research question.

Period	*Description*		*Analysis/Explanation*			
	Context (Q 2)	Decision making (Q 1)	Influence perceptions on strategies (Q 3)	Institutio- nalization (Q 4)	Learning (Q 5)	Influence context on perceptions and strategies (Q 6)
1967- 1985 (*Rounds I-IV*)	4	5.2	5.3.1	5.3.2	5.3.3	5.3.4
1985- 1993 (*Rounds V-IX*)	4	6.2	6.3.1	6.3.2	6.3.3	6.3.4
1993- 1997 (*Rounds Xa-XIIb*)	4	7.2	7.3.1	7.3.2	7.3.3	7.3.4

Part II:
The Scheldt Case

CHAPTER 4

CONTEXT OF DECISION MAKING ON INTERNATIONAL SCHELDT ISSUES

4.1 Introduction

The context may contain important explanatory variables for the development of a decision making process. Although in Section 3.3.1 it was argued that in a study of decision making on international river issues it is hardly possible to distinguish clearly between context and phenomenon, this chapter attempts to describe and analyze several dimensions of the context of decision making on international Scheldt issues. With that, it presents the answer to the second research question. The contextual analyses presented in this chapter will be used for the analyses of the relation between context and decision making process, which will be presented in the Chapters 5 to 7.

Section 4.2 begins with a concise hydrographic description of the Scheldt basin, and a description of its main functions. In Section 4.3 the main Scheldt issues, which were placed on the international agenda between 1967 and 1997, are introduced. On the basis of their intellectual coherence, two issue-areas are distinguished: the issue-area of the maritime access to the port of Antwerp, and the issue-area of water and sediment pollution.

In the Sections 4.4 to 4.8 the five dimensions of the context distinguished in Section 2.4 are described and analyzed. These sections successively address the underlying hydrological structure of the issues at stake, the international context of decision making on international Scheldt issues, the decision making cultures of the Scheldt basin states and regions, the *intra*national developments in the Scheldt basin states and regions during the case study period, and the history of conflict and cooperation over the Scheldt river basin.

Finally, in Section 4.9 the main conclusions of the analysis of the context are summarized.

4.2 The Scheldt basin

4.2.1 Hydrographic description[1]
Together with the Rhine, Meuse, and Ems, the Scheldt is one of the four rivers flowing through the Dutch delta to the North Sea (See Appendix 7). The Scheldt river rises near

---

1. Detailed hydrographic descriptions of the Scheldt basin can be found in Ovaa (1991a and b), IPR (1991a), ISG (1994 a and b), and ICBS/CIPE (1997).

Saint-Quentin in northern France, and flows through Belgium and the Netherlands to the North Sea (See Appendix 8). The catchment area stretches out over France (31%), the Belgian regions Wallonia (17%), Brussels (1%), and Flanders (43%), and the Netherlands (8%). The Scheldt basin is bounded by the North Sea and the Yser basin in the west, a number of small coastal river basins in the southwest, and the Rhine, Meuse and Oise basins in the northeast, east, and southeast. The river, which is about 350 km long, has a catchment area of 21.863 km². The average annual flow, which is controlled by rainfall, is 120 m³/s. Table 4.1 shows that with these characteristics the Scheldt is relatively small compared with other European rivers, such as the Rhine or Meuse, not to mention rivers, such as the Amazon and the Nile.

Table 4.1 *Catchment area and average annual discharge of international rivers* (After Jansen *et al.*, 1979).

River	Catchment area (10^3 km²)	Average annual discharge (m^3s^{-1})
Amazon	7,000	100,000
Nile	2,900	3,000
Rhine	170	2,200
Meuse	33	250
Scheldt	21	120

Because the fall between the source and the mouth of the river is only 100 meters, the Scheldt and her tributaries are lowland river systems, which are characterized by relative low current velocities and meandering. The Scheldt estuary is that part of the Scheldt where tidal influence is present, and extends from the weirs in Ghent and several tributaries to Flushing (See Appendix 9). The estuary can be divided into the fluvial estuary, which is the part of the estuary upstream of Rupelmonde, the Upper Estuary or Lower Sea Scheldt, which extends from Rupelmonde to the Belgian-Dutch border, and the Lower Estuary or Western Scheldt, which extends from the Belgian-Dutch border to Flushing. The Scheldt is a heavily modified river. Appendix 5 contains an overview and description of the various subbasins of the Scheldt, and the artificial connections that were made between the Scheldt basin, the North Sea and adjacent basins. A study of the geography of the Scheldt basin learns that, with the exception of the Leie, which partly is a boundary river, all international Scheldt tributaries are transboundary rivers.

4.2.2 River functions
Like most river systems in developed countries, the Scheldt basin is exploited intensively by the riparian states, who have assigned several user functions to the river. Besides the socio-economic user functions, however, the river still has an important nature function. In the following paragraphs, the main functions of the Scheldt river are described shortly.[2]

---

2. Detailed descriptions of the Scheldt river functions can be found in Ovaa (1991a and b), IPR (1991a), ISG (1994a and b), and ICBS/CIPE (1997).

Discharge of water and sediments

Because it is so obvious, this important river function is mostly forgotten. The discharge function of the river, however, enables to drain agricultural and urban areas, and contributes to the prevention of floods along the river. A considerable part of the fresh water discharge of the river Scheldt is diverted to the North Sea via several canals (See Appendix 5), to improve navigation possibilities on these canals, and for industrial water use. The fertile sediments transported to the Scheldt estuary are an important constituent of the downstream marsh lands.

Navigation

Although there is always subjectivity in the appraisal of river functions, nobody will deny that navigation is one of the most important, if not the most important function of the river Scheldt. The Scheldt estuary is the maritime access to the port of Antwerp, which is one of the biggest ports of the world. Together with some smaller ports in the Scheldt basin, such as those of Ghent, Flushing and Terneuzen, the port of Antwerp is situated in the Rhine-Scheldt-Delta, which belongs to the most prosperous economic areas in Europe. Table 4.2 produces some economic data for the ports in the Rhine Scheldt Delta.

Nature

In spite of the intensive human exploitation of the river Scheldt, some parts of the basin still have high ecological values. Together with the estuary of the river Ems and the Wadden Sea, the Scheldt estuary is one of the remaining real estuaries in the Netherlands.[3] The brackish tidal water areas and marsh lands, such as the *Verdronken land van Saefthinge*, and the fresh tidal water areas in the upper estuary are unique.[4]

Discharge of domestic, industrial and agricultural waste loads

Whether one likes it or not, the use of the river for waste discharge is an important socio-economic function of the river Scheldt. The pollution of water and water beds in the Scheldt river basin may be explained by the high population density, the high level of economic development of the Scheldt basin states and regions, the relative low discharge of the river, and the relative late development of water quality policies. Although it was stated in the previous section that the Scheldt is a relative small international river basin, Bijlsma has calculated that once the Gross National Products in the main international river basins are divided by their discharges, the Scheldt belongs to the ten most important river basins in the world.[5] Main discharges of pollutants in the Scheldt basin stem from the industrialized areas around Lille and Roubaix in northern France, the conurbations Brussels and Antwerp[6], and the industrial zone along the Canal Ghent-Terneuzen.

---

3. Hoogweg and Colijn (1992). Because of the implementation of the Delta Act, and the construction of large dams, the natural conditions of the (mixed) estuaries of the rivers Rhine and Meuse have changed fundamentally.

4. An attractive description of the Scheldt ecosystems with many illustrations can be found in Meire *et al.* (1995).

5. Bijlsma (1990, p. 10).

6. When this thesis is completed the Brussels Capital Region does not yet have operational waste water treatment plants. Two waste water treatment plants, north and south of the city of Brussels, are planned, and the construction of the southern plant has started.

Table 4.2 *Economic data ports Rhine Scheldt Delta* (O-RSD, 1994). The first column lists the ports situated in the Rhine Scheldt Delta (RSD). The ports of Antwerp, Ghent, Terneuzen, and Flushing are situated in the Scheldt basin. The second column produces information on the transfer of goods in 1990. The third column presents index numbers of the transfers in 1990 compared with the transfers in 1982. The fourth column gives information about the employment in port-related industries. In the final column the Gross Regional Product (GRP) for the region in which the port is situated is compared with the average GRP in the European Union.

Ports in RSD	Transfer 1990 (10³ ton)	Transfer 1990 (1982 = 100)	Employment related to the port	GRP (Average EU = 100)
Antwerp	102,000	121	66,000	142
Ghent	24,400	107	18,000	121
Zeebrugge	30,300	322	1,700	108
Oostende	4,600	107	1,800	100
Rotterdam	300,000	116	53,338	115
Terneuzen/ Flushing	17,800	129	12,605	110

Recreation
Recreation in the Scheldt basin mainly concerns river side recreation. In the Dutch Province of Zeeland recreation along the mouth of the river Scheldt is rather important. Marinas along the river are situated in Antwerp, Terneuzen, Breskens, and Flushing. Because of the intensive professional navigation to the port of Antwerp, yachting is concentrated in the western part of the Western Scheldt. Recreational fishing takes place on those places in the catchment area where river pollution has not reduced the total number of fishes too much.

Use of river water for agriculture, industry, and drinking water production
Scheldt water is used for agriculture and several industrial purposes. Only in the upstream part of the Scheldt basin, and on a small scale, river water is used for drinking water production. Because the estuary of the river Scheldt contains salt or brackish water, this water is not used for drinking water production. The Dutch part of the Scheldt basin, and the agglomerates Brussels and Antwerp mainly use Meuse water for drinking water production. The Dutch province of Zeeland receives its drinking water from the water works of Rotterdam whose basins are situated in the Dutch Biesbosch. The Brussels region buys Walloon drinking water, which is diverted from the river Meuse, and transported to Brussels via an immense pipeline. Finally, the city of Antwerp extracts water from the Albert canal, which is fed with Meuse water.

4.3   International Scheldt issues

In this section the main international Scheldt issues, which were placed on the agenda between 1967 and 1997, are introduced shortly. Agenda-setting processes, and decision making on these issues are not addressed, because these processes are described and analyzed in the Chapters 5 to 7. Furthermore, the discussion of international Scheldt issues is restricted to the bilateral Belgian-Dutch, and the multilateral Scheldt issues. Bilateral French-Belgian, and the many interregional issues in the Scheldt basin are not addressed in this research.

Because of the interrelations between the different components of a river system, most international river issues are related to each other somehow. Nevertheless, because the intellectual coherence between some issues is more intense than between others, for the sake of clarity two main international issue-areas are distinguished in the Scheldt basin: the issue-area of the maritime access to the port of Antwerp, and the issue-area of water and sediment pollution.

4.3.1 The issue-area of the maritime access to the port of Antwerp
In Section 4.2.2 it was argued that navigation belongs to the most important functions of the river Scheldt. The Scheldt estuary forms the maritime access to the port of Antwerp. Because the Western Scheldt is situated on Dutch territory, the Belgian and later Flemish government depends on the Dutch as regards the maintenance and improvement of this important navigation channel. Several issues, which were placed on the international agenda in the case study period, show an intellectual coherence, since they are related to the improvement or maintenance of the maritime access to the port of Antwerp.

As regards the improvement of the maritime access to the port of Antwerp, the main international issues were the construction of the Baalhoek canal, the straightening of the bend near Bath, and the deepening of the navigation channel of the Western Scheldt. The first issue, which was placed on the international agenda by the Belgian government in 1967, concerns the construction of a canal between the left bank of the river Scheldt near Antwerp and the Dutch village of Baalhoek (See Appendix 10). This issue still is on the international agenda, but presently has a rather low priority. The second issue, which was placed on the international agenda by the Belgian government together with the previous issue, concerns the construction of a by-pass near the Dutch village of Bath (See Appendix 11). The final issue concerns the implementation of a deepening programme of the navigation channel in the Western Scheldt. In 1984, this issue was placed on the international agenda by the Belgian government, whereas at the same time the issue of the straightening of the bend near Bath was removed from the agenda. In 1995 the Flemish and Dutch governments reached an agreement on the Convention on the deepening of the navigation channel in the Western Scheldt.[7] Implementation of the deepening programme, among other things, involved the issuing of a number of permits by Dutch government agencies.

Also the maintenance of the navigation channel of the Western Scheldt was an important issue in the case study period. Ever since 1906 the Belgian government has to apply yearly for a dredging permit to be issued by the Dutch government. Since 1985 the Belgian government also has to apply for a permit for dumping dredged material in the Western Scheldt, which is based on the Dutch Surface Water Pollution Act, the so-called WVO-permit.[8]

Finally, during the past decade the ecological aspects of dredging and dumping in the Scheldt estuary were placed on the international agenda. Because of the water and sediment

---

7. Verdrag inzake de verdieping van de vaarweg in de Westerschelde (Trb. 1995 Nr. 51).

8. WVO =Wet Verontreiniging Oppervlaktewateren.

pollution in the Scheldt basin, dredging and dumping caused a further dissemination of pollutants in the Scheldt estuary. Especially the transport of polluted sediment from the eastern to the relative clean western part of the Western Scheldt became an important topic of discussion. This is an important intellectual relation between the two issue-areas distinguished in this study. Besides the dissemination of pollutants, awareness of the geomorphological disturbances caused by the dredging works increased. In the nineties dredging strategies and compensation of nature losses became important international issues.

4.3.2 The issue-area of water and sediment pollution in the Scheldt basin

The second cluster of interrelated issues distinguished in this study are the issues related to water and sediment pollution in the Scheldt basin. Unlike the issues that were discussed in the previous section, most pollution-issues are multilateral issues involving all Scheldt basin states and regions. In Section 4.2.2 it was argued that discharge of pollutants was and still is an important socio-economic function of the river Scheldt. In all parts of the basin discharge of industrial and domestic waste water, and land-based pollution negatively influence the water quality of the river Scheldt. Part of the pollutants adheres to fine sediment fractions. Deposition of these sediments causes river bed pollution. The mixing of fresh and salt water in the Lower Sea Scheldt shapes a specific physical condition, indicated a turbidity maximum, which causes high sedimentation rates near the port of Antwerp.

Main pollution-related issues, which were placed on the international agenda in the case study period, were the construction of the Tessenderlo pipeline, the formulation of international water quality objectives, the development of an alarm system for cases of accidental pollution, the extraction of polluted sediments from the Lower Sea Scheldt, and the ecological rehabilitation of the river Scheldt. The Dutch had the opportunity to place pollution issues on the international agenda in 1967. Until 1992 Belgium and the Netherlands discussed Scheldt water quality issues, and several draft conventions for the Protection of the Scheldt against pollution bilaterally. Only since 1992 the upstream riparian state France was involved in decision making on pollution issues. Since the conclusion of the Convention on the protection of the Scheldt against pollution in 1994[9], all basin states are involved in the preparation of the first Scheldt Action Programme. Furthermore, they discussed and reached an agreement on the development of an alarm system for cases of accidental pollution.

Since 1985 the Belgian and Dutch governments discuss bilaterally the conditions the Dutch government formulates to the Belgian permit for dumping dredged material in the Western Scheldt. The last WVO permits issued to the Flemish government contain the condition that the Flemish government has to extract specified amounts of (polluted) sediments from the Lower Sea Scheldt, because this would reduce the transboundary load of pollutants at the Belgian-Dutch border.

Because the issue of water and sediment pollution is strongly related to the ecology of the river Scheldt, the multilateral discussions on the development of joint water quality policies were accompanied by discussions concerning the ecological rehabilitation of the river Scheldt.

---

9. Trb. (Treaty series of the Netherlands) 1994 Nr. 150. Verdrag inzake de bescherming van de Schelde. The Flemish Region signed this convention in 1995.

4.4   Underlying hydrological structure of the issues at stake

The underlying hydrological structure of an international water issue influences the distribution of interests and resources among the basin states, and consequently their perceptions of (inter)dependence. Therefore, the underlying hydrological structure of an international water issue may be an important explanatory variable for the development and outcome of decision making on that issue. In Chapter 2 several types of international water issues, and their underlying hydrological structure were discussed. It was argued that for some issues, such as the exploitation or pollution of a transboundary acquifer, transboundary lake, or boundary river, the underlying hydrological structure causes a rather symmetric distribution of resources and interests among the parties. For most issues in transboundary rivers, however, the underlying hydrological structure causes an asymmetric distribution of interests and resources. In the following, the underlying hydrological structure of the international Scheldt issues introduced in the preceding section will be analyzed.

The underlying hydrological structure of most issues in the issue-area of the maritime access to the port of Antwerp causes a clear asymmetric distribution of interests and resources between the basin states Belgium (Flanders) and the Netherlands. Belgium, which is situated upstream of the Netherlands, perceives a dependence on the Dutch as regards the maintenance and improvement of the maritime access to the port of Antwerp. Therefore, navigation issues are clear upstream-downstream issues. As regards the ecological issues related to the improvement and maintenance of the navigation channel to Antwerp, however, upstream-downstream asymmetries are less predominant.

Because of their multilateral character the underlying hydrological structure of most issues related to water and sediment pollution does create a less unambiguous distribution of interests and resources among the Scheldt basin states and regions. The Scheldt and her international tributaries all are transboundary waters. Table 4.3 produces information on the relative hydrologic positions of the Scheldt basin states and regions. France is situated upstream of the Flemish and Brussels Capital Regions, and the Netherlands, and consequently has a relative power advantage in the issue-areas of water and sediment pollution. The underlying hydrological structure of the pollution issues, however, causes a more symmetric distribution of interests and resources between France and the Walloon region, because for the tributaries Haine and Scarpe both parties do have a relative downstream and upstream position (See Appendix 8). The Walloon region is situated upstream of the Flemish and Brussels Capital Regions, and the Netherlands, and therefore has a relative power advantage compared to these basin state and regions. The Brussels Capital region is entirely situated within the Flemish region, and consequently has a downstream as well as an upstream position in relation to the Flemish region. Finally, the Brussels Capital and Flemish regions are situated upstream of the Netherlands, and therefore have a relative power advantage in relation to the downstream basin state as regards international pollution issues. Because upstream discharges of pollutants do have an impact on downstream ecosystems, the ecological issues related to water and sediment pollution are characterized by the same power asymmetries.

An interesting result of the analyses of the underlying hydrological structure of international Scheldt issues is that the Belgian (Flemish)-Dutch power asymmetry in the issue-area of water and sediment pollution is the inverse of the power asymmetry between these parties in the issue-area of the maritime access to the port of Antwerp.

Table 4.3 *Relative hydrologic positions of the Scheldt basin states and regions*[10] The table produces information on the relative hydrologic positions of the Scheldt basin states and regions. It indicates the parties having a relative upstream position. Fl = Flemish Region, W = Walloon Region, B = Brussels Capital Region, Fr = France, X = no hydrologic relation between parties, Equal = both parties do have a relative upstream and downstream position.

	France	Walloon Region	Brussels Capital Region	Flemish Region	The Netherlands
France	X	Equal (Scarpe)	X	Fr	Fr
Walloon Region	Equal (Haine)	X	W	W	W
Brussels Capital Region	X	W	X	Equal	B
Flemish Region	Fl	W	Equal	X	Fl
The Netherlands	Fr	W	B	Fl	X

4.5   International context

In an analysis of decision making on international river issues the total context of international relations between the basin states should be taken into account. First, the underlying hydrological structure of other water management issues between the basin states may influence the distribution of interests and resources among them, and consequently the course of their joint decision making processes. The same applies to other international issues at stake. Secondly, decision making in international organizations on more general water policies may influence decision making on more specific international river issues. Finally, the strategic behaviour of the basin states in other (joint) decision making processes shape their mutual reputations, and consequently the degree of trust between them. The Sections 4.5.1 and 4.5.2 successively address decision making on the main other international water management issues between the Scheldt basin states and regions, and decision making on water policies in the EU, the North Sea region, and the UN-ECE.

4.5.1 Decision making on other international water management issues
Besides decision making on international Scheldt issues, two or more of the Scheldt basin states and regions were involved in decision making on several international Meuse and Rhine issues in the period between 1967 and 1997.

---

10. The author is grateful to L. Santbergen who suggested to include this table in the thesis.

International Meuse issues

The river Meuse is situated east of the Scheldt basin (See Appendix 7), and just like the river Scheldt rises in France, and flows through Belgium and the Netherlands to the North Sea. Part of the Meuse basin is situated in Luxembourg and Germany. Whereas the biggest part of the Belgian Scheldt basin is situated in the Flemish region, the biggest part of the Belgian Meuse basin is situated in the Walloon region.

In the period between 1967 and 1997 several international Meuse issues have been placed on the Belgian-Dutch agenda. Because of the many linkages that were made between Scheldt and Meuse issues, decision making on the international Scheldt issues discussed in Section 4.3 cannot be understood without an analysis of decision making on the international Meuse issues. Moreover, the organizations and persons involved in decision making on Scheldt and Meuse issues often are the same. Because of the strategic coherence between international Scheldt and Meuse issues, decision making on international Meuse issues can hardly be conceived of as a context, but rather is an integral part of the decision making process that will be described and analyzed in the Chapters 5 to 7. Therefore, this section is confined to a short introduction of the international Meuse issues, and an analysis of their underlying hydrological structure.

A first important Meuse issue concerns the water distribution between the basin states. The natural discharge of the river Meuse is rather fluctuating. During periods of a low river discharge, it is impossible to satisfy all demands that are made by the various users of Meuse water. The water distribution between the Meuse basin states has been an important issue throughout Belgian-Dutch history.[11] In 1967 the Dutch had the opportunity to place the issue high on the Belgian-Dutch agenda. First, the discussion concentrated on the construction of storage reservoirs in the Walloon Ardennes. Later, the discussion focused on possibilities to reduce water use, and on a distributive code for cases of a low river discharge. In 1994 Flanders and the Netherlands reached an agreement on a Convention concerning the flow of the water of the river Meuse.[12]

Also the other extreme, namely high water levels, causes severe international problems. Since the river floodings of 1993 and 1995 the issue of flood alleviation is placed on the international agenda. The floods in the river Meuse are a perfect illustration of the influence disasters may have on processes of agenda setting and decision making. Until 1993 the Meuse basin states did hardly address water quantity issues, whilst presently the basin states have organized permanent deliberations concerning the water quantity management of the river Meuse.[13]

A final main cluster of issues are the issues related to water and sediment pollution in the Meuse basin. In Section 4.2.2 the main user functions of the river Scheldt were discussed. An interesting conclusion from this overview was that the main population centres Brussels and Antwerp do not use Scheldt but Meuse water for the production of their drinking water. Also in the Netherlands the drinking water function of the river Meuse is of the utmost importance, because the southern part of the Randstad and the Province of Zeeland mainly use Meuse water for drinking water production.[14] This is the main reason for the Dutch interest in the water quality of the river Meuse. Especially the accidental pollution in the

---

11. See also the historic overview presented in Section 4.8.

12. Verdrag inzake de afvoer van het water van de Maas (Trb. 1995 No. 50).

13. See Section 2.3.2.

14. The Randstad is the urbanized western part of the Netherlands.

upstream parts of the Meuse basin, which forced Dutch drinking water companies to stop the intake of Meuse water several times, was an important topic of discussion among the Meuse basin states. There is, however, a second main reason for the Dutch interest in a clean-up of the river Meuse. Whilst in the Scheldt basin the port of Antwerp has to cope with river bed pollution, in the Meuse basin the Dutch port of Rotterdam has to cope with contaminated water beds, and consequently with high costs of the storage of polluted dredged material. Finally, the Dutch always wanted to improve the water quality of the Meuse for ecological reasons. In 1994 the Meuse basin states reached an agreement on a Convention on the protection of the Meuse against pollution, and a provisional International Commission for the Protection of the Meuse against pollution (ICPM) was installed. The Meuse convention and the provisional ICPM resemble the Scheldt convention and the provisional International Commission for the Protection of the Scheldt against pollution (ICPS).

The underlying hydrological structure of most international Meuse issues in many respects is similar to the underlying hydrological structure of the Scheldt quality issues. The upstream riparian states do have a relative power advantage compared to the Netherlands, and the latter perceives a dependence on the former actors as regards the discharge of a sufficient amount of water of a reasonable quality. Part of the river Meuse, however, forms the border between Flanders and the Netherlands (See Appendix 7). Therefore, in that part of the basin the underlying hydrological structure causes a more symmetric distribution of resources and interests, and the parties are mutual dependent as regards the quantitative, qualitative, and ecological water management.

International Rhine issues[15]
In the previous section it was argued that because of the many linkages decision making on international Scheldt issues can hardly be understood without an understanding of decision making on international Meuse issues. In addition to the river Meuse, another international river is of relevance for the developments in the Scheldt basin, namely the river Rhine (See Appendix 7). The Rhine is situated next to the Meuse basin, which can even be considered as a large subbasin of the Rhine basin as the Rhine basin is the largest of the two river basins, and both rivers end up in the same estuary. Due to the geographical nearness of these basins some Scheldt and Meuse basin states also are Rhine basin states. The Netherlands is situated in the delta of all three rivers, and the sources of the rivers Scheldt and Meuse, and a considerable part of the Rhine basin are situated in France. In the past France and the Netherlands had some severe conflicts about salt discharges by potassium mines in the French Elzas. After this main dispute had been settled, however, international cooperation between the Rhine basin states went rather smooth, and is generally considered as very successful. As long ago as 1950 the International Commission for the Protection of the Rhine against Pollution (ICPR) was established. In 1963 the ICPR got its legal base according to international law.[16] The first joint activities concerned the development of a monitoring system and several studies to identify the different types of pollution and their impacts. In 1976 the first two water quality conventions, against chemical and chloride pollution, were signed. Only after a disastrous fire in the chemical enterprise Sandoz near Basel in 1986,

---

15. For a description and analysis of Rhine river cooperation, see Durth (1996), de Villeneuve (1996a), and Dieperink (1997).

16. Overeenkomst nopens de Internationale Commissie ter bescherming van de Rijn tegen verontreiniging (Trb. 1963, No. 104) (ICPR agreement).

however, rapid progress could be made. The accident increased the political willingness to clean up the river Rhine, and formed the immediate cause for the development of the Rhine Action Programme (RAP), which was formulated in 1987.[17] The developments in the Rhine are relevant for the Scheldt case for a completely different reason than the developments in the Meuse. In the case analysis presented in the Chapters 5 to 7 it will be shown that because of the rapid development of the intensity and scope of international cooperation in the Rhine basin, the ICPR often is used as an example for the rivers Scheldt and Meuse. Therefore, decision making on international Rhine issues is an important contextual variable for the study of decision making on international Scheldt issues.

4.5.2 Decision making on general water management policies

During the case study period the Scheldt basin states also were involved in decision making on water policies of the EU, of the North Sea basin states, and of the UN-ECE.

EU policies

All Scheldt basin states are member states of the European Union. Because, unlike other international organizations, the EU is a supranational organization, its policies are binding, and of the utmost importance for the development of water management policies in its member states. In the case study period the EU has developed a considerable number of emission and water quality directives[18], which had to be implemented by its member states. Most of these directives had a sectoral character, and did address ground water or surface water, or various types of water pollution.

During the completion of this thesis, the European Commission is preparing a new water directive, which aims at integrating several existing ones.[19] Because the draft of this directive contains several ideas about European river basin management, this directive may also have consequences for the international institutional arrangements for the management of the Scheldt river basin. In Chapter 9 of this thesis some remarks and recommendations concerning the content of this directive will be made.

North Sea policies

The Scheldt is one of the rivers flowing into the North Sea and indirectly the Northeast-Atlantic Ocean. Therefore, the Convention on the protection of the marine environment in the Northeast-Atlantic Ocean[20], and the agreements that were made at the North Sea

---

17. See Section 2.3.2.

18. The combined approach has led to several emission and water quality directives. The most important emission directives are the directives for the discharge of hazardous substances in the aquatic environment (76/464/EEC), the protection of groundwater against pollution caused by hazardous substances (80/068/EEC), waste water discharge from urban areas (91/271/EEC), and for the protection of water against nitrogen pollution from agriculture (91/676/EEC). The most important water quality directives are the directives for the quality of surface water for drinking water abstraction (75/440/EEC), the quality of bathing water (76/160/EEC), the procedure for the exchange of information on the quality of fresh surface water (77/795/EEC), the quality of fresh water for supporting fish life (78/659/EEC), the quality of fresh water for supporting shellfish growth (79/923/EEC), and the quality of drinking water for human consumption (80/778/EEC).

19. Commission proposal for a council directive establishing a framework for community action in the field of water policy. Draft 25/02/97 and several amendments made afterwards. For a discussion on the possible function and contents of this directive, see de Villeneuve (1996b).

20. Paris-Oslo Convention (1992).

ministerial conferences are relevant for the Scheldt water management as well. All three Scheldt basin states participated in the North Sea Ministerial Conferences and the OSPAR-commission. The four Ministerial North Sea Conferences were concluded with Ministerial declarations, which contain important agreements concerning emission reduction by the North Sea basin states.[21] The Convention on the protection of the marine environment in the Northeast Atlantic Ocean aims at a reduction of land based pollution in the North-East Atlantic Ocean and the North Sea. Because in 1997, the Scheldt basin states and regions have not yet reached an agreement on Scheldt water quality policies[22], the EU and North Sea policies are the main international water quality policies that apply to the river Scheldt.

UN-ECE river policies
All Scheldt basin states are member of the United Nations-Economic Commission for Europe (UN-ECE), and were involved in the negotiations on the Convention on the protection and use of transboundary watercourses and international lakes, which was signed in 1992.[23] This convention, which contains a framework for bilateral and multilateral cooperation on the protection and use of international waters, came into force on 8 October 1996. The Netherlands ratified the convention on 14 March 1995. Belgium has not yet ratified the UN-ECE convention, because the Belgian regions have not yet been able to reach an agreement on that issue. When this thesis was finished, France has not yet ratified the convention as well.

4.6   Cultural context

4.6.1 Introduction
In Chapter two, it was argued that any reasonable explanation of international decision making should include the cultural aspects of the international relationship. This section contains a description and comparison of the decision making cultures of the Scheldt basin states and regions, which will be used for the analysis of decision making on international Scheldt issues. It is based on an in depth study of cultural differences between countries, the Dutchman Hofstede carried out between 1973 and 1978[24], and several other studies addressing cultural characteristics of the Netherlands, Belgium, and France. Although the research of Hofstede is rather dated, the research results are interesting for at least three reasons. First, this thesis research addresses decision making in the period between 1967 and

---

21. At the second North Sea Ministerial Conference, which was held in London, the North Sea Action Programme (NAP) was adopted. According to this Programme the flow of dangerous substances to the North Sea had to be reduced in 1995 with approximately 50% in comparison with 1985. On the third North Sea Ministerial Conference in The Hague in 1990, the North Sea co-states reached an agreement on a further pollution reduction, namely a minimum reduction of 50% for 36 prior substances and at least 70% for mercury, cadmium, lead and dioxin. On the fourth North Sea Ministerial Conference, which was organized in Esbjerg (Denmark) in 1995, hardly new objectives were formulated (Elek, 1996).

22. The Scheldt water quality convention concluded in 1994 marked the beginning of the preparation of joint water quality policies for the Scheldt basin, but did not yet contain concrete policy objectives or policy programmes.

23. United Nations Economic Commission for Europe, Convention on the Protection and Use of Transboundary Watercourses and International Lakes, Geneva, E/ECE./1267, 1992.

24. Hofstede (1980).

1997. Secondly, it is well known among cultural experts that cultures do only change in the long run. Thirdly, Hofstede's research has been replied, and the results of this replication were very similar to the research results of Hofstede.[25]

The next section contains a brief description of the cultures of the Scheldt basin states, and discusses the main differences between them. In the Sections 4.6.3 to 4.6.5 the decision making cultures of the Netherlands, Belgium, and France are discussed more into detail. Finally, in Section 4.6.6 some trends in the development of the decision making cultures in the Scheldt basin states and regions are addressed.

4.6.2 Cultural differences between Scheldt basin states and regions

Hofstede's research on cultural differences between nation states concerns an investigation among 100,000 IBM employees in 53 countries. His research results are used in many studies of international cooperation.[26] The national cultures are described with scores on four dimensions[27]:

- Power Distance (PDI);
- Uncertainty Avoidance (UAI);
- Individualism (IDV);
- Masculinity (MAS).

The first of the four dimensions of national culture is called Power Distance. The basic issue involved, to which different societies have found different solutions, is human inequality.[28] Inequality may be experienced between ordinary workers and employers, between citizens and authorities or between consumers and media 'stars'. Societies can be more or less equal in these respects.

The second dimension of national culture is labelled Uncertainty Avoidance. Uncertainty about the future is a basic fact of human life with which human beings have to cope through the domains of technology, law, and religion.[29] Tolerance for uncertainty varies considerable among nations. A high score on this dimension is accompanied with nervous tensions, a need for predictability, and formal and informal rules.

The third dimension of national culture is called Individualism. It describes the relationship between the individual and the collectivity which prevails in a given society.[30] In individualistic societies everybody is expected to take care of him- or herself and for his or her family. A society is more collectivistic if individuals from their birth are part of a strong, cohesive group offering them whole-life protection in exchange for unconditional loyalty.[31]

The fourth dimension is called Masculinity with its opposite pole Femininity. The issue is whether the biological differences between the sexes should or should not have

---

25. Hoppe (1990) cited in Hofstede (1994).

26. e.g. Soeters (1993), Soeters *et al.* (1995), and Van Vugt (1994).

27. Hofstede (1980, p. 315).

28. Ibid., p. 92.

29. Ibid., p. 153.

30. Ibid., p. 213.

31. Ibid.

implications for their roles in social activities.[32] In a masculine society the social roles of the sexes are clearly separated. Men are expected to be assertive and hard, and directed to material success. Women, however should be modest and tender, and be directed toward quality of life. A society is feminine if social roles are overlapping, and both men and women are expected to be modest, tender, and directed toward quality of life.

Table 4.4 produces the scores on these four dimensions by each of the Scheldt basin states. Because Belgium is a bilingual state, Table 4.4 also values for the two Belgian language groups. As regards the scores indicated in this table, an important remark should be made. The scores characterize a *collectivity*, namely a state or region. When interpreting these scores, one should keep in mind that individuals are able to back out of the characteristics of this collectivity. For example, although the Belgian scores on the dimensions Power Distance and Uncertainty Avoidance are much higher than the Dutch ones, there may be Dutch individuals that have much higher scores on these dimensions than some Belgians.

Table 4.4 *French, Belgian, and Dutch scores on dimensions of national culture*[33] Scores on Power Distance, Uncertainty Avoidance, Individualism, and Masculinity for the Scheldt basin states and regions.

	Power Distance (PDI)	Uncertainty Avoidance (UAI)	Individualism (IDV)	Masculinity (MAS)
The Netherlands	38	53	80	14
Belgium total	65	94	75	54
Belgium Dutch	61	97	78	43
Belgium French	67	93	72	60
France	68	86	71	43

On the basis of the scores presented in Table 4.4 two main conclusions can be drawn. The first conclusion is that the French culture resembles the Belgian culture, and that the Dutch culture is very different from both the French and Belgian cultures. The second conclusion is that cultural differences between the Dutch and French speaking parts of Belgium are rather small.

France and Belgium have relative high scores on the cultural dimensions PDI and UAI. In Hofstede's research France and Belgium belong to a cluster of countries with a more developed Latin culture.[34] Other countries in this cluster are Argentina, Brazil, Spain and

---

32. Ibid., p. 261.

33. Ibid., p. 337.

34. Hofstede (1980, p. 336). Hofstede distinguished between more developed and less developed Latin cultures. The more developed Latin cultures do have medium to high scores on the cultural dimension IDV, whilst the less developed Latin cultures do have low scores on the IDV dimension.

Italy.[35] The Latin culture is characterized by a rather unequal society with a tendency to avoid problems or - if unavoidable - to respond to them in an inflexible manner.[36] The historical explanation of the relative high PDI- and UAI-scores in the European Latin countries is the cultural inheritage of the Roman Empire. In this Empire, the emperor had absolute authority and stood above the law, which may explain the high PDI-scores in these countries.[37] The effective system of formal control of the roman territories, and the unified legal system may explain the uncertainty-avoiding pattern.[38]

The Netherlands belong to a cluster of countries with a Nordic culture.[39] Other countries in this cluster are the Scandinavian countries Denmark, Finland, Norway, and Sweden. The Nordic culture is characterized by a rather equal and feminine society, and a low score on the cultural dimension UAI.

The conclusion concerning the internal Belgian cultural differences perhaps is more surprising, because language generally is seen as one of the most important cultural characteristics of a society. Cultural differences between the Flemish- and French speaking parts of Belgium, however, are much less than differences between the Dutch speaking part of Belgium and the Netherlands. According to Hofstede, there even are no two neighbouring countries speaking the same language, which do have such different cultural characteristics as Belgium and the Netherlands.[40] The common French culture of the two language areas in Belgium can be explained by their common history, and by the fact that since Belgium split from the Netherlands in 1831, French was the language of government, the upper classes, and secondary and higher education for more than 100 years.[41] That the language split has been a 'hot' political issue in Belgium for long, may be explained by the monolithic political structure the country had until 1988.[42]

The only similarity between on the one hand the French and Belgian cultures, and on the other hand the Dutch culture is the score on the dimension IDV. This may be explained by the strong correlation between the scores on this dimension and economic welfare. Rich countries do generally have rather individualistic cultures.[43]

A comparison of the scores on the four dimensions of national culture learns that there are significant cultural differences between the Scheldt basin states, which cannot be neglected in a study of international cooperation and decision making in this basin. Therefore, in the next sections the decision making cultures of the Scheldt basin states are discussed in more detail.

---

35. For an interesting discussion of the many similarities between Belgium and Italy, see Deschouwer *et al.* (1996).

36. Soeters *et al.* (1995).

37. Hofstede (1980, p. 127).

38. Hofstede (1980, p. 179), See also NRC, 25-2-1995.

39. Hofstede (1980, p. 336).

40. Hofstede in: NRC Handelsblad, 25-2-1995, See also Couwenberg (1997, p. 21).

41. Hofstede (1980, p. 337).

42. Hofstede (1980, p. 337). Cultural differences in Switzerland, where there are different linguistic groups as well are much bigger than in Belgium. That language is not a hot issue may be explained by the federal state structure of Switzerland.

43. Ibid., p. 214.

4.6.3 Dutch decision making culture

A first and perhaps main characteristic of Dutch decision making is the consensus culture. In the Netherlands, decision makers often try to reach a consensus between the parties involved in a decision making process. Consultation and involvement of the public and interest groups, negotiations and compromising are essential elements of decision making in almost any policy sector. The typical Dutch consensus culture may be explained by the relative low scores on the PDI- and MAS-dimensions. The relative low score on the former dimension explains why all citizens and interest groups do have opportunities to express their opinion, and to participate in the decision process. The score on the latter dimension may explain the strong legal protection of citizens that are grieved by governmental policies, and the highly democratic character of most decision making processes.[44]

Advantages of consensual approaches are the social and administrative support of public policies, which decrease the chance on the occurrence of implementation gaps. The lengthy decision making procedures to reach a consensus, however, may be perceived as disadvantageous. Research on the relationship between characteristics of national cultures, and the length of decision making processes on the construction of motorways has learned that because of the lower Dutch PDI- and MAS-scores decision making processes take much longer in the Netherlands than in Belgium.[45]

The consensus culture has influenced the institutional arrangements for policy making. In the Netherlands, there are numerous institutional arrangements for consultation and participation of NGOs and citizens. In addition, policy making needs a lot of 'Overleg', which Van der Horst describes expressively as "[...] *a form of group communication which aims not so much at reaching a decision as giving the parties involved the opportunity to exchange information. [...] At the end of the 'overleg' everyone has the idea of what the other wishes to achieve"*. "*[...] Often the chairman will conclude with satisfaction that 'de neuzen weer in dezelfde richting wijzen' (all noses are pointing in the same direction), or -in typically Dutch fashion- 'de klokken zijn gelijk gezet' (the clocks are all showing the same time)." Such 'overleg' may lead to practical decisions being made, but more often than not results in little more than an exchange of information.*"[46]

Although the Dutch traditionally aim at consensus-building, the Dutch Ministry of Transport, Public Works and Water Management, which probably is the most important Dutch actor in decision making on international Scheldt issues, for a long time had the reputation of a 'state within the state'. According to critics the Ministry was so powerful that it was able to take decisions without taking into account the opinion of parties that would be bothered by the planned (infrastructure) projects of the Ministry. In recent times, however, the Ministry is leading as regards interactive or participatory ways of policy-making, and open planning processes. These latest 'fashions' in policy making seem to fit very well in the Dutch consensus culture.

The Dutch consensus culture often is associated with the Dutch 'polder model' as well.[47] This socio-economic model is characterized by a high degree of consensus among the Dutch government, employer's organizations, and trade unions on the need of wage restraints. The Dutch consensus culture might explain why employers and employees could reach an

---

44. Soeters (1995, p. 305).

45. Reitsma (1995), Ministerie van Verkeer en Waterstaat (1994).

46. Van der Horst (1996, p. 158).

47. Van Ingen and de Ruiter (1997).

agreement on wage restraints, whilst these restraints, in turn, would explain the relative high economic growth in the Netherlands in the 1990's.

A second characteristic of Dutch decision making is the planning culture.[48] Government traditionally plays an important role in Dutch society, and interferes in almost any aspect of life.[49] Traditionally, the Dutch do not have much confidence in the market mechanism. Especially in the policy fields of physical planning, and water, nature and environmental management advanced planning systems were developed. Because of these planning systems Dutch bureaucracies produce numerous policy documents, and to coordinate these policies new policy documents are made. Most policy documents contain an extensive overview of related documents, and the formal relations between them. Because of the complexity of the Dutch planning system, only (some) professional bureaucrats are able to get an overview of all plans relevant to a certain policy issue.

A third major characteristic of Dutch decision making is the relative important role of professionals in decision making. The power distance between politicians and civil servants is relatively small, and the recruitment and career perspectives of civil servants depend on their professional capacities, and mostly not on their political preferences. Because of the relative small power distance, civil servants and negotiators do have large mandates, and ample possibilities to anticipate on political decision making.[50]

A fourth cultural characteristic of the Dutch is their directness. The Dutch are used to express what they think, which may be explained by their relative low score on the UAI-dimension. Observers from other countries may get a rather arrogant impression of the Dutch, and may easily perceive the Dutch directness as an overestimation of the value of their ideas. Soeters gives an interesting example of this.[51] The example concerns the Dutch succession of Luxembourg as chair of the European Union in 1991. Although Luxembourg had prepared a draft treaty for European cooperation, and the main parts of this draft had already been approved by the EU Council of Ministers, the Dutch decided to start all over again, and to develop their own draft treaty. Later, however, the other member states of the EU refused to discuss the Dutch proposal. Consequently, the Dutch had to change their strategy, and to continue work on the draft treaty its predecessor as chairman of the EU, Luxembourg, had prepared.

A fifth cultural characteristic of the Dutch is their flexibility, which also may be explained by the relative low score on the cultural dimension Uncertainty Avoidance. The Dutch do have a rather flexible bureaucracy, which generally is better able to cope with complex issues involving many uncertainties, such as most environmental issues, than bureaucracies of states with higher UAI-scores.[52]

Finally, the Dutch have a real feminine culture, which implies that they tend to value quality of life more than material success. The femininity of the Dutch culture may explain the relative early awareness of environmental issues, and development of environmental policies in the Netherlands.[53] Applied to the growth-versus-environment controversy, states

---

48. Soeters (1991, p. 26).

49. Soeters (1995, p. 306).

50. Soeters (1993, p. 650; 1995, p. 302).

51. Soeters (1995, p. 304).

52. Soeters (1993, p. 651).

53. Hofstede (1980, p. 297), Soeters (1995, p. 304).

with a more feminine value position will put higher priority on environmental conservation, whilst states with a more masculine value position will put priority on economic growth.[54]

4.6.4 Belgian decision making culture

In Section 4.6.2 it was shown that the cultures of the Dutch and French-speaking parts of Belgium are much alike. Therefore, in this section discussing the Belgian decision making culture, these main linguistic groups are not distinguished. Only the score on the dimension Masculinity makes it useful to discriminate between the two linguistic groups.

A first characteristic of the Belgian culture is the relative high PDI-score, indicating a relative unequal and stratified society. The distrust ordinary citizens have of the elites and the government may be explained historically. Until the mid 19th century Belgium always had been suppressed by foreign powers, and until the sixties of the 20th century the Flemish region felt suppressed by the Walloon (French-speaking) part of Belgium. For these reasons most of the Belgian population perceives government not so much as its own government, but as an externally imposed regime, they have to cope with.[55] The inequality of Belgian society may explain why there are only few checks and balances in the Belgian administrative system, and subordinates do have few possibilities to correct imperfect behaviour of the elites. Concentration of power with the elites is enhanced further by the acceptance of combining public functions.[56] Decision making often takes place outside parliament in so-called 'conclaves' of the elites, and therefore seems be less democratic and more hierarchical than in, for example, the Netherlands.[57]

A second important characteristic of the Belgian decision making culture is the relative large influence of politics on decision making. Unlike the Netherlands, where professional bureaucrats are rather influential, decision making in Belgium has a very political character. This characteristic of Belgian decision making is reinforced by structural characteristics, because in the Belgian political system each minister has the competence to select a group of advisers, who constitute the cabinet of the minister.[58] These advisers generally are member of the same political party as the minister, and do have much influence on decision making. As a consequence, top level professional bureaucrats of the several ministries are less influential. Because of the existence of political cabinets, a change of the political colour of government may have a considerable impact on current policy programmes. The influence of politics on decision making is enhanced further by the ways in which civil servants are selected and their careers are determined. Career perspectives within a Ministry often depends on the political colour of the minister who is in charge.[59] Because of the high power distance, and the large influence of politics on decision making, bureaucrats generally do have less mandates than their Dutch colleagues, and have to consult their superiors more often.[60] Another characteristic of Belgian politics is the clientilism, which means that politicians do have close relations with their electorate. If politicians are able to arrange

---

54. Ibid.

55. Van Vugt (1994, p. 11).

56. Soeters (1995, p. 301).

57. Ibid.

58. Ibid., p. 302.

59. Ibid.

60. Soeters (1995, p. 302; 1991, p. 26).

things that are beneficial to some parts of their electorate (their clients), they can count on their votes.[61] Because of this clientilism Belgian politics has a rather personal character.[62]

A third characteristic of Belgian culture is the relative high score on the cultural dimension Uncertainty Avoidance (UAI). This may explain why Belgians generally feel uncomfortable with complex, unstructured problems, why they are in favour of detailed regulations, and tend toward formalism.[63] Because of the high UAI-score decision making on complex issues, such as most environmental issues, is postponed easily, or shifted to other echelons.[64] The high UAI-score may also explain the rather extensive Belgian delegations to international negotiations comprising representatives of several administrative echelons.[65] Finally, the high UAI-score may explain why Belgians, unlike the Dutch, do not express their personal opinions easily, do hardly tell what they really think, and generally are conceived very polite.[66]

The high Belgian MAS-score may explain why environmental issues appeared on the agenda relatively late, and institutional arrangements and policies for environmental protection only developed recently. The higher MAS- score of the French-speaking part of Belgium may explain why in this part environmental issues appeared on the political agenda later than in the Flemish part of Belgium.

4.6.5 French decision making culture
In Section 4.6.2 it was argued that like Belgium France has a Latin culture. Because the Belgian and French scores on the four cultural dimensions are almost similar, most of what has been written about the Belgian decision making culture applies to France as well.

France does have an even higher PDI-score than Belgium, which implies a relative unequal society. Decision making tends to be more hierarchical than in the Netherlands, since the French have a lot of confidence in hierarchy as an effective means of problem solving.[67] According to Hofstede: "*in low PDI countries, power is something of which power holders are almost ashamed and which they will try to underplay*". In his book he writes: "*I heard a Swedish university official state that in order to exercise power, he tries not to look powerful. This theory definitely does not hold in Belgium or France. I once met the Dutch prime minister with his caravan on a camping site in Portugal; I could not very well see his French or Italian colleague in that situation.*"[68] French decision making in many respects is the opposite of Dutch consensus decision making. The French are used to functional and open conflicts, in which conflicting or antagonistic points of view of the parties become apparent. In a negotiation process involving mutual concessions they have to reach an agreement.[69] In case they are unable to solve their conflicts, the parties may demand

---

61. Soeters (1995, p. 302).

62. Van Vugt (1994, p. 12).

63. Soeters (1995, p. 303).

64. Soeters (1995, p. 303; 1993, p. 651;1991, p. 26).

65. Soeters (1995, p. 303; 1993, p. 651).

66. Soeters (1995, p. 303).

67. Hofstede (1980, p. 320).

68. Ibid., p. 121.

69. d' Iribarne (1989, pp. 28-29).

arbitration by a hierarchical superior.[70] *"On est très loin du consensus Néerlandais, et des réunions paisibles qui permettent de le mettre en oeuvre."* (*"They are very distant from the Dutch consensus culture, and the meetings aiming at pacification, which make it possible to reach this consensus."*)[71] In France, conflicts are seen as an inevitable consequence of society, and therefore are perceived as normal.[72] To solve conflicts, however, parties are expected to make mutual concessions and to show understanding for the points of view of the other parties. These moderating mechanisms explain why most conflicts are solved eventually. An interesting example of the open French conflicts concerns the massive strike of French truck drivers, and their blockades in 1997.

Although the literature clearly indicates that France has a hierarchical decision making culture, Appendix 6 describing the administrative organization of water management in the Scheldt basin states and regions shows that the administrative organization of French water management is not that hierarchical, and that on the basin level the water users are directly involved in decision making.

Cultural and structural characteristics of French society are clearly interrelated. The hierarchical decision making culture is reflected in the hierarchical organization of French public administration.

The second cultural characteristic Belgium and France do have in common is the high score on the UAI-dimension. This characteristic may be illustrated with the acting of the French delegation in the negotiations on the development of a regime for the river Rhine: *"the French delegation was often described as showing a tendency to approach issues by concentrating on "principles", procedures, and legal aspects and to back positions by elaborate and obstinate argumentation. In contrast, the Dutch often were described as more direct, pragmatic, and result oriented."*[73]

4.6.6 Trends in the development of the decision making cultures in the Scheldt basin states

Cultures do only change in the long run, i.e. during a period of a few decades or more. Nevertheless, it is worthwhile to discuss some trends in the development of the decision making cultures in the Scheldt basin states. Both international and intranational developments may have caused gradual cultural changes.

The most eye-catching international development is the continuing European integration, and the increasing influence of the European Union on decision making. In Section 4.5.2 the EU water policies were discussed shortly. These and other EU-policies largely influence national policies and institutional arrangements, such as decision making procedures. Although it cannot yet be proven, cultural experts do expect that these changes will cause a convergence of the decision making cultures of the EU member states.[74] Partly because of the ongoing European integration, the intensity of cooperation between the Dutch-speaking part of Belgium (Flanders) and the Netherlands is increasing.[75] The same pattern of an

---

70. Ibid.

71. Ibid, p. 29.

72. Ibid., p. 32.

73. Dupont (1993, p. 109).

74. Delmartino and Soeters (1994, p. 251).

75. Soeters (1995, p. 306).

increasing intensity of cooperation can be observed between Wallonia and France.

Besides international developments, also intranational developments in the Scheldt basin states may change their decision making cultures. During the past decades intranational developments in Belgium probably were most eye-catching. Because of the economic growth in the Flemish region, and the economic decline in Wallonia, the Dutch-speaking part of Belgium was able to continue its emancipation process.[76] During the successive stages of the Belgian state reforms, Flanders and Wallonia became relative autonomous regions, which enabled them to develop their own decision making cultures. Interesting development in Flanders, and to a lesser extent in Wallonia is the increasing resistance against the political influence on career perspectives of bureaucrats. Political appointments are more and more criticized, and bureaucrats are judged on the basis of their merits rather than on their political preferences.[77] Secondly, the Flemish government started to copy the ideas of rational planning from the Netherlands, and in the policy fields of land-use and environmental planning the first plans have been developed.[78] It generally takes a long time until these structural changes do have an impact on cultural characteristics. Although there remains a large cultural gap between Flanders and the Netherlands, research has shown some evidence for a convergence of their cultures during the past decades.[79] Cultural experts do expect that the relative autonomous Belgian regions will develop different decision making cultures, and that the internal Belgian cultural pattern will be going to show similarities with the pattern found in Switzerland, where the various linguistic groups have clearly separated cultures.[80]

Although the Dutch culture seems to have been less dynamic during the past decades, there are some changes that may have contributed to a convergence between the Flemish and Dutch decision making cultures. First, because of a gradually withdrawing government, influence of government on society is gradually decreasing. In the eighties and nineties successive Dutch governments cut government budgets, and stimulated privatization of government agencies. Dutch society tends to rely more on the market mechanism.[81] Secondly, Dutch government tried to reduce the lengthy decision making procedures, for example by the development of the Tracé and NIMBY Acts[82], and the Delta Act for the large rivers. An interesting observation is that the explanatory memorandum of the Tracé Act refers explicitly to the situation in Belgium and France.[83] Finally, there are indications that the number of political appointments of top level bureaucrats in the Netherlands is increasing gradually.[84]

Also developments in France tend to cause a gradual convergence of the decision making cultures of the Scheldt basin states and regions. The French decentralization rounds in the eighties may lead to a less hierarchical decision making culture in the long run (See also Appendix 6).

---

76. Ibid.

77. Delmartino and Soeters (1994, pp. 247-248), Soeters (1995, p. 308).

78. Delmartino and Soeters (1994, p. 247-248), Soeters (1995, p. 310).

79. Soeters *et al.* (1995).

80. Delmartino and Soeters (1994, p. 251).

81. Delmartino and Soeters (1994, pp. 249-250), Soeters (1995, p. 309).

82. NIMBY = Not In My Backyard.

83. Delmartino and Soeters (1994, pp. 247-248), Soeters (1995, p. 310).

84. Soeters (1995, p. 309).

4.7   Intranational context

The perceptions of the actors involved in decision making on international river issues, and their willingness to develop international institutional arrangements and policies, may be influenced by *intra*national developments in the basin states as well. Therefore, in this section the main *intra*national developments in each of the Scheldt basin states and regions are discussed shortly. The primary focus is on the development of water management policies. These developments, however, take place in a context of other *intra*national developments, such as administrative reorganizations, and economic developments. The Sections 4.7.1 to 4.7.3 successively point out the main *intra*national developments in French, Belgian, and Dutch water management during the case study period. Section 4.7.4 concludes this section with a comparison and analysis of these developments. Appendix 6 contains a description of the administrative organization of water management in the Scheldt basin states and regions in 1997.

4.7.1 Developments in French water management

Because of the industrialization and the absence of an adequate waste water treatment infrastructure, in the beginning of the 1960s, like most waters in the other Western European countries, the French waters were severely polluted. Therefore, a national water act was developed. The Water Act of 1964 (*Loi sur l' Eau 1964*) did not only contain water quality regulations, but changed the French water management organization fundamentally as well.[85] The water management organization introduced by this Act is largely inspired by the river basin approach. The Act provided for the installation of a national water committee, and six basin committees (*Comités de bassin*), which have a river catchment area as administrative delimitation. One of these committees, the basin committee Artois-Picardie, is involved in the management of the waters in the French part of the Scheldt basin. The main task of the newly created basin committees was the formulation of water management policies. Besides the basin committees, six basin agencies (*Agences financières de Bassin*[86]) were established. Their main tasks are to levy charges and to finance the policies that are developed by the basin committees. The basin committees develop five-year action programmes for their respective territories. To clean-up the French waters, investments were raised continuously. In the VIth programme of the water agencies, which was valid from 1992 to 1996, for example, the budget was doubled compared with the preceding programme, which was valid from 1987 to 1991.[87]

The economic situation in Nord-Pas de Calais, the biggest region in the Artois-Picardie basin, is very bad. Nord-Pas de Calais is an industrialized region, with mainly mining, steel and textile industries. During the case study period, many industries were closed, whilst others do still have hard times. As an example, in 1952 about 200,000 people were employed in the mining industry, whereas nowadays no mining industry is left at all. Therefore, the unemployment rate is the highest of France, and the income per capita the lowest.[88] Because of this bad economic situation, industries and citizens are not inclined to increase their

---

85. Loi sur l'Eau 1964, Gustafsson (1990, p. 1).

86. Later the name of the *Agences financières de Bassin* is changed into *Agences de l' Eau*.

87. Ministère de l' Environnement, pp. 6-7.

88. Interview with M. Grandmougin.

contribution to the development and implementation of new environmental (water quality) policies.

In 1992 the Water Act of 1992 (*Loi sur l' Eau 1992*) was issued[89], which introduced a new planning system in French water management. According to this law, each of the basin committees has to develop a strategic water management plan, called SDAGE (*Schéma Directeur d' Aménagement et de Gestion des Eaux*). On the regional and local levels the SDAGE is elaborated in SAGE-plans (*Schéma d' Aménagement et de Gestion des Eaux*). In the SDAGE to be developed for the Artois-Picardie basin the new water policies that apply to the French part of the Scheldt basin will be formulated.

4.7.2 Developments in Belgian water management

During the case study period, the water management organization in Belgium changed fundamentally. In 1967, when the case study starts, Belgium was a unitary state. At that time the Belgian central government had almost all water management competencies, and the exclusive competence to conclude international conventions. In 1997, at the end of the case study period, Belgium has become a full federal state, and almost all water management competencies, including the competence to conclude international conventions, have been attributed to the Walloon, Flemish, and Brussels Capital Regions. In the following sections, the successive stages of the Belgian state reforms are discussed first. Subsequently, the main development of water policies in Belgium, and within the three separate regions are discussed briefly.

The Belgian state reforms
The state reforms that transformed the Belgian unitarity state into a full federal state comprised several constitutional amendments. The first constitutional amendment took place in 1970. By this amendment a restricted cultural autonomy was given to the Dutch-, and French-speaking communities[90], which, among other things, received competencies to develop and implement acts. The next reforms concerned the constitutional amendments of 1980 and 1983, and several acts to implement these amendments. In 1980 the Belgian regions Wallonia and Flanders were established, which received competencies for the development and implementation of regional acts, and in 1983 the small minority of German-speaking Belgians received cultural autonomy as well.[91] The last Belgian state reforms concerned the constitutional amendments of 1988 and 1993. In 1988, besides the Walloon and Flemish Regions, the Brussels Capital Region was created, and the three regions received most water management competencies that had not yet been attributed to them in 1980.

Since 1988 Belgium is a full federal state with three regions and three communities, which each do have a directly elected parliament, and a federal state level with a relative small parliament. The constitutional amendment of 1993 concluded the latest (last?) stage of the Belgian state reforms. Although most water management competencies had already been attributed to the three regions in 1980 and 1988, the amendment of 1993 still caused a fundamental change, because the competencies to conclude international conventions were attributed to the regions and communities as well. Since than the regions and communities are able to start international negotiations on policies for which they have the competencies.

---

89. Loi sur l' Eau 1992.

90. Alen *et al.* (1990, p. 178).

91. Ibid.

For the coordination of the Belgian standpoint in international organizations, such as in the EC, a special coordination commission was installed.

Two main reasons for the rather complicated organization of the Belgian state, and the establishment of both communities and regions are that Belgium has a German-speaking minority, and that the city of Brussels is bilingual, i.e. has a French- and Dutch-speaking part.[92] Therefore, the architects of the Belgian federal state designed two types of administrative units: the communities and the regions. The former units, the French, Flemish and German Communities, are responsible for person related issues, such as family and immigration policies. The latter ones, the Walloon, Flemish and Brussels Capital Regions, are responsible for territory-related issues, such as land-use planning, environmental, water, and nature management, and transport.[93] Consequently, for the study of decision making on international Scheldt issues, the Belgian regions are most relevant. In the Flemish part of Belgium, the parliaments, governments, and ministries of the Flemish Community and Region are united. Consequently, there is only one Flemish parliament, government, and one Ministry of the Flemish Community.[94] This sometimes is called the asymmetry of the Belgian federal structure.[95]

Development of Belgian water management
Like in France, also in Belgium the awareness of the severe water pollution caused by domestic and industrial waste discharges increased in the sixties. Therefore, in 1971 the Act on the protection of surface waters was adopted.[96] According to this Act, which has been modified several times since then, the discharge of waste water into surface water is prohibited, unless the Belgian government has granted a waste water permit. In 1983 the Act concerning the water quality objectives for surface waters was adopted.[97] This Act mainly aimed at the implementation of several EC water quality Directives. According to this Act, the assignment of functions to the waters is the responsibility of the Belgian regions.

Development of water management in the Flemish Region
Because of the successive stages of the Belgian state reforms, the Belgian regions became the most important actors in Belgian water management. To implement the national Act of 1983, the Flemish government developed a General Waste water treatment Programme (AWP).[98] This programme is the basis for the operational water quality policies in the Flemish region. Anticipating the constitutional amendment of 1988, which attributed the remaining water management competencies to the regions, in 1987 the Flemish government decided that all surface waters in the Flemish region should have the basis quality in 1995.[99] In addition, the region started with the implementation of several EC-water quality directives, and the

---

92. Van Istendael (1989).

93. TK 1993-1994, 23 536, Nr. 1, p. 1.

94. Ibid., p. 4.

95. Ibid.

96. Wet op de bescherming van het oppervlaktewater tegen verontreiniging (26-3-1971).

97. Wet betreffende de kwaliteitsobjectieven van het oppervlaktewater (24-5-1983).

98. Ovaa (1991, p. 87), AWP=Algemeen Waterzuiveringsprogramma.

99. Ovaa (1991, p. 86), Paepe (1991, p. 211).

Flemish Agency for Waste water Treatment (VMZ) developed ambitious investment plans.[100] In spite of these ambitious plans, the quality of the Flemish surface waters did hardly ameliorate. The main Belgian problems concerned the lack of waste water treatment infrastructure, and sewer systems in the main urban areas. According to Pallemaerts, in 1989 only 15 % of the about 14,000 polluting industries did have a valid permit.[101] Cappaert estimated that about 42 % of the 8,600 industries that discharge waste water into surface waters did have a permit at that time.[102]

In 1990 a combined Environmental Policy and Nature Development Plan (MINA-plan) was issued[103], which aims at a sustainable development of the Flemish region. This policy plan is elaborated in an environmental policy plan, and a nature development plan. To finance the planned environmental policies, the MINA-fund was established, which receives its revenues from, among others, charges on waste water discharges. After a reorganization of the Flemish administration, the Flemish Environment Agency (VMM) received the task to develop the AWPs[104], and the private company AQUAFIN, which is responsible for the implementation of these plans, was established. The investments made by AQUAFIN are financed by the MINA-fund. In 1991 the VLAREM regulation, which aims to streamline environmental permitting in the Flemish region, is introduced.[105]

Finally, since 1992 special attention is given to the development of several basin committees (*Bekkencomités*) in the Flemish region, which indicates that policy makers in the Flemish region are inspired by the water systems and river basin approach.[106] With these new institutional arrangements and policies the Flemish region tries to eliminate the backlog it has compared with the surrounding Western European countries.

Development of water management in the Walloon Region
After the Belgian constitutional amendment of 1980, like the Flemish region also the Walloon region received most water and environmental management competencies. The region used these competencies for the development of regional water quality regulations that largely replaced the national Belgian water quality Act of 1971. The most important regulations introduced by the Walloon government were the regulation concerning the protection of surface waters against pollution of 7 October 1985[107], and the regulation concerning the charges on the discharge of industrial and urban waste water of 30 April 1990.[108] The first Act aims at the protection of the water quality, and regulates the issuing of permits for waste discharges, and the formulation of water quality objectives. The general rule is that the discharge of industrial waste water is prohibited, unless the Walloon administration has issued a permit. The second Act introduces a system of charges on the discharge of waste water. In spite of these regulations, in 1991 only about 800 of the 3,000 industries that would

---

100. Cappaert (1988), VMZ =Vlaamse Maatschappij voor Waterzuivering.

101. Pallemaerts (1991) cited in van Ast and Korver-Alzerda (1994, p. 51).

102. Cappaert (1990) cited in van Ast and Korver-Alzerda (1994, p. 51).

103. MINA =Milieu- en Natuurbeleidsplan Vlaanderen.

104. VMM =Vlaamse Milieu Maatschappij.

105. VLAREM =het Vlaamse Reglement betreffende Milieuvergunningen.

106. De Wel (1994). For more information on these committees, see also Appendix 6.

107. Décret du 7 octobre 1985 sur la protection des eaux de surface contre la pollution.

108. Décret instituant une taxe sur le déversement des eaux industrielles et domestiques, 30-4-1990.

have to apply for a permit, did actually have a permit.[109] Like the other Belgian regions, also the Walloon region had to develop a new administration, and had to cope with a lack of personnel capacity. In 1995 an environmental management plan was issued.[110] Since a few years, for several (sub) basins in the Walloon regions so-called river contracts (*contrats de rivière*) are concluded.[111]

Development of water management in the Brussels Capital Region
Unlike the Walloon and Flemish Regions, which had received most water management competencies already in 1980, the Brussels Capital Region received these competencies only in 1988.[112] Therefore, a new administration had to be established.[113]

The water quality of the Senne, the tributary of the river Scheldt that flows through the city of Brussels, is very bad.[114] A report of the Brussels government of 1995 states that : "*As regards the treatment and monitoring of the water quality, and waste discharges, Brussels has a considerable backlog.*" ("*Wat de zuivering en het toezicht op de kwaliteit van het water en de controle op de lozingen betreft, vertoont Brussel een aanzienlijke achterstand.*")[115] Two large waste water treatment plants, which are planned north and south of the city of Brussels, however, will improve water quality of the river Senne considerably.[116]

4.7.3 Developments in Dutch water management[117]
The two decades after the second world war were characterized by high economic growth rates. The average annual economic growth between 1960 and 1969 was 5.5 %. This economic growth was accompanied by a rationalization of agriculture and increasing industrial activities. In the first decennia after the second world war the emphasis in water management was put on flood protection, and the control of water levels for agricultural purposes. In 1968 the first national policy document on water management was issued, which focused on the user functions, and the expected increase of water use by industries and households. In the sixties, the negative effects of the economic growth and industrialisation became apparent, and awareness of water pollution increased. Acts and regulations to deal with these emerging problems effectively, however, were lacking.

Whilst the sixties can be characterized by economic growth and prosperity, the seventies and early eighties were characterized by rather low economic growth rates. Awareness of environmental issues increased further. The PAWN-study contributed to the knowledge of the interrelations between the many interests or user functions, and for the first time explicitly addressed the nature function.[118] The sanitation of water resources started, and the

---

109. Stichting Reinwater (1992, p. 77).

110. Gouvernement wallon (1995); Plan d' Environnement pour le Développement durable.

111. For more information on these river contracts, see Appendix 6.

112. BIM (1995, p. 276).

113. For a description of the present administrative organization in the Brussels Capital Region, See Appendix 6.

114. BIM (1995, p. 117).

115. Ibid., p. 289.

116. Ibid., p. 135.

117. This brief overview largely draws on the work of Grijns and Wisserhof (1992).

118. PAWN =Policy Analysis of Water management for the Netherlands.

necessary legal-administrative instruments were developed. In 1970 the Surface Water Pollution Act (WVO) came into force. This Act is based on the polluter pays principle, and introduced a permit system and charges on the discharge of waste water. The revenues were used to finance the construction of waste water treatment plants. In 1971 the waste water treatment capacity in the Netherlands was 5 million inhabitant equivalents (i.e.). In 1984 this was already 20 million i.e. In 1985 85% of the waste water discharges by households were treated. In addition to emission policies immission policies were developed. The water quality policy plan IMP Water 1980-1984 contained water quality objectives for the Dutch water systems.[119] In 1984 the second national policy document on water management, which largely draws on the PAWN-study, was issued. The second policy document focused on water quantity issues, because water quality issues were addressed in the IMP-Water. In the second policy document on water management the interrelations between the water quality and water quantity of both groundwater and surface waters were emphasized, and the water beds were addressed as well. In this policy document the problems of transboundary water pollution in the rivers Rhine, Meuse, and Scheldt were mentioned explicitly.

The late eighties and early nineties were characterized by high economic growth and an increasing consensus on the need for more environmental friendly policies and sustainable development. Environmental issues got a high place on the political agenda. The report 'Living with Water' (Omgaan met Water) of 1985 introduced the water systems approach and emphasized the need for an integrated approach of water management problems. Also in the IMP-Water 1985-1989 the interrelations between the compartments of the environment were recognized. In 1989 the third policy document on water management was issued. This policy document elaborated the concepts of water systems approach and integrated water management, which had been introduced in the report 'Living with Water', and put the emphasis on the internal functional interrelations (between water quantity and quality of groundwater and surface water), and the external functional interrelations (between water management, environmental and nature management, and land-use planning). It specifically addressed international water management issues, and aimed at the development of integrated river basin policies coordinated by international river commissions.[120] In 1989 the Water Management Act came into force, which aimed at the coordination of all sectoral water management policies.[121] This Act introduced a new planning system. In this system the Dutch provinces do have an important coordinating role.

When this thesis is finished, the fourth policy document on Water Management is issued.[122] This document mainly aims to intensify the policies that have been formulated in the third policy document on water management.

4.7.4 Comparison and analysis of developments in the Scheldt basin states and regions

Within the Scheldt basin different stages of development can be observed as regards the development of water policies. The Netherlands was the first basin state that developed huge investment plans for waste water treatment. Subsequently, investments in waste water treatment were increased in the French and Flemish parts of the Scheldt basin. More

---

119. IMP = Indicatief Meerjarenprogramma.

120. V&W (1989).

121. Wet op de Waterhuishouding.

122. V&W (1997).

recently, also in the Walloon and Brussels region investments in waste water treatment were increased. Four supplementary explanations for the different stages of development of water quality policies in the Scheldt basin states and regions are the upstream-downstream relationships in the basins of the rivers Scheldt and Meuse, the Belgian federalization process, the different socio-economic circumstances in the basin states, and the different scores on the cultural dimensions Femininity and Uncertainty Avoidance.

A first reason for the relative early development of water quality policies in the Netherlands may be that the Netherlands is situated downstream of the other basin states and regions. Because of its downstream position the Netherlands was confronted first and most with pollution problems. To have a strong position in international negotiation processes with its neighbours, the Dutch had to clean-up their own waters first. Only in second instance the other basin states and regions could be demanded to reduce the transboundary load of pollutants. Later, a similar development can be observed in the Flemish region. This region first had to develop plans to clean-up the waters in the Flemish region, before it was able to make a credible demand on the Walloon and Brussels Capital Regions, and France to reduce the transboundary load of pollutants flowing to the Flemish region.

A second reason why water quality policies did develop later in Belgium than in the Netherlands may be the Belgian federalization process. The successive constitutional amendments caused a permanent redistribution of competencies and responsibilities among the federal state and the regions, including those concerning the development of environmental policies. The absence of clearly defined competencies and responsibilities seem to have hindered the development of effective environmental policies for a long time.[123]

Thirdly, differences in economic prosperity may explain partly the different stages of development of environmental policies in the basin states and regions. The structural (long lasting) socio-economic problems in the upstream parts of the Scheldt basin, i.e. in the Walloon region and northern France create a situation in which political priorities are with the fight against unemployment and with economic development, more than with the development of environmental policies.

Finally, the cultural differences between the Scheldt basin states discussed in Section 4.6 may be a deep cause of the different stages of development of environmental policies. It was argued that in feminine countries environmental issues generally did appear on the agenda earlier than in more masculine countries. Therefore, the femininity of the Dutch culture may be one explanation for the relative early development of water quality policies in the Dutch part of the Scheldt basin. The different scores of the Dutch- and French-speaking parts of Belgium on the cultural dimension Masculinity (MAS) may explain why environmental policies in Flanders did develop earlier than in Wallonia. Besides the MAS-scores also the different scores on the dimension Uncertainty Avoidance (UAI) may explain partly the different stages of development. In Section 4.6 it was argued that low UAI-countries generally are better able to deal with unstructured and complex problems, such as most environmental issues, than high UAI-countries, and the Netherlands has a much lower UAI-score than the other basin states.

---

123. Bedet (1993, p. 27).

4.8   Historical context

The Scheldt basin states have a long history of conflict and cooperation over the Scheldt river, which mainly concerns the navigation on that river. In the next sections this history is described shortly. The overview draws on several secondary sources of information. Interesting historical studies in Dutch were made by Smit[124], Strubbe[125], Van den Heuvel and Ligtermoet[126], and Somers.[127] Overviews in French were made by d'Argent and Planchar.[128] Finally, Bouchez made a detailed historical overview in English.[129]

Antwerp prosperity
In early times the main downstream part of the river Scheldt was the present Scheldt eastern branch. Only in 838 a storm surge brought about a connection between the river Scheldt and the "Honte", the eastern part of the present Scheldt western branch or Western Scheldt.[130] This event caused an increase of the tidal volume in the Scheldt, and the development of the present estuary of the river Scheldt.[131] These geophysical changes improved the maritime access to the harbour of Antwerp, which in turn enabled a steady development of this harbour. Other storm surges in the 14th and 15th centuries increased erosion of the Western Scheldt, and improved navigation possibilities further. Partly because of these advantageous developments, economic development of Antwerp reached its height in the 16th century. At that time the city had 82,000 inhabitants and was, after Paris, the second largest city of Europe.[132] The Dutch Province of Zealand also took advantage of the economic prosperity of the harbour of Antwerp, because it took a toll on commodities transported to Antwerp via the Eastern and Western Scheldt.[133]

Blockade of the Scheldt
The religious wars of the 16th century brought an end to Antwerp prosperity. Antwerp joined the rebels against the Spanish rule, commanded by prince William of Orange.[134] The King of Spain, Philips II, appointed Alexander Farnese, the duke of Parma, as governor of the Netherlands with the main task to recapture the Netherlands. In 1584 Farnese started to lay siege to Antwerp. A blockade of the river Scheldt was a crucial part of this strategy. Farnese ordered to build forts on the left and right bank of the river Scheldt, and to link together 32 ships between these forts.[135] In March 1585 the blockade of the Scheldt was completed, and

---

124. Smit (1966).

125. Strubbe (1988b).

126. Van den Heuvel and Ligtermoet (1989).

127. Somers (1992).

128. d'Argent (1997), Planchar (1993).

129. Bouchez (1978).

130. Strubbe (1988b, p. 237).

131. For an overview of the development of the course of the river Scheldt, see Franssen and Schuurman.

132. Strubbe (1988b, p. 237).

133. Ibid.

134. Ibid.

135. Ibid.

after violent fights Antwerp capitulated on 17 August 1585.[136] During the war between Spain and the northern Netherlands the Scheldt was not reopened. Although commercial navigation was allowed, the loads had to be transshipped twice. The first transshipment had to take place in one of the Zeeland harbours. Subsequently, Zeeland ships transported the loads to Lillo, where they were transshipped once more, and the northern Netherlands took a toll. Obviously, these obligations were beneficial to Zeeland harbours, such as Middelburg and Flushing.[137] They did, however, cause a population decline and economic recession in Antwerp.[138]

In 1609 the war between the northern Netherlands and Spain was concluded temporarily, and Antwerp expected the Western Scheldt to be re-opened. The region of Zeeland, however, supported by the Province of Holland (including the harbour of Amsterdam) continued to prohibit direct navigation to Antwerp.[139]

In 1648 the Treaty of Münster concluded the war between Spain and the northern Netherlands once and for all.[140] It provided for the separation between Spain and the northern Netherlands. According to the Treaty the northern Netherlands maintained the right to prohibit direct navigation to the harbour of Antwerp.[141] This was a real disillusionment for Antwerp. Even though the war was ended, direct navigation to Antwerp still was impossible. Was this the reward for their decision to side with the northern Netherlands, and to support the rebellions commanded by Prince William of Orange?

The Republic of the Netherlands, however, benefited from the Treaty, and got its golden age.[142] Because the Netherlands was one of the most powerful nations of the world, it was able to disregard objections against the Treaty of Münster, which were made by other countries, and to continue the blockade for more than a century.[143]

Discontinuance of the blockade
In 1795 the French army invaded the Republic of the Netherlands. The French were inspired by nature law and were of opinion that rivers are a common property of mankind. A restriction of the freedom of navigation was interpreted as a violation of human rights.[144] On 16 May 1795 France and the Batavian Republic concluded the Peace Treaty of The Hague.[145] According to this Treaty, navigation on Rhine, Meuse and Scheldt is free for all riparian states.

After the abdication of Napoleon the Paris Peace Treaty of 30 May 1814 regulated the Peace between France and the Allies. It stipulated that navigation on the international rivers has to be part of a definitive regulation. The General Act of the Vienna Congress of 1815

---

136. Ibid.

137. Ibid., p. 238.

138. Ibid.

139. Ibid.

140. Treaty of Münster, 30 January 1648.

141. Strubbe (1988b, p. 238).

142. 'Golden Age' is another name of the 17th century in the northern Netherlands. This was a century of unprecedented economic prosperity and a prime of arts and science.

143. Strubbe (1988b, p. 238).

144. Ibid.

145. The Batavian Republic was the name of the Netherlands under French rule.

provided this regulation. It laid down the principle of freedom of navigation on the large international rivers.[146] Because the southern and northern Netherlands were re-united, and until 1830 navigation on the Scheldt was free, a flourish of commerce started in Antwerp.[147]

The re-union between the northern and southern Netherlands would not last long. On 4 October 1830, after the Belgian rebellion in Brussels, a provisional Belgian government declared Belgians independence. As a reaction the Dutch King William I closed the Scheldt. France and England, however, forced the northern Netherlands to discontinue the blockade of the Scheldt in 1831, and decided to separate the northern and southern Netherlands.[148]

The statute of the Western Scheldt

The separation between Belgium and the Netherlands was regulated in the Separation Treaty of 19 April 1839.[149] Article 9 of this Peace Treaty contains the statute of the Western Scheldt. In Section 1 of Article 9 the provisions of Articles 108 to 117 inclusive of the General Act of the Vienna Congress were applied to all rivers that form or cross the border between the two countries. These articles regulate, among other things, that the basin states have to regulate navigation issues 'de commun accord'.[150] Furthermore, they state that commercial navigation is free[151], and that each state is obliged to implement the necessary works to maintain the navigability.[152] To compensate the maintenance costs caused by these obligations the basin states were allowed to levy duties on transported loads.[153]

Section 2 of Article 9 of the Peace Treaty provides for a joint supervision on piloting and buoyage of the Western Scheldt by a Permanent Commission for supervising Scheldt navigation (Permanente Commissie voor Toezicht op de Scheldevaart). The statute of the Western Scheldt is worked out in more detail in the Belgian-Dutch Treaty of 1842[154], and a Belgian-Dutch agreement of 1843.[155] Because the statute of the Western Scheldt did allow the Dutch to take tolls, and the Belgian government did not want to hinder the development of the harbour of Antwerp, the Belgian government decided to pay the necessary tolls for commercial navigation.[156]

On 12 May 1863 the Netherlands and Belgian reached an agreement on the abolition of tolls on Scheldt navigation. The Belgian government had to redeem the Dutch by paying an amount of 36.278.566 BF.[157] The State of Belgium, however, did not have to pay the whole sum, because other maritime superpowers, among which England, showed solidarity and

---

146. Strubbe (1988b, p. 238).

147. Ibid.

148. Ibid.

149. Stb. (Official Law Journal of the Netherlands) 1839 No. 26.

150. General Act of Vienna of 9 June 1815, Article 108.

151. Ibid., Article 109.

152. Ibid., Article 113.

153. Ibid., Article 111.

154. Stb. 1843 No. 3.

155. Stb. 1834 No. 43.

156. Strubbe (1988b, p. 239).

157. Ibid.

contributed to the redemption as well.[158] Simultaneous to the agreement on the abolition of tolls on Scheldt navigation Belgium and the Netherlands reached agreement on a Treaty concerning the use of Meuse water.[159]

Different interpretations of the statute of the Western Scheldt
Halfway the 19th century the Dutch developed plans to construct a railway between the Province of Brabant and Flushing. Implementation of these plans made it necessary to dam the Scheldt eastern branch, which was completed in 1867.[160] To provide for an alternative navigation channel connecting the Scheldt western branch to the eastern branch, the Dutch completed the canal through South Beveland in 1866. Belgian authorities objected against the Dutch engineering works, because they were not consulted. According to the statute of the Western Scheldt, however, the Dutch would be obliged to discuss these issues with Belgium.[161]

In 1889 the Belgian government informed the Dutch government for the first time that a bar in the Western Scheldt hindered commercial navigation. The Belgian government added that according to the statute of the Western Scheldt the Dutch government was obliged to carry out the necessary dredging works.[162] The Dutch, however, replied that the Belgian demand was not appropriate, and the problem could be solved by the installation of warning lights. Although the Dutch were afraid that Belgium would not accept their reaction, the Belgian government decided not to reply.[163]

Some years later Belgium authorities observed another bar in the navigation channel on Dutch territory. Once again, they informed the Netherlands that they would like the Dutch to carry out the necessary dredging works. This time, however, the Belgian authorities added that they were willing to finance the necessary dredging works.[164] Later, the requested dredging works turned out to be less urgent than the Belgian government had expected.

In 1905 Belgian authorities intended to carry out dredging works on Dutch territory. Main objective of these dredging works, however, was not the maintenance of the navigation channel in the Western Scheldt. The Belgian government aimed at sand mining for extending the port of Antwerp.[165] Because the Dutch wanted to safeguard their sovereignty over the Western Scheldt, they informed the Belgian authorities that they would need a dredging permit to implement these dredging works.[166] The first dredging permit for the Belgian government was issued in 1906.

---

158. Ibid.

159. Strubbe (1988b, p. 241). This example shows that the frequent use of the strategy of tactical issue-linkage between Scheldt and Meuse issues in the negotiations on the so-called water conventions between 1967 and 1995 is nothing new under the sun.

160. Strubbe (1988b, p. 239).

161. Ibid.

162. Van den Heuvel and Ligtermoet (1989, p. 219).

163. Ibid.

164. Ibid.

165. Ibid.

166. Ibid, p. 216.

Attempts to revise the statute of the Western Scheldt

After the first World War Belgium aimed at a revision of the Separation Treaty of 1839. It wanted to annex Zeeuws-Vlaanderen and Dutch Limburg, and to have absolute sovereignty over the Western Scheldt and the Canal Ghent-Terneuzen.[167] Although the super powers showed understanding for the Belgian request, they did not support the Belgian annexation plans.[168] Negotiations on a revision of the Separation Treaty were started in 1919, and were concluded with a draft Treaty in 1925. This draft Treaty contained several important amendments of the statute of the Western Scheldt.[169] First, it stated that the navigation channel of the Western Scheldt always had to meet the needs of modern navigation. Secondly, each country would have to bear the costs of necessary infrastructure and maintenance works on its territory. Thirdly, Belgium would receive co-sovereignty over the Western Scheldt. Finally, the Treaty provided for the construction of two canals connecting Antwerp to the Rhine.[170] The draft Treaty of 1925 was part of a package deal enabling the Dutch Minister of Foreign Affairs, Van Karnebeek, to arrange a number of territorial issues, among others in Limburg, with Belgium.[171] Whilst the Dutch Second Chamber of parliament approved the draft convention, the First Chamber was of the opinion that the proposed revision of the statute of the Western Scheldt did not serve the Dutch interests sufficiently, and sent the Dutch Minister of Foreign Affairs away in 1927.[172]

The connection between Rhine and Scheldt[173]

In 1948 the Benelux Ministers decided to install the Technical Scheldt Commission (TSC). Its task is to advise and carry out research on technical problems concerning navigation on the Western Scheldt.

In 1949 Belgium and the Netherlands installed a commission to study problems concerning waterways and harbours of common interest to the two states.[174] One of the main issues discussed in this commission was the improvement of the connection between the Rhine and the Scheldt. As mentioned in the previous section, Antwerp had demanded for such a connection after the first World War. It wanted to dig a canal between Antwerp and the 'Hollands Diep', the Moerdijk canal. Negotiations, however, went laboriously, and the delegations were unable to reach an agreement. Major issues preventing a solution acceptable to the two countries were that: (1) Belgium insisted on the construction of the Moerdijk canal in spite of the Delta project undertaken by the Netherlands as a result of the flood which

---

167. Strubbe (1988b, p. 239).

168. Ibid.

169. Ibid.

170. Ibid. The two canals were the Moerdijk canal connecting Antwerp to the Hollands Diep via Roosendaal, and the Ruhrort canal connecting Antwerp to the Rhine crossing Dutch Limburg.

171. Strubbe (1988b, p. 239).

172. Van den Heuvel and Ligtermoet (1989, p. 220). According to one respondent this event has influenced the Dutch strategic choices in the negotiations on the water conventions between 1967 and 1995, because the Dutch delegation wanted to prevent a similar event.

173. For a detailed overview of the negotiations on the Treaty concerning the connection between Rhine and Scheldt, see Smit (1966).

174. Bouchez (1978, p. 269).

afflicted the country, and in particular the province of Zeeland, on 1 February 1953[175]; and (2) the financial support granted by the Belgian government to its Rhine boats.[176] These problems were finally settled by the Treaty concerning the Connection between the Scheldt and the Rhine that was signed on 13 May 1963.[177] This Treaty provided for the construction of a canal between the harbour of Antwerp and the Krammer-Volkerak. Article 16 of the Treaty states that Dutch fresh water losses in the Scheldt-Rhine connection have to be compensated by Belgium somewhere along the Belgian-Dutch border. Furthermore, the 1963 Treaty confirmed the principle of freedom of navigation, and Belgium waived all claims relating to the maintenance and quality of the navigation routes connecting the Scheldt with the Rhine derived from treaties in force at the time of the signing of the 1963 Treaty.[178] For that reason, at that time international Scheldt issues seemed to be settled. The case study, however, will show that several new issues will appear on the international agenda, many of them related to the navigation to and from Antwerp.

Pollution reduction, a new issue on the Belgian-Dutch agenda
Another Belgian-Dutch navigation issue was the improvement of the Canal between Ghent and Terneuzen. The Treaty concerning the improvement of this canal was signed on 20 June 1960.[179] Most interesting is that this treaty was the first Belgian-Dutch agreement concerning the prevention of pollution in the Scheldt estuary[180], because it contains specific provisions governing pollution prevention.

In Table 4.5 the crucial events in the history of conflict and cooperation over the Scheldt river, and their implications are listed.

4.9   Conclusions

The description and analysis of the context of decision making on international Scheldt issues presented in this chapter focused on the underlying hydrological structure of the issues at stake, the international context, the cultural context, the intranational developments within each of the basin states, and the historical context. Hereafter, the main conclusions of the analysis of each of these dimensions of the context are presented.

The intellectual coherence between some international Scheldt issues made it possible to distinguish between two main issue-areas: the issue-area of the maritime access to the port of Antwerp, and the issue-area of water and sediment pollution. Both issue-areas do contain ecological issues. The issues related to the maritime access to the port of Antwerp mostly are

---

175. The Delta project encompassed, *inter alia*, the reinforcement of the dikes along the Rotterdam waterway and the western branch of the Scheldt, and closing the outlets to the sea between Walcheren and North Beveland, between North-Beveland and Schouwen, between Schouwen and Goeree-Overflakkee, between Goeree-Overflakkee and Voorne. This Project was approved by the Act of 8 May 1958.

176. Bouchez (1978, p. 269).

177. Trb. 1963 Nr. 78. Verdrag betreffende de verbinding tussen de Schelde en de Rijn.

178. Bouchez (1978, p. 269).

179. Trb. 1960 Nr. 105. Verdrag betreffende de verbetering van het kanaal van Terneuzen naar Gent en de regeling van enige daarmede verband houdende aangelegenheden.

180. Bouchez (1978, p. 272).

Table 4.5 *Crucial events in the history of conflict and cooperation over the Scheldt river, and their implications*

Year	Event	Implications
1585	Blockade of the Scheldt	Capitulation of Antwerp Spanish suppression
1648	Treaty of Münster	Separation between Spain and the northern Netherlands Scheldt remains closed
1795	French suppression of the Netherlands	Discontinuance blockade of the Scheldt
1815	General Act of Vienna Congress	Regulation of freedom of navigation on international rivers
1839	Separation Treaty (including Scheldt Statute)	Separation of Belgium and the Netherlands Application Articles of Act of Vienna to rivers and waterways that form or cross the Belgian-Dutch border
1863	Redemption of Scheldt toll	Unconditional freedom of navigation to Antwerp
1906	Issuing of first dredging permit	Belgium needs permission for dredging activities on Dutch territory
1948	Installation of Technical Scheldt Commission	Permanent deliberations concerning technical Scheldt issues
1961	Treaty concerning the improvement of the Canal Ghent-Terneuzen	First Belgian-Dutch agreement concerning pollution prevention in the Scheldt estuary
1963	Treaty concerning the connection between Scheldt and Rhine	Issue of inland navigation between Scheldt and Rhine is settled

bilateral Belgian (Flemish)-Dutch issues, whereas the issues related to water and sediment pollution mostly are multilateral issues involving all Scheldt basin states and regions. Most international Scheldt issues are typical upstream-downstream issues. The underlying hydrological structure of most issues in the issue-area of the maritime access to the port of Antwerp causes a clear distribution of interests and resources between Belgium (Flanders) and the Netherlands. The former perceives a dependence on the latter as regards the maintenance and improvement of the maritime access to the port of Antwerp. The underlying hydrological structure of most issues in the issue-area of water and sediment pollution causes a more complicated pattern of relations of (inter)dependence, since five main parties are involved in decision making on pollution-related issues. The party situated downstream in the basin of the Scheldt or one of its tributaries perceives a dependence on the parties situated upstream in the basin as regards the supply of a reasonable amount of water of a reasonable quality (See Table 4.3).

Beside the decision making on international Scheldt issues, two or more of the Scheldt basin states and regions were involved in decision making on international issues in the adjacent basin of the river Meuse, and the Rhine basin. All Scheldt basin states and regions were involved in decision making on international Meuse issues, which mainly concerned water and sediment pollution, and the distribution of Meuse water in periods of low flow. The underlying hydrological structure of these issues influenced the distribution of interests and resources among the Scheldt (and Meuse) basin states and regions, and consequently their perceived (inter)dependencies as well. France and the Netherlands were (are) involved in decision making on the clean-up of the river Rhine, which generally is considered to be a successful case of international river cooperation. The Scheldt basin states were (are) involved in decision making on general water policies, such as European water policies, North Sea policies, and the river policies of the UN-ECE, as well. Most of these policies apply to the river Scheldt.

Two main conclusions from the cultural analysis are that there is a cultural gap between on the one hand the Latin cultures of Belgium and France, and on the other hand the Nordic culture of the Netherlands, and that cultural differences between the French- and Dutch-speaking parts of Belgium are rather small. France and Belgium are rather unequal societies with a relatively low tolerance for uncertainty, whereas the Netherlands is a relatively equal society with a relatively high tolerance for uncertainty. Unlike France and Belgium, the Netherlands has a rather feminine culture.

The most eye-catching *intra*national development are the Belgian state reforms. During the case study period the Belgian unitary state was transformed into a full federal state. The successive stages of the Belgian state reforms entailed a continuous redistribution of competencies between the federal state and the three regions. Furthermore, comparing the Scheldt basin states and regions, different stages of development can be observed as regards the development of water and environmental policies. The Netherlands was the first party that developed huge investment plans for waste water treatment. Subsequently, investments in waste water treatment were increased in the French and Flemish parts of the Scheldt basin. More recently, also in the Walloon and Brussels Capital regions investments in waste water treatment infrastructure have been increased. Four supplementary explanations for the different stages of development of environmental policies in the Scheldt basin states and regions are their different relative hydrologic positions in the Scheldt and Meuse basins, the Belgian federalization process, the different socio-economic circumstances in different parts of the basin, and the different scores on the cultural dimensions Femininity and Uncertainty Avoidance.

Although all Scheldt basin states and regions are Western European market economies characterized by a high standard of living, in the period studied in this thesis, there was a considerable difference between the economic prosperity in the northern and southern parts of the basin. In the Dutch and Flemish parts of the basin economic growth is relatively constant, whilst the Walloon and French parts of the basin have to cope with economic recessions and high unemployment rates.

The Belgian-Dutch history concerning the management of the river Scheldt mainly relates to the navigation on the river, and can be characterized by conflict better than by cooperation. During history the two basin states did have fundamentally different interests. The southern Netherlands or Belgium were interested in the maintenance or improvement of the maritime access to the harbour of Antwerp. The northern Netherlands, however, wanted to develop the Zeeland harbours and the harbours of Amsterdam and Rotterdam.

CHAPTER 5

PREPARATION OF THREE DRAFT CONVENTIONS AND A TOTAL DEADLOCK (1967-1985)

5.1   Introduction

In the preceding chapter the first part of the case study was presented. In that chapter the second research question was answered, and the context of decision making on international Scheldt issues between 1967 and 1997 was described and analyzed. The second part of the case study comprises a reconstruction and analysis of decision making on international Scheldt issues in the period indicated. This chapter focuses on the period between 1967 and 1985. Decision making in the periods between 1985 and 1993, and between 1993 and 1997 will be addressed in the Chapters 6 and 7.

For the reconstruction of the decision making process the rounds model of decision making introduced in Chapter 2 is used. In Section 5.2 the selection of crucial decisions, and the distinction between four decision making rounds are addressed. Next, the strategic interactions between the Scheldt basin states and regions in each of the rounds distinguished are described. By that the first research question is answered for the period between 1967 and 1985. In Section 5.3 the decision making process is analyzed and explained. In this section the explanatory research questions three to six are answered for the period studied in this chapter. These research questions successively address the influence of the perceptions of the actors involved in the decision making process on their strategies, institutionalization and learning processes, and the influence of the context on the perceptions and strategies of the actors. Finally, in Section 5.4 the main conclusions of the first 18 years of the case study are summarized.

5.2   Reconstruction of the decision making process (1967-1985)

Selection of crucial decisions, and distinction between decision making rounds
The reconstruction of decision making between 1967 and 1985 draws on written material. Reports of the Dutch Parliament, files and minutes of meetings of the Technical Scheldt Commission (TSC), and secondary literature were the main data sources consulted. The analysis of these written data made it possible to inventory the decisions taken by the actors involved in the decision making process. Appendix 12 contains this list of decisions. In the series of decisions, five crucial decisions which influenced the composition of the international arena or the international issues discussed, could be traced and therefore four decision making rounds could be distinguished. Table 5.1 produces information on the crucial

decisions marking the beginning and/or end of the decision making rounds and the main actors participating in these rounds. In the following, the strategic interactions preceding the crucial decisions will be described.

Table 5.1 *Decision making rounds, crucial decisions, and major actors (1967-1985)*

Round	Start situation	Major actors	End situation
I (1967-1968)	Belgian request to start negotiations on: 1. the construction of the Baalhoek canal 2. the straightening of the bend near Bath	Belgian government Dutch government	Belgian-Dutch agreement to start negotiations on: 1. the construction of the Baalhoek canal 2. the straightening of the bend near Bath 3. the water quality of the Scheldt 4. the water quality of the Meuse 5. the water quantity of the Meuse
II (1969-1975)	Installation of Belgian-Dutch negotiation commission	Belgian delegation Dutch delegation	Agreement between Belgian and Dutch delegations on three draft water conventions dealing with the issues 1 to 5
III (1975-1983)	Three draft conventions	Dutch government Belgian government Walloon and Flemish politicians	Walloon-Flemish disagreement Belgian government does not want to sign the conventions
IV (1983-1985)	Dutch decision to link negotiations on the 48'/43'/38' deepening programme to the negotiations on the issues 3 to 5	Belgian government Dutch government Belgian and Dutch Ministers of Foreign Affairs	Declaration of intent of Belgian and Dutch Ministers of Foreign Affairs stating that joint solutions will be searched for: 1. the water quantity of the Meuse 2. the water quality of the Meuse 3. the 48'/43'/38' deepening programme 4. the Baalhoek canal

Round I: Dutch linkage of Scheldt and Meuse issues (1967-1968)

The completion of the Scheldt-Rhine connection, which has been made possible by the Treaty concerning the connection between the Scheldt and the Rhine of 13 May 1963[1], shortens the navigation distance from Antwerp to the Rhine with 35 kilometres[2], and brings the port of Antwerp at the same distance to German cities, such as Leverkusen, Frankfurt and Ludwigshafen, as the port of Rotterdam.[3] Another project improving the access to the port of Antwerp is the construction of the Zandvliet sluice in 1967.[4] Due to this sluice the access to the harbour is moved downstream of the river Scheldt. The implementation of these infrastructure projects, and other Belgian policies aimed at an expansion of the port of Antwerp causes spatial planning problems because the expansion of the harbour on the right bank of the river has reached its limits. Therefore, Belgian government agencies develop plans to expand the harbour to the left bank of the river Scheldt[5], which raises the problem of the access to these harbours. The Belgian government intends the main access to the harbours to be on the left bank of the river Scheldt near the village Baalhoek on Dutch territory, and is developing plans to construct a canal between Baalhoek and the left bank of the river Scheldt.[6] Appendix 10 shows the planned alignment of this canal. Since the sluice and canal have to be constructed on Dutch territory, the Belgian government has to negotiate these plans with the Dutch government.

Beside the issue of the access to the harbours on the left bank of the river Scheldt, Belgium faces another problem, which needs to be discussed with the Dutch. Near the Belgian-Dutch border a sharp bend in the Scheldt exists, the bend of Bath. Because several shipping accidents had taken place near Bath, the Antwerp harbour developed plans to straighten this bend and by that to ameliorate the navigation channel from the North Sea to the port of Antwerp.[7] The construction of an artificial navigation channel would shorten the navigation distance from the North Sea to Antwerp with three kilometres.[8] Appendix 11 shows the planned alignment of the by-pass to straighten the bend near Bath.

In September 1967, the Belgian government informed the Dutch government that it would like to start negotiations on these two infrastructure projects. The Dutch government replied that it is prepared to discuss these issues and to be neighbourly, as long as no essential Dutch interests will be harmed, i.e. if the construction of the Baalhoek canal will not (1) cause planning or water management problems, (2) disturb competition between Belgian and Dutch sea ports, or (3) cause transboundary water and air pollution.[9] Furthermore, the Dutch government proposes to discuss the two issues simultaneously with three issues in which the Dutch government has an interest, namely the water quality and water quantity of the river Meuse, and the water quality of the river Scheldt.

---

1. Trb. (Treaty Series of the Netherlands) 1963 No. 78. Verdrag betreffende de verbinding tussen de Schelde en de Rijn.

2. Bouchez (1978, p. 270).

3. Suykens (1995, p. 229).

4. Ibid.

5. Ibid.

6. Ibid.

7. Ibid.

8. Strubbe (1988a, p. 152).

9. Ibid., p. 151.

Belgian-Dutch negotiations on the water quantity of the Meuse had been going on since 1963, and were aimed at a modification of two Meuse conventions of 1863 and 1873 dealing with the distribution of Meuse water between Belgium and the Netherlands.[10] Between 1930 and 1939 Belgium had constructed the Albert Canal thus connecting the inland port of the city of Liège to the port of Antwerp (See map in Appendix 5).[11] The water for this canal was diverted from the river Meuse. According to the Dutch, during dry periods the flow of the river Meuse at the Belgian-Dutch border would be too small to feed the Juliana canal on Dutch territory.[12] In addition, a new Meuse agreement would have to implement the obligation provided for in article 16 of the Treaty concerning the connection between the Scheldt and the Rhine.[13] This article states that Dutch fresh water losses in the Scheldt-Rhine connection have to be compensated by Belgium somewhere along the Belgian-Dutch border (See map in Appendix 5).[14] The losses are caused by the use of the Kreekrak sluices, through which brackish water from the Antwerp harbour flows into the Dutch freshwater lake "Zoommeer".[15] Later, Belgium and the Netherlands agreed that Belgium would compensate the Dutch fresh water losses in the Scheldt-Rhine connection by guaranteeing the Dutch a minimum flow of the river Meuse.[16]

The main reason for the Dutch interest in the water quality of the Meuse is the use of Meuse water for drinking water production. Although the Dutch are interested in an improvement of the water quality of the Scheldt as well, they have less economic interest in that issue. In 1968, the Belgian and Dutch governments decided to install a negotiation commission and to start negotiations on:
1. the construction of the Baalhoek canal;
2. the straightening of the bend near Bath;
3. the water quality of the Scheldt;
4. the water quality of the Meuse;
5. the water quantity of the Meuse.

Round II: Preparation of three draft conventions (1969-1975)
The linkage of issues in 1967 marks the beginning of more than 25 years intermittent negotiations on the so-called Belgian-Dutch water conventions. The first negotiations start on 14 May 1969, and last six years.[17] For reasons discussed above, the Dutch are interested most in regulations concerning the management of the water quantity and quality of the river Meuse. They, however, also emphasise the need to formulate water quality objectives for the

---

10. Stb. (Official Law Journal of the Netherlands) 1863 No. 18, and Stb. 1874 No. 23; Traité entre le Royaume des Pays-Bas et le Royaume de Belgique au sujet du partage et de la qualité des eaux de la Meuse, Article 42.

11. Strubbe (1988a, p. 149).

12. Ibid.

13. Trb. 1963 No. 78, Article 16. See also the historic overview presented in Section 4.8.

14. Bouman (1996, p. 161).

15. Strubbe (1988a, p. 149).

16. Ibid.

17. Maes (1990, p. 6).

river Scheldt in the Baalhoek and Bath conventions.[18] Eventually, all Scheldt water quality issues are regulated in the Bath convention. In addition, the Dutch want to develop regulations to prevent a distortion of competition between Belgian and Dutch Sea ports. The Belgian government is mainly interested in the construction of the two infrastructure projects to improve the maritime access to the port of Antwerp. On 19 June 1975, the delegations reach an agreement on three draft conventions dealing with the five issues[19]:

1. A Convention on the distribution and the quality of the water of the river Meuse[20];
2. A Convention on the construction of the Baalhoek canal[21];
3. A Convention on the improvement of the navigation channel in the Scheldt near Bath, which also contains regulations on the water quality of the river Scheldt.[22]

In the following, the contents of each draft convention will be summarized.

Draft Convention on the distribution and the quality of the water of the river Meuse
The Convention on the Meuse contains provisions for the water quantity and the water quality of the river. The waters of the Meuse are defined as: "*The waters of the river Meuse and her tributaries, as well as the waters connected to the Meuse and her tributaries, including the groundwater basins within the basin of the river Meuse and her tributaries, which are related hydrologically to the Meuse and her tributaries.*" ("*les eaux de la Meuse et de ses affluents ainsi que les eaux communiquant avec la Meuse et ses affluents, y compris les eaux souterraines qui, dans les limites du bassin de la Meuse et de ses affluents, sont, du point de vue hydrologique, en communication avec le fleuve et ses affluents.*")[23] As regards the water quantity Belgium and the Netherlands agree that from 1 January 1982 the minimum flow of the river Meuse upstream of Monsin, where the Albert canal starts, will be 50 m³/s.[24] Furthermore, it is agreed that Belgium will not use more than 22 m³/s for the feeding of the Albert canal and some smaller canals, namely the Kempen canals.[25] This implies that the minimum flow of the river Meuse at the Belgian-Dutch border will be 28 m³/s. The flow of the river Meuse will be regulated with storage reservoirs, which were already planned in the Belgian Ardennes.[26] The convention also contains provisions for monitoring the discharge of

---

18. Minutes meetings of the Technical Scheldt Commission (TSC) of 10 March 1970 in The Hague, and 2 March 1971 in Antwerp.

19. TK (Reports of discussions in the Second Chamber of the Parliament of the Netherlands) 1977-1978, Aanhangsel van de Handelingen, 854.

20. Traité entre le Royaume des Pays-Bas et le Royaume de Belgique au sujet du partage et de la qualité des eaux de la Meuse.

21. Traité entre le Royaume des Pays-Bas et le Royaume de Belgique au sujet de la construction du canal de Baalhoek.

22. Traité entre le Royaume des Pays-Bas et le Royaume de Belgique au sujet de l'amélioration de la voie navigable dans l' Escaut près du goulet de Bath.

23. Traité entre le Royaume des Pays-Bas et le Royaume de Belgique au sujet du partage et de la qualité des eaux de la Meuse, Article 2 Sub a.

24. Ibid., Article 6.

25. Ibid., Articles 7 and 10.

26. Ibid., Article 6.

the river Meuse on specified locations.[27]

Beside the agreements on the distribution of the water of the river Meuse, the draft convention contains detailed regulations concerning the water quality of the river. Objectives are formulated for the water quality at the Belgian-Dutch border (Eysden). The parties agree on a stand still for the first five years[28], and a significant amelioration of the water quality of the river Meuse in the periods between 5 and 15, and 15 and 25 years after ratification of the convention. The water quality objectives are described in an appendix to the convention.[29] The convention contains provisions for monitoring the water quality on specified locations[30] and the development of a sanitation plan to reach the water quality objectives formulated in the appendix to the convention.[31] In addition, the parties agree that they will inform each other about the planning of new industries or the expansion of existing ones if implementation of these plans may entail a deterioration of the water quality of the river Meuse.[32] The convention provides for the installation of a Belgian-Dutch Meuse commission[33], which consists of three to five representatives of the basin states[34], and meets at least twice a year.[35] The tasks of the commission are[36]: (1) to promote the implementation of the convention, (2) to answer questions of the respective governments, (3) to advise governments concerning the implementation of the convention, (4) to discuss and solve problems concerning the implementation of the convention, and (5) to perform all other tasks given to the commission in the convention. The decision making rule in the commission is unanimity.[37] The convention enables the contracting parties to modify the technical details of the convention if this is necessary because of new technological developments.[38] Furthermore, the parties agree that if after ratification of the convention Belgium and the Netherlands become contracting parties of a multilateral convention which contains more strict water quality policies, these policies will be integrated in the Meuse convention.[39] Finally, the convention contains provisions for conflict resolution.[40]

Draft Convention on the construction of the Baalhoek canal
The Baalhoek convention contains provisions for the construction of the Baalhoek canal. It describes the technical infrastructure that is necessary to construct the Baalhoek canal, such

---

27. Ibid., Article 16.
28. Ibid., Article 20 sub 1a.
29. Ibid., Appendix C.
30. Ibid., Articles 22 and 27.
31. Ibid., Articles 22 and 23.
32. Ibid., Article 25.
33. Ibid., Article 31.
34. Ibid., Article 32.
35. Ibid., Article 33.
36. Ibid., Article 37.
37. Ibid., Article 38.
38. Ibid., Article 20.3.
39. Ibid., Article 26.3.
40. Ibid., Chapter V (Articles 39-41).

as the Baalhoek sluice, the canal itself, the dockyards, and several dams.[41] All projects have to be financed by the state of Belgium.[42] In addition, the Baalhoek convention contains detailed economic regulations.[43] The considerans of the convention, among other things, contains the following statement: "*Desiring that the possibility for economic development of the left bank of the river Scheldt does not hinder the harmonious development of the other ports of the two countries, and in particular will not cause a distortion of competition between the sea ports of the two countries.*" ("*Désiraut que l'ouverture de la rive gauche de l' Escaut au développement n' entrave pas le développement harmonieux des autres zones portuaires maritimes de Leurs deux pays et que notamment ne soient pas perturbées les conditions de concurrence entre les ports de mer de Leurs deux pays.*")[44] The convention contains several regulations which should prevent a distortion of competition between Belgian and Dutch ports. First, the convention prohibits the granting of investment subsidies to new industries on the left bank of the river Scheldt.[45] Secondly, the Belgian and Dutch authorities have to inform each other about their negotiations with industries which would like to make investments in a specified zone along the planned canal.[46] Thirdly, the convention sets a minimum price for the ground which is needed for the construction of the canal.[47] Beside these economic regulations, the convention contains provisions to prevent air pollution.[48] Finally, the convention provides for an arbitration commission and mechanisms for conflict resolution.[49]

Draft Convention on the improvement of the navigation channel in the Scheldt near Bath (including regulations concerning the water quality of the river Scheldt)
The Bath convention provides for the straightening of the bend near Bath in the Western Scheldt, and contains regulations concerning the water quality of the river Scheldt. It describes the technical infrastructure which is necessary for the construction of the canal, such as the canal itself, dikes and dams.[50] All infrastructure projects will be financed by Belgium.[51] In addition to the regulations concerning the construction of the infrastructure projects to improve the navigation channel in the Western Scheldt, the convention contains detailed regulations concerning the water quality of the river Scheldt.[52] Among other things,

---

41. Traité entre le Royaume des Pays-Bas et le Royaume de Belgique au sujet de la construction du canal de Baalhoek, Chapitre II (Articles 2-4).

42. Ibid., Article 24.

43. Ibid., Chapitre X (Articles 35 to 43).

44. Traité entre le Royaume des Pays-Bas et le Royaume de Belgique au sujet de la construction du canal de Baalhoek, considerans.

45. Traité entre le Royaume des Pays-Bas et le Royaume de Belgique au sujet de la construction du canal de Baalhoek, Article 37.

46. Ibid., Article 38.

47. Ibid., Article 39.

48. Ibid., Chapitre XI (Articles 44-51).

49. Ibid., Chapitre XV (Articles 64-66).

50. Traité entre le Royaume des Pays-Bas et le Royaume de Belgique au sujet de l'amélioration de la voie navigable dans l' Escaut près du goulet de Bath, Chapter II (Articles 2 and 3).

51. Ibid., Article 13.

52. Ibid., Chapter VI (Articles 23-41).

the delegations agreed that Belgium will create a waste water treatment capacity of at least 2,500,000 inhabitant equivalents (i.e.) before 1 January 1981[53], a capacity of at least 5,900,000 i.e. before 1 January 1987[54], and a capacity of at least 6,600,000 i.e. before 1 January 1990.[55] These objectives are elaborated in detail and specific targets are formulated for the conurbations of Antwerp, Brussels, and Ghent, and the basin of the Lys.[56] An appendix to the convention contains water quality standards at the Belgian-Dutch border, which have to be achieved in 1987.[57] Furthermore, the Bath convention states that from 1 May 1979 onwards Belgian industries are no longer allowed to discharge untreated waste water in the Scheldt basin.[58] The parties also agree that they will inform each other about the planning of new industries or the expansion of existing ones if implementation of these plans may entail a deterioration of the water quality of the river Scheldt.[59] The Bath convention provides for the installation of a Belgian-Dutch Scheldt Water Commission (SWC) similar to the Meuse commission described in the Meuse convention. The main difference is that the scope of the SWC is limited to water quality issues.[60] Like the Meuse convention, also the Bath convention enables the contracting parties to modify the technical details if this is necessary because of new technological developments.[61] Furthermore, the parties agree that if after ratification of the convention Belgium and the Netherlands become contracting parties of a multilateral convention which contains more strict water quality policies, these policies will be integrated in the Bath convention.[62] Finally, like the other two conventions, also the Bath convention contains provisions for conflict resolution.[63]

Round III: Internal Belgian disagreement (1975-1983)

Shortly after the delegations reached an agreement on the three draft conventions, the Dutch national government expressed its willingness to sign the Bath and Baalhoek conventions if the Belgian government would sign the Meuse convention and contribute to an improvement of the water quality of the Scheldt.[64] Lower level governments in the Dutch province of Zeeland and environmental NGO's, however, opposed the construction of the Baalhoek canal and the by-pass near Bath, because of the consequences for the spatial planning in Zeeland and the negative impacts of these projects on the environment.[65]

---

53. Ibid., Article 24.1 Sub a.

54. Ibid., Article 24.1 Sub b.

55. Ibid., Article 24.1 Sub c.

56. Ibid., Article 24.2 and 24.3.

57. Ibid., Article 31.

58. Ibid., Article 27.

59. Ibid., Article 36.

60. Ibid., Chapitre VIII (Articles 45-52).

61. Ibid., Articles 24.1, and 30.5.

62. Ibid., Article 30.3.

63. Ibid., Chapitre XI (Articles 60-62).

64. Letter from the Dutch Minister of Transport, Public Works, and Water Management, Westerterp, to the Chairman of the Second Chamber of the Dutch Parliament of 1 October 1975; TK 1977-1978, Aanhangsel van de Handelingen, 854.

65. Anonymous, 12-2-1979.

The Belgian government does not approve the draft water conventions. The main reason for this are the objections of the Walloon region. In 1975, it appears that the Belgian delegation had hardly consulted representatives of the Walloon region during the negotiations on the water conventions.[66] According to the Walloon region "*the two Antwerp projects, the Bath and Baalhoek convention, formed in fact a real trilogy because the closing element - a third convention on the water of the river Meuse - was presented to the Walloon region as a 'fait accompli', real exchange for the desires of the Netherlands and Antwerp.*" ("*les deux projets anversois, Traité Bath et Traité Baalhoek, comportaient donc en fait une véritable trilogie dont la clé de voûte - un troisième traité dit "des eaux de la Meuse" - fut présenté à la région wallone comme un fait accompli, véritable soulte à des exigences ou hollandais ou anversoises.*")[67] On 8 September 1975 the Regional economic council of Wallonia states that the Belgian government had "[...] *let the Walloon region completely uninformed about the negotiations on the conventions, even though the law of 1 August 1974 had introduced a regionalized water policy.*" ("[...] *tenu la Région wallone dans l'ignorance la plus complète de la négociation de ces traités, alors que la loi de 1er août 1974 avait instauré la régionalisation de la politique de l'eau*")[68] According to the Walloon politicians[69]: (1) Wallonia is not interested in a modification of the Meuse conventions of 1863 and 1873, (2) Wallonia does not use Meuse water for drinking water production, the water quality of the Meuse is sufficient for the needs of the Walloon region, and Wallonia will decide itself whenever and how to clean up the rivers Meuse and Scheldt, and (3) Wallonia does not want to build storage reservoirs on its territory.

In addition to the objections to the convention on the Meuse, the Walloon region opposed the Bath and Baalhoek conventions. According to the Walloon politicians the two infrastructure projects stimulate the industrial expansion of Antwerp, but have to be paid by the Belgian state and consequently at least partly by the Walloon region. On 16 September 1975, the Regional economic council of Wallonia protested once more and declared that: "*The two conventions aimed at an improvement of the access to the port of Antwerp will cause a stagnation of the industrialization in the Meuse basin. Since the law of August 1974 on the regionalization provided for a regionalized water management, the Belgian government did not have the competence to negotiate and initiate a convention that would bind Wallonia for a long time.*" (*Les deux traités ayant pour objet l'amélioration de l'accessibilité du port d'Anvers auront pour effet d'arrêter l'industrialisation du bassin mosan. La Loi du 1er août 1974 sur la régionalisation ayant prévu la régionalisation des problèmes de l'eau, le gouvernement belge n'avait aucune qualité pour négocier et parapher un projet de traité engageant lourdement et à long terme la Wallonie.*")[70]

Unlike the Walloon politicians, the Flemish pass a positive judgement on the draft conventions. The construction of the Baalhoek canal and the straightening of the bend of Bath in the Western Scheldt would create the preconditions for an expansion of the port of Antwerp. The Flemish region only opposed the economic regulations to prevent a distortion

---

66. Planchar (1993, p. 92).

67. Ibid., p. 93.

68. Statement of the Conseil économique régional wallon on 8 September 1975. Cited in: Planchar (1993, p. 95).

69. Suykens (1995, p. 230).

70. Statement of the Conseil économique régional wallon on 16 September 1975. Cited in: Planchar (1993, p. 96).

of competition between Dutch and Belgian ports.[71]

A joint Walloon-Flemish statement concerning the three draft conventions was necessary for ratification of the conventions or a restart of Belgian-Dutch negotiations on the water conventions. Several attempts were made to reach an agreement between Flanders and Wallonia, but they all failed. In January 1978, a commission of civil servants and representatives of the ministerial cabinets is established, which is chaired by the Belgian Secretary-General of the Ministry of Foreign Affairs. This commission assembled eleven times, and during these meetings the texts of the draft conventions were modified according to Belgian preferences.[72] The work of this commission, however, was hindered by the attribution of water management competencies to the Belgian regions[73] because of the constitutional amendment of 1980. On 7 July 1980, a second attempt is made to come to a Walloon-Flemish consensus and a new working group with representatives of several cabinets is established. On 23 July 1981, after 40 meetings, this group issues a study report. The Walloon government, however, does not agree with the results, mainly because the report contains a proposal to ensure a minimum discharge of the river Meuse, namely 50 m^3/s in Monsin (Liège).[74] A third attempt is made by an *ad hoc* commission of ministers which assembles from 1982 until 1984. This commission, with representatives of the Belgian regions, once more tried to reach a Walloon-Flemish agreement. The study report, which was finished in 1981, forms the input for the discussions in this commission.[75] Also in this third commission no consensus is reached.

Infrastructure and maintenance dredging works in the navigation channel to Antwerp
Because of the increasing dimensions and draught of modern ships which are needed for the increasing tonnage and the desirability of low transportation costs, Belgium would like to deepen the navigation channel of the Western Scheldt. In August 1977, Belgian government agencies issued a report, which describes the 48'/43'/38' deepening programme.[76]

---

71. Suykens (1995, p. 230).

72. Ibid.

73. Ibid.

74. Ibid.

75. Ibid.

76. Minutes of TSC-meeting 22, on 3 October 1978 in Brussels.
Without implementation of a deepening programme, possibilities of navigation on the Western Scheldt were the following (TSC, 1984a, p. 6):
Ships with a draught of 44' (13.42 m.) could reach the harbour during one tide.
Ships with a draught of 48' (14.65 m.) could reach the harbour during two tides, but only in cases of favourable spring tide.
Ships with a draught of 34' (10.37 m.) could sail down independent of the tide.
Ships with a draught up to 40' (12.20 m.) could sail down during one or two tides, dependent on the tidal level.
    Implementation of the 48'/43'/38' deepening programme in the Western Scheldt would create the following navigation possibilities (Ibid., pp. 6-7):
A ship for bulk goods with a draught of 48' (14.65 m.) could reach the harbour during one tide.
A ship for bulk goods with a draught of 50' (15.25 m.) could reach the harbour during two tides.
A container ship with a draught of 41' (12.50 m.) could sail down during one tide.
A container ship with a draught of 42'8" (13.00 m.) could sail down during one tide.
A ship for bulk goods with a draught of 41' (12.50 m.) could sail down during one tide.
A ship with a draught of 38' (11.60 m.) could reach the harbour or sail down the river during an average-low-

Implementation of this programme entails additional dredging works on several bars in the navigation channel of the Western Scheldt. In the Technical Scheldt Commission (TSC) informal discussions on this issue took place, and the Dutch delegation expressed its concerns about the consequences implementation of the deepening programme may have for the water level in the estuary during storm surges and for the banks along the Western Scheldt.[77] At a bilateral ministerial meeting on 20 November 1979, the Dutch Minister of Transport, Public Works, and Water Management, Tuijnman, announced that the Dutch government would like to have more information on the consequences of the deepening programme first, including the consequences for the environment. He also stated that he will consult a Dutch advisory body, the *Raad van de Waterstaat*. His Belgian colleague, Chabert, expects resistance from those concerned with navigation if decision making on the implementation of the deepening programme would take too long, and therefore pleas for a quick decision making procedure.[78] The ministers decide to jointly carry out a study on the impacts of the 48'/43'/38' deepening programme on the Western Scheldt.[79] This study should also comprise an evaluation of the need to straighten the bend near Bath if Belgium and the Netherlands would decide to implement the deepening programme.[80] To carry out this study the TSC installed the Subcommission for the deepening of the Western Scheldt.

Round IV: Dutch linkage of the 48'/43'/38' deepening programme to Meuse water management issues (1983-1985)

The Dutch linkage and the TSC report of 1984
On 28 October 1983, the Dutch Minister of Transport, Public Works, and Water Management, Smit-Kroes, informs her Belgian colleague, the Minister of Public Works, Olivier, that the deepening of the navigation channel of the Western Scheldt requires a convention to be approved by both the Dutch and Belgian parliaments.[81] Furthermore, she links the Dutch approval of this convention to the approval of the other water conventions.[82] Although the Dutch Minister is pleased with the installation of a Technical Meuse Commission (TMC)[83], she stresses once more the need to conclude an international convention on the water management of the Meuse.

On 15 June 1984, the TSC finished its research on the impacts of the 48'/43'/38' deepening programme. The main conclusion of this study is that the deepening programme can be implemented without major changes of the Scheldt regime, without a significant

---

low water-spring.

77. Minutes TSC-meeting 22, 3 October 1978, Brussels.

78. Minutes TSC-meeting 24, 20 June 1980, Bokrijk (Belgian Limburg).

79. Ibid.

80. Ibid.

81. Letter from the Dutch Minister of Transport, Public Works, and Water Management, Smit-Kroes, to the Belgian Minister of Public Works, Olivier, of 28 October 1983.

82. Ibid.

83. Because the Convention on the distribution and the quality of the water of the river Meuse had not yet been signed, at a bilateral Belgian-Dutch ministerial meeting it was decided to install a Technical Meuse Commission (TMC) in June 1983. This commission met for the first time on 20 October 1983, and mainly discussed transboundary pollution issues (TK 1985-1986, Aanhangsel van de Handelingen, 122).

impact on the ecological values of the Western Scheldt, and without increased risks for navigation, the population, and the environment.[84] Consequently, the TSC concludes that implementation of the deepening programme is acceptable.[85] The Dutch Minister Smit-Kroes requested the *Raad van de Waterstaat* to advise on the deepening program.[86] In the participation rounds organized by this advisory body, governments and NGOs formulated several objections to the proposed deepening programme.[87] Dutch Water Boards demand financial compensation for the costs of the protection of the banks along the Western Scheldt against erosion. The Province of Zeeland equally states that the Dutch national government should bear the financial responsibility for the implementation of the necessary works to protect the banks along the Western Scheldt. The provincial executive for the environment, Boersma, criticizes the Dutch Minister, because the deepening programme is mainly linked to the water quality of the Meuse instead of the water quality of the Scheldt.[88] The municipal executives of Flushing and Oostburg are of the opinion that the planned displacement of polluted sediments from the eastern to the relatively clean western part of the Scheldt estuary is unacceptable. The executive of Flushing also states that dumping polluted sediments on Flushing territory is not allowed, unless the municipality grants a permit for this activity.[89] The regional department of the Ministry of the Environment and Public Health puts the emphasis on the risks of the transport of dangerous substances on the Western Scheldt, and proposes to link the Dutch cooperation on the implementation of the deepening programme to the implementation of waste water treatment programmes on Belgian territory. Whereas various Dutch government agencies and NGOs formulate a wide range of objections and conditions to the deepening programme, the municipal executive of the city of Antwerp is pleased with the proposal to deepen the navigation channel in the Western Scheldt, and emphasizes that implementation of the deepening programme is beneficial for the Dutch harbours of Flushing and Terneuzen as well.

Beside government agencies also environmental NGOs use the opportunity to express their concerns. The Zeeland Fishery Association (*Zeeuwse Visserij Belangen, ZEVIBEL*) expressed its worries about the geomorphological disturbances, and the changes of the water and sediment quality if the deepening programme would be implemented. The Zeeland Landscape Foundation (*Stichting het Zeeuws Landschap*) states that the implementation of the deepening programme is unacceptable, because this will have a very negative impact on the nature values of the Western Scheldt, for example because of the loss of tidal areas, the distribution of pollutants, and the increased pollution in the western part of the Western Scheldt. The International Scheldt working group (*Internationale Scheldewerkgroep*), an association of Flemish and Dutch environmental NGOs, demands an EIA procedure to assess carefully the impacts of the implementation of the deepening programme on the tidal areas and the pollution of the Scheldt estuary. This working group also proposed to insert the water

---

84. TSC (1984b, p. 172).

85. Ibid., p. 174.

86. Letter from the Dutch Minister of Transport, Public Works, and Water Management to the chairman of the 'Raad van de Waterstaat' of 2 November 1984.

87. Report of rounds of participation organized by the Dutch advisory council, the 'Raad van de Waterstaat'. In files of TSC-meeting 29, 12 July 1985, Delft.

88. PZC, 13-11-1984.

89. This Permit is based on the Dutch Act on Land-use planning.

quality regulations formulated in the Bath convention in the Convention on the deepening of the navigation channel in the Western Scheldt. The Zeeland Environment Federation (*Zeeuwse Milieufederatie, ZMF*) proposed to link the water quality of the Scheldt to the deepening programme, and to make an EIA.

On 13 July 1984, the Belgian Minister of Foreign Affairs informed the Dutch government that Belgium is prepared to restart negotiations on the water conventions.[90] Furthermore, he states that Belgium would like to place the issue of the deepening of the navigation channel in the Western Scheldt on the Belgian-Dutch agenda and to remove the issue of the straightening of the bend near Bath from the agenda, because implementation of the deepening programme would make the straightening of this bend unnecessary.[91]

At a bilateral meeting between the Ministers Smit-Kroes and Olivier on 18 July 1984 in Brussels, the Belgian Minister declared that the Dutch linkage between the deepening of the Western Scheldt and other bilateral issues, in particular the Meuse issues, is a violation of the freedom of navigation on the Scheldt, which is regulated in the Scheldt Statute.[92] The Dutch Minister Smit-Kroes did not agree with this point of view, and declared that the preparation, modification, and partial replacement (straightening of the bend near Bath) of the draft conventions between the Netherlands and Belgium should take place simultaneously with the deliberations between France and Belgium on the construction of a barrage in the Houille, a transboundary tributary of the river Meuse.[93] With that the Dutch Minister confirmed the linkage between the negotiations on Scheldt infrastructure projects and the water quantity of the Meuse. Whilst the Ministers do not agree on the linkage between the deepening programme and the Meuse issues, they are able to reach an agreement on the installation of the Belgian-Dutch Scheldt Water Commission (SWC) provided for in the draft Bath convention of 1975.[94]

After the conclusion of the research on the 48'/43/'38'-deepening programme, and after the 48'/43'/38' deepening programme has been placed on the international negotiations agenda, Belgium wants to continue research in the TSC on the implementation of a 50'-deepening programme, which is also called the second stage of the deepening programme. The Dutch do not want to start this research immediately, because research on a further deepening of the Western Scheldt could be used as exchange in the negotiations on a Meuse convention.[95] The Dutch Minister of Transport, Public Works, and Water Management, however, proposes to start negotiations on the protection of the banks along the Western Scheldt.[96]

---

90. Suykens (1995, p. 230). On 4 July 1984 the Belgian Ministerial Committee for the water conventions had decided to place the issue of the deepening of the Western Scheldt on the international negotiation agenda, and to remove the issue of the straightening of the bend near Bath from this agenda (d' Argent, 1997, p. 17).

91. Ibid.

92. Dutch report of the meeting between the Dutch Minister of Transport, Public Works, and Water Management, Smit-Kroes, and the Belgian Minister of Public Works, Olivier, on 18-7-1985 in Brussels. For information on the Scheldt Statute, see the historic overview presented in Section 4.8.

93. Ibid.

94. The Dutch had proposed to install this commission at the TSC-meeting of 24 June 1983.

95. Files Dutch Ministry of Transport, Public Works, and Water Management, TSC-meeting 28, internal Dutch document.

96. Letter from the Dutch Minister of Transport, Public Works, and Water Management to the Belgian Minister of Public Works of 12 July 1985.

Installation of the Scheldt Water Commission, and the issue of the Tessenderlo pipeline
The SWC met for the first time on 13 June 1985 in The Hague. At a Dutch preparatory meeting for the first SWC meeting it is decided to be flexible as regards the installation of working groups and the sequence of activities of the commission, mainly because the Dutch delegation expects that the Belgian delegation is bound more by their administrative and decision making rules.[97] At the first SWC meeting it is stated that between 1975 and 1985 there actually have been no Belgian-Dutch deliberations concerning the water quality of the river Scheldt. The main issue discussed in the SWC is the use of the Belgian pipeline Tessenderlo.[98] The Belgian Minister responsible for water and environmental policies in Flanders declares that he would like to exploit a collector of sewage water near Antwerp, which receives its waste water from the area around Tessenderlo in Belgian Limburg. The Dutch delegation strongly opposes these Belgian plans. The Dutch Ministry of Foreign Affairs stated that the use of the pipeline Tessenderlo is not in accordance with European directives and is not neighbourly. The environmental NGOs *'Benegora'*, Save the Voorkempen (*'Redt de Voorkempen'*), and the International Scheldt working group protest against the bringing into use of the collector.[99] Also the Antwerp region opposes the bringing into use of the collector. *"All Antwerp efforts to ameliorate the quality of the Scheldt water, and to disprove the Dutch objections against the further deepening of the Scheldt, are threatened to be undone."*[100] In the SWC, the Dutch also pay attention to illegal discharges into the Scheldt and ask several times for more information on waste discharges in the upstream parts of the basin.[101]

The Belgian-Dutch declaration of intent of 1985
On 7 October 1985, the Dutch Minister of Foreign Affairs, Van den Broek, and the Belgian Minister of Foreign Affairs, Tindemans, reach an agreement on a declaration of intent. In this declaration both parties state that they want to solve the Scheldt and Meuse issues as

---

97. Report preliminary talks of Dutch delegation on 13-5-1985 in The Hague.

98. TK 1985-1986, Aanhangsel van de Handelingen, 909.
 In the period between 1930 and 1938 Belgian government agencies had developed plans to construct a pipeline (waste water collector) along the newly constructed Albert canal to protect this canal against industrial waste discharges without hindering economic development in the region the 'Kempen'. On 6 January 1939 the Belgian Minister of Public Works demanded the Department for the Albert canal and the Department for the Treatment of Waste Water to carry out a study on pollution prevention in the Albert canal. In June 1958 the Belgian Ministers of Public Health and Public Works initiated research on the construction of a waste water collector. On 24 October 1961 a Belgian Ministerial committee asked the Minister of Public Health to design a pipeline between Tessenderlo and Antwerp, including a pipeline from Herentals-Mol. In 1965 this design is approved by the Minister of Public Health, and in 1967 construction works were started. Because of technical and administrative problems the constructed collector was never used. In 1982, however, the Belgian government decided to use the collector to transport brackish waste water stemming from fertilizer industries in the region of Tessenderlo to the Lower Sea Scheldt. In July 1983 the Flemish government decided to test the collector, and to investigate the conditions for these waste water discharges (Report of the Flemish region for a discussion in the meeting of the SWC on 16 January 1986). A major argument to use the pipeline was the protection of the water quality in the basins of the Laak and the Nete in the Kempen region (SWC-meeting 3, 23 October 1986, Renesse).

99. Benegora Leefmilieu, Redt de Voorkempen, Belt, Internationale Scheldewerkgroep (1985).

100. De Standaard, 28-8-1985.

101. TK 1986-1987, Aanhangsel van de Handelingen, 597.

soon as possible in a way that is acceptable for both parties.[102] First, they express their willingness to start negotiations on the distribution of the water of the Meuse during periods of a low flow. To ensure a minimum flow of the river Meuse, the ministers propose to construct one or two storage reservoirs in Wallonia.[103] The Convention on the flow of the river Meuse should also provide for the Belgian compensation for Dutch fresh water losses which had been agreed upon in the Treaty concerning the connection between the Scheldt and the Rhine of 1963.[104] Secondly, the Ministers agree that a Meuse convention should contain regulations concerning the water quality of the Meuse at the Belgian-Dutch border. The Belgian Minister of Foreign Affairs declares that the Belgian regions will receive financial support in case implementation costs of planned water quality regulations in the conventions would exceed the regional budgets for waste water treatment and sewerage.[105] Thirdly, the declaration confirms that the issue of the straightening of the bend near Bath is skipped from the negotiation agenda, and that the 48'/43'/38' deepening programme for the Western Scheldt is placed as a new issue on the Belgian-Dutch negotiation agenda. The Convention on the deepening of the navigation channel in the Western Scheldt should contain the technical aspects of the implementation of the deepening programme, regulations concerning environment and nature protection, and regulations concerning the prevention or restoration of damages to the banks of the Scheldt.[106] A final negotiation issue mentioned in the declaration is the construction of the Baalhoek canal. In the declaration the Convention on the deepening of the navigation channel in the Western Scheldt and the one on the construction of the Baalhoek canal are linked to the Meuse Convention.[107] The declaration does not address the issue of the water quality of the river Scheldt.

The WVO Permit for Belgium
Maintenance of the navigation channel in the Western Scheldt requires continuous maintenance dredging works on the bars in the Western Scheldt. According to the Scheldt statute, Belgium and the Netherlands are responsible for the maintenance of the navigation channel on their respective territories. Each year, the Belgian state receives a permit for the implementation of the necessary dredging works on Dutch territory.[108]

On 7 June 1985, the Dutch Minister of Transport, Public Works and Water Management, Smit-Kroes, informs the Belgian Minister of Public Works, Olivier, that from 1 May 1985 Belgium needs a permit for the dumping of dredged material in the Western Scheldt as well.[109] Legal basis of this permit is the Dutch Surface Water Act (WVO).[110] To give the

---

102. TK 1985-1986, 19 200 No. 12.

103. Ibid.

104. Trb. 1963 No. 78.

105. TK 1985-1986, 19 200 No. 12.

106. Ibid.

107. Ibid.

108. See also the historic overview presented in Section 4.8.

109. Letter from the Dutch Minister of Transport, Public Works, and Water Management, Smit-Kroes, to the Belgian Minister of Public Works, Olivier, concerning the Belgian obligation to apply for a WVO permit for dumping dredged material in the Western Scheldt of 7 June 1985.

Belgian government enough time to prepare for the application for the permit, the Dutch Minister decides that Belgium should have the permit from 1 May 1986. Unlike the dredging permit, the issuing of a WVO permit entails a procedure for public participation according to the Dutch General Environmental Act (Awbm).[111] On 19 November 1985 the Belgian government applies for two permits: a first one for the dumping of dredged material from the dredging works on the bars in the Lower Sea Scheldt in the Western Scheldt, and a second one for the dumping of dredged material from the dredging works on the bars in the Western Scheldt on other locations in the Western Scheldt.[112]

5.3   Analysis and explanation of the decision making rounds I to IV

In this section, the strategic interactions in the four decision making rounds distinguished in the period between 1967 and 1985 will be analyzed and explained. In Section 5.3.1 the influence of the perceptions of the actors involved in the decision making process on their strategies are analyzed, and with that research question 3 will be answered. Subsequently, in Section 5.3.2 the strategic interactions and institutionalization processes are discussed, and with that research question 4 will be answered. In Section 5.3.3 the strategic and cognitive learning processes are analyzed and with that research question 5 will be answered. To conclude, in Section 5.3.4 the influence of the context, which was described and analyzed in Chapter 4, on the perceptions and strategies of the actors involved in the decision making process is analyzed and research question 6 will be answered.

5.3.1 Influence of perceptions of actors involved in the decision making rounds I to IV on their strategies

Perceptions and strategies of the Scheldt basin states and regions
Table 5.2 produces information on the main strategies of the Scheldt basin states and regions in each decision making round. The typology of strategies used in this table was introduced in Section 2.2.3. In the following, the strategies of these actors are related to their perceptions of (inter)dependence. This analysis sheds light on the action rationality of the individual actors, and therefore contributes to a better understanding of the course and outcomes of the decision making process.

The Belgian government, demanding party for infrastructure projects
Belgium is situated upstream of the Netherlands, and perceived a dependence on the downstream riparian state as regards the implementation of three infrastructure projects for the improvement of the maritime access to the port of Antwerp: the construction of the

---

110. Stb. 1969 No. 536; Wet Verontreiniging Oppervlaktewateren; in fact, a permit for the dumping of dredged material had been obligatory since an amendment of the Dutch Surface Water Act in December 1974, but only since 1985 this amendment was implemented for the main waters in the Netherlands. Ministerie van Verkeer en Waterstaat, Hoofddirectie van de Waterstaat, Hoofdafdeling Bestuurlijke en Juridische zaken (1988); *WVO-vergunning voor België*, Nr. RJW 51525, p.5.

111. Awbm =Algemene wet bepalingen Milieuhygiëne.

112. Ministerie van Verkeer en Waterstaat, Hoofddirectie van de Waterstaat, Hoofdafdeling Bestuurlijke en Juridische zaken (1988), *WVO-vergunning voor België*, Nr. RJW 51525, and *WVO-vergunning voor België*, Nr. RJW 51526.

Table 5.2 *Main strategies of Scheldt basin states and regions in decision making rounds I to IV*

Round	Actor	Type of strategy	Strategic actions
I (1967-1968)	Belgian government	Offensive interactive	Places the construction of the Baalhoek canal and the straightening of the bend near Bath on the Belgian-Dutch agenda
	Dutch government	Reactive interactive, Issue-linkage	Links the two infrastructure projects to the water quality of Scheldt and Meuse and the water quantity of the Meuse
II (1969-1975)	Belgian delegation	Reactive interactive, Issue-linkage,	Links water management issues to infrastructure issues
		Negotiation	Approves draft conventions
	Dutch delegation	Reactive interactive, Issue-linkage,	Links infrastructure issues to water management issues
		Negotiation	Approves draft conventions
III (1975-1983)	Dutch government	Interactive	Approves draft conventions
	Walloon politicians	Reactive aimed at maintaining autonomy	Reject draft conventions
	Flemish politicians	Interactive	Judge positively the draft conventions
	Belgian government	Reactive aimed at maintaining autonomy	Rejects draft conventions
IV (1983-1985)	Belgian government	Offensive interactive	Places the 48'/43'/38' deepening programme on the Belgian-Dutch agenda
	Dutch Minister	Reactive interactive, Issue-linkage	Links deepening programme and Baalhoek canal to Meuse issues
		Offensive, Increasing (control of) resources	Introduces obligatory WVO Permit for dumping of dredged material in the Western Scheldt
	Belgian Minister	Reactive interactive Financial compensation	Accepts Dutch linkages Promises to help regions financially with clean-up of Scheldt and Meuse

Baalhoek canal, the straightening of the bend near Bath, and the 48'/43'/38' deepening programme. This dependence is caused by the distribution of resources among the two basin states. The Dutch did have the authority to approve the implementation of the three projects. To reach its objectives the Belgian government used offensive interactive strategies and placed these issues on the Belgian-Dutch agenda. Because of the Dutch linkages, the Belgian government had no choice but to start negotiations on Scheldt infrastructure and Scheldt and Meuse water management issues simultaneously. The Belgian government aimed at a package deal including all Belgian-Dutch Scheldt and Meuse issues. During the negotiations, however, the Belgian government hardly consulted the Walloon region, and did not take its interests into account. Therefore, the Walloon region opposed the proposed package deal and the Belgian government was unable to approve the draft conventions the Belgian and Dutch delegations had reached agreement on in 1975. The Belgian government was of the opinion that the Dutch linkage between the 48'/43'/38' deepening programme and the Meuse issues was a violation of the Scheldt Statute. Because the Dutch did not give in and Belgium had a big interest in the Scheldt infrastructure projects, the Belgian Minister of Foreign Affairs did accept the Dutch linkage eventually (See Round IV: Declaration of intent of 1985). In addition, the Belgian national government understood that the problems of the Walloon Region were mainly linked to the financial aspects of the Scheldt and Meuse water quality regulations and therefore expressed its willingness to compensate the regions financially (See Round IV: Declaration of intent of 1985).

The Dutch government, the linking actor
The Dutch were mostly interested in the two Meuse issues: the improvement of the water quality and the distribution of the water of this river. Meuse water pollution endangered the use of Meuse water for drinking water production in the Netherlands. In addition, it caused severe water bed pollution and consequently high dredging and storage costs, mainly in the port of Rotterdam. During low flows in the river Meuse water quantity problems emerged. The Dutch government, however, did not possess the necessary resources to influence the water quality and quantity in the upstream parts of the Scheldt and Meuse basins and consequently perceived a dependence on the upstream riparian states as regards the supply of a reasonable amount of water of a reasonable quality. Therefore, the Dutch did not want to start negotiations on the Scheldt infrastructure projects unconditionally, but made several linkages across Scheldt and Meuse issue-areas. Table 5.3 lists the linkages that were made in the period between 1967 and 1985. The Dutch government used its relative power advantage in the issue-area of the maritime access to the port of Antwerp to force Belgium to cooperate on the Meuse water management issues. This linkage is in accordance with the theory that in cases where decision making is based on unanimity and there are heterogenous preference intensities across issue-areas, the probability is high that movement away from the status quo will involve tactical issue-linkage.[113] The same applies to the Dutch linkage of the deepening programme of the Western Scheldt to the Meuse water management issues.

The Dutch also tried to reduce their dependence on the upstream basin states by the introduction of a new resource: the WVO permit for dumping dredged material in the Western Scheldt. Later, the Dutch would use this resource to make a linkage between the dumping of sediments in the Western Scheldt, and Belgian policy measures to improve the water quality of the river Scheldt. The negotiations between 1969 and 1975 on the Baalhoek convention clearly show that the Dutch did not only perceive an interest in the Scheldt and Meuse water management issues, but also were concerned with the competitive position of

---

113. See Section 2.3.2 and Figure 2.2.

the Dutch ports, in particular the port of Rotterdam. To prevent a distortion of competition between Belgian and Dutch ports, the Dutch wanted to include several economic conditions in the Convention on the construction of the Baalhoek canal.

Table 5.3 *Issue-linkages in the negotiations on the water conventions (1967-1985)*

Year	Actor placing issue on the agenda	Issues placed on the agenda	Linking actor	Linked issues
1967	Belgian government	1. Construction of the Baalhoek canal 2. Straightening of the bend near Bath	Dutch government	3. Water quality of the Meuse 4. Water quantity of the Meuse 5. Water quality of the Scheldt
1983	Belgian government	6. 48'/43'/38' deepening programme	Dutch government	Issues 3 and 4
1985	Belgian Minister	Issues 1 and 6	Dutch Minister	Issues 3 and 4
	Dutch Minister	Issues 3 and 4	Belgian Minister	Issues 1 and 6

The Walloon region, maintainer of autonomy

The Walloon region did not perceive an interest in any of the international Scheldt and Meuse issues at stake. Consequently, the region did not perceive a dependence on the Netherlands, and disapproved the Belgian-Dutch draft conventions. The statements of the Walloon regional economic council were clear (See section 5.2, Round III). According to the Walloon politicians, Wallonia would have to pay the price for the 'Flemish' infrastructure projects. In addition, the Walloon politicians argued that because of the institutional reforms in Belgium, the Walloon region should have been involved in the negotiations with the Netherlands. For these reasons the region used a strategy aimed at maintaining autonomy. Because in the period between 1975 and 1985, the Belgian state and the Netherlands did not change their strategies basically, the Walloon region continued to protest against the proposed package deal.

The Flemish region, demanding party for infrastructure projects

The three infrastructure projects that were part of the negotiations on the water conventions would increase economic welfare primarily in the Flemish region, since these projects would stimulate the development of the Flemish port of Antwerp. Therefore, unlike the Walloon region the Flemish region was very interested in a package deal with the Dutch and formulated a rather positive judgement on the conventions drafted in 1975. Because of the linkages made by the Dutch, the Flemish region perceived a strong dependence on the Walloon willingness to cooperate on the international Meuse water quantity and quality issues.

An interesting observation is that because of the Dutch linkage some Flemish parties seem to have stimulated waste water treatment in the Antwerp region (See Round IV). Possibly, these parties wanted to build up a reservoir of goodwill they could draw on in the negotiations on the deepening of the Western Scheldt. This strategy may easily be interpreted as a strategic learning process. The region had experienced that it is dependent on the Netherlands, and therefore gives in to the Dutch objections concerning the water quality of the Scheldt.

Perceptions and strategies of lower level governments and NGOs
Beside the international arena, which consists of (representatives of) the Belgian and Dutch national governments, there is an important regional arena in the Dutch province of Zeeland, which consists of regional directorates of national ministries, regional and local governments, and NGOs. Since decision making in the international and regional arenas is linked, decision making may be conceived of as a two-level game. In the period between 1967 and 1985 the participation round concerning the 48'/43'/38' deepening programme organized by the Dutch advisory body *Raad van de Waterstaat* formed the main link between the international and regional arenas. In this participation round regional actors were given the opportunity to formulate their concerns. Several issues introduced by these actors, such as the issues of the displacement of polluted sediments, the safety of navigation on the Scheldt, and the morphological disturbances caused by dredging works and the dumping of dredged material, showed an intellectual coherence with the implementation of the deepening programme and would get a place on the international negotiation agenda. The lower level governments and NGOs, however, also used the opportunity to introduce issues that were less related intellectually to the implementation of the deepening programme, such as the water quality of the river Scheldt and the Belgian waste water treatment capacity. Like the Dutch national government, the regional government agencies and NGOs also had specific preferences concerning the linkages to be made in the international negotiations. Generally, the actors in the province of Zeeland emphasized the linkage between the deepening of the Western Scheldt and the water quality of the Scheldt and opposed the policy of the Dutch national government to link the deepening programme to the Meuse issues only.

Some Dutch municipalities were able to reduce their dependence on the national governments. With the competence to grant a permit for dredging and dumping activities on their territories, they got an indispensable resource for the implementation of the deepening programme. Later, these municipalities would use this resource to exert influence on the national governments of the Scheldt basin states.

5.3.2 Strategic interaction and institutionalization processes

Development of international institutional arrangements and policies
In the decision making rounds I en II the prospects on the development of ambitious international institutional arrangements and policies were good. The three draft conventions presented in 1975 provided for many international institutional arrangements, such as the installation of international river commissions for Scheldt and Meuse, provisions for monitoring and provisions for conflict resolution. In addition, the draft conventions contained detailed policies. They provided for the construction of two main infrastructure projects, detailed regulations concerning the water quality of the rivers Scheldt and Meuse, policy programmes to achieve the formulated objectives and detailed regulations concerning the distribution of the Meuse water. Some interesting observations as regards the content of the draft conventions are that: (1) although a significant part of the Scheldt and Meuse basins is

situated in France, France was not a contracting party, (2) the conventions contain water quality objectives at the Belgian-Dutch border and therefore only contain detailed obligations to improve the water quality for the contracting party Belgium, (3) the regulations concerning the water quantity of the river Meuse do not fix a minimum flow at the Belgian-Dutch border, but do specify the distribution of Meuse water in Belgium. These characteristics of the regulations on the water management of the rivers Scheldt and Meuse clearly show that these regulations were primarily used as an exchange in the negotiations on the Scheldt infrastructure projects.

Because the draft conventions were not approved, the ambitious international institutional arrangements and policies could not develop. The reason for this is as simple as it is clear: the Dutch and Belgian governments had not created a 'win-win situation', i.e they had not created a situation in which all parties, including the ones on the sub-national level, having indispensable resources for the implementation of the conventions, perceived that they would be better off with the conventions than without them. Whereas for the implementation of the Meuse convention cooperation of the Walloon region was indispensable, this region did not perceive benefits from any of the proposed conventions. Because the perceptions of the Walloon politicians did not change fundamentally and neither the Dutch nor the Belgian government compensated the Walloon region, and created a win-win situation, there was a deadlock in the international negotiation process between 1975 and 1985 and the Belgian-Dutch relations concerning the management of their main shared natural resources may be characterized by conflict rather than by cooperation.

Distrust

The Belgian-Dutch history of conflict on the management of the navigation channel in the Western Scheldt and some other bilateral water management and infrastructure issues[114], caused the growth of a culture of distrust between the two basin states. In first instance, this distrust mainly concerned the relationship between the Netherlands and (the Antwerp region of) Flanders. The Dutch decision to make a linkage between several infrastructure projects aimed at an improvement of the maritime access to the port of Antwerp and Meuse water management issues, confirmed the Dutch reputation of being unwilling to cooperate on the further development of the port of Antwerp and the economic development of the Flemish region. The Dutch unwillingness to cooperate on the further development of the port of Antwerp was frequently related to the competition between the port of Antwerp and the Dutch port of Rotterdam. A Walloon member of the Belgian delegation of the Commission Biesheuvel[115], Planchar, concludes in his study of the negotiations on Meuse and Scheldt that: *"The problem of the Scheldt, the problem of the Meuse and the linkage between them are in the first place problems of competition between Antwerp and Rotterdam"* (*"Le problème de l'Escaut, puis celui de la Meuse et la liaison des deux, sont, avant tout, des problèmes de concurrence Rotterdam-Anvers."*)[116]

---

114. See also the historic overview presented in Section 4.8.

115. See Section 6.2.

116. Planchar (1993, p. 105); The negotiations on the Baalhoek convention illustrate that the Dutch were afraid of a distortion of competition between Belgian and Dutch Sea Ports at least. Whether the competitive position of the port of Rotterdam played a role in intranational decision making in the Netherlands on the strategies to be used in the negotiations on the water conventions in the period between 1975 and 1985 could not be assessed. It was, however, argued before that not 'objective facts', but feelings and perceptions of the

The Dutch linkage between Scheldt infrastructure and Meuse water management issues, the issue of the construction of storage reservoirs in Wallonia and the detailed water quality policies in the Meuse convention, contributed to the development of a culture of distrust between Wallonia and the Netherlands, and between Wallonia and Flanders as well.

The Belgian plans to use the pipeline of Tessenderlo and the continuous refusal to sign the Scheldt and Meuse water management conventions, gave the Belgian state and the Belgian regions the reputation to be unwilling to clean-up the rivers Meuse and Scheldt and to have one of the less well developed water quality policies in Western Europe. Summarizing, between 1975 and 1985, mutual bad reputations were reinforced or developed and the Belgian-Dutch relationship concerning the management of the rivers Scheldt and Meuse may be characterized as one of profound distrust.

5.3.3 Learning processes

Strategic learning
In the decision making rounds I to IV examples of strategic learning can hardly be found. For 10 years the basin states were unable to overcome the deadlock in the negotiations after the Walloon disapproval of the draft water conventions in 1975, because neither of the basin states or regions changed its objectives or strategies fundamentally. The only interesting example of strategic learning concerns the Antwerp objections to the use of the pipeline Tessenderlo (See Section 5.3.1).

Cognitive learning
The parties recognized the intellectual coherence between several issues related to the deepening of the Western Scheldt. First, the TSC research had indicated clearly that implementation of the deepening programme would cause erosion of the banks along the Western Scheldt. Other related issues introduced in the decision making process were the displacement of polluted sediments, geomorphological disturbances caused by the dredging works and the dumping of dredged material, and the safety of navigation on the Western Scheldt. Because of the recognition of the intellectual coherence between these issues, these issues were placed on the (international) agenda. Because research had also indicated that the straightening of the bend near Bath would no longer be necessary if the deepening programme would be implemented, the construction of a by-pass near Bath was removed from the international agenda.

5.3.4 Influence of the context on the perceptions and strategies of the Scheldt basin states and regions

The analysis of the perceptions and strategies of the actors involved in the decision making process and the strategic interactions between them, made it possible to trace some contextual factors and developments that have influenced the course and outcomes of the decision making process.

---

actors involved in a decision making process influence their strategic behaviour. Consequently, for this study the Antwerp, Flemish or Walloon *perception* that the competition between Antwerp and Rotterdam *did* play a role in the negotiations on the water conventions is more interesting than, for example, 'objective' information on the actual influence the port of Rotterdam has had on decision making in the Netherlands on the strategies to be used in the negotiations on the water conventions.

Underlying hydrological structure of the issues at stake
The underlying hydrological structure of the issues at stake seems to have strongly influenced the actors' perceptions of (inter)dependence and their strategies. Because of the upstream-downstream relations in the rivers Scheldt and Meuse, the downstream basin state the Netherlands perceives a dependence on the upstream basin state Belgium as regards the discharge of a reasonable amount of water of a reasonable quality. In the issue-area of the maritime access, however, the upstream basin state Belgium perceives a dependence on the downstream basin state the Netherlands as regards the improvement or maintenance of the maritime access to the port of Antwerp. The strategies used by the Scheldt basin states and regions make it very plausible that the underlying hydrological structure of the issue-areas were a very important, if not the most important contextual factor that influenced the perceptions and strategies of the basin states and regions and consequently the course and outcome of the decision making process.

International context
Because of the linkages made across Scheldt and Meuse issue-areas, decision making on international issues in the adjacent Meuse basin formed the most important international context of decision making on international Scheldt issues (See Table 5.3). Because of these linkages no agreement on the Scheldt infrastructures issues could be reached unless an agreement would have been reached on the water quantity and quality issues in the river Meuse basin.

Cultural context
The case study provides some evidence for the influence of the cultural context on the perceptions and strategies of the actors involved in the decision making process. A first example concerns the Belgian reaction on the statement of the Dutch Minister Tuijnman that the Dutch advisory body, the *Raad van de Waterstaat*, will be consulted on the deepening programme. Whilst the Dutch Minister intends to consults the advisory body and with that to start a participation round, the Belgian Minister Chabert pleas for quick decision making (See Round III). This different approach may be explained partly by the difference between the Dutch consensus culture, which is characterized by lengthy decision making processes and ample possibilities of participation, and the more hierarchical Belgian decision making culture.[117] The Dutch government was of the opinion that it needed the commitment of the region. The difference, however, may also be explained by the different interests of the basin states in the implementation of the deepening programme. The Belgian government, which is interested most, pleaded for quick decision making, whilst the Dutch government, which does not have that much interest in the deepening programme, prefers to carry out research, to consult an advisory body and to organize a participation round.

A second example concerns the installation of the Scheldt Water Commission (SWC). The preliminary talks of the Dutch delegation for the first meeting of this commission clearly show that some Dutch representatives perceived cultural differences between Belgium and the Netherlands. According to them, at the first SWC meeting the Dutch would have to be flexible as regards the institutional design of the commission and the sequence of the topics to be addressed in the commission, mainly because they expected the Belgian delegation to be bound more by their administrative and decision making procedures (See Round IV). This

---

117. See Section 4.6.4.

example clearly illustrates that some Dutch perceived that they had more mandate than their Belgian colleagues.[118] These two examples show that, although most likely less influential than the underlying hydrological structure of the issues at stake and the international context, the cultural context of the actors did influence their perceptions and strategies.

Intranational context

Undoubtedly, intranational developments in Belgium were most influential. Although in the period between 1967 and 1985 Belgium was a unitary state, there actually were four instead of two main parties involved in decision making on international Scheldt and Meuse issues: the Netherlands, the state of Belgium, and the Walloon and Flemish regions. The Act of 1 August 1974 had given the regions some competencies to develop regional policies which were used by the Walloon region to oppose the draft conventions presented in 1975. According to the Walloon politicians, the Walloon region was not involved in the negotiations on these conventions, which would have a significant impact on the water management in the Walloon region (See Round III). Because the Flemish and Walloon regions did have rather different perceptions of the issues at stake and more competencies were attributed in 1980 to the Belgian regions, the three Belgian parties, i.e. the national government and the two regions, were hardly able to formulate one joint Belgian standpoint for the negotiations with the Netherlands.

Beside the Belgian constitutional amendments, the different stages of development of water quality policies in Belgium and the Netherlands may explain partly the perceptions and strategies of the national governments. After the adoption of the Dutch Surface Water Act in 1970, which enabled the Dutch government to clean up the Dutch waters, the Dutch got a lead on Belgium as regards the development of water quality policies. Beside the economic interest in the water quality of the Meuse, this may explain the Dutch emphasis on the need to conclude international conventions to clean up the rivers Scheldt and Meuse. In Belgium, however, water quality policies had not yet been developed and in particular the Walloon region did not want to formulate strict emission policies to reduce industrial and domestic waste discharges. The Walloon perceptions should be related to the economic situation in the Walloon region. Whilst the Flemish economy was flourishing, the Walloon economy was declining. According to the Walloon region, the water quality regulations in the Meuse and Bath conventions would oblige the Walloon region to impose costly measures on its industries, whilst the Bath and Baalhoek projects would have stimulated the Flemish economy. Consequently, a Belgian approval of the water conventions would have increased economic differences between the Belgian regions even more.

5.4   Conclusions

In this chapter decision making on international Scheldt issues between 1967 and 1985 is described, analyzed, and explained. The use of the rounds model of decision making made it possible to generate a long list of decisions taken by the actors involved in the decision making process, to trace five so-called crucial decisions and to distinguish between four decision making rounds. The crucial decisions marking the beginning and/or end of the decision making rounds were the decision of the Dutch government to link the Scheldt

---

118. Ibid.

infrastructure issues to Meuse and Scheldt water management issues in 1967, the Belgian-Dutch decision to start negotiations on the 'water conventions' in 1969, the approval of three draft conventions by the Belgian and Dutch delegations in 1975, the Dutch linkage of the 48'/43'/38' deepening programme to the Scheldt and Meuse water management issues in 1983 and the declaration of intent of the Belgian and Dutch Ministers of Foreign Affairs to restart negotiations on the water conventions of 1985.

Belgium and the Netherlands had different perceptions of the international Scheldt and Meuse issues at stake. The Belgian government perceived a dependence on the Dutch as regards the implementation of infrastructure projects to improve the maritime access to the port of Antwerp, whereas the Dutch government perceived a dependence on Belgium as regards the distribution of the Meuse water and the amelioration of the water quality of the rivers Scheldt and Meuse. Within the state of Belgium, the regions Wallonia and Flanders also perceived different problems. The Flemish region was the demanding party for the three infrastructure projects, whilst the Walloon region did not perceive an interest in any of the international Scheldt or Meuse issues at stake. The Dutch linkage across the Scheldt and Meuse issue-areas is in accordance with the theory that in cases where decision making is based on unanimity and where there are heterogenous preference intensities across issue-areas, the probability that movement away from the status quo involves issue-linkage is high. The Dutch linkage, however, did not contribute to international agreement, because the Walloon region did not perceive the proposed package deal as beneficial and used a reactive strategy aimed at maintaining autonomy. In spite of the total deadlock between 1975 and 1985, neither of the parties fundamentally changed its objectives or strategies, and strategic learning processes hardly developed. Because the parties were unable to create a win-win situation, hardly any institutional arrangements and no international policies were developed. The continuous disagreement between Belgium and the Netherlands and their rigid mutual strategies, however, reinforced the Belgian-Dutch distrust which had grown during the history of conflict on the management of the river Scheldt.

The upstream-downstream power asymmetries caused by the underlying hydrological structure of the Scheldt and Meuse issue-areas seem to have strongly influenced the perceptions of (inter)dependence of the basin states and therefore to have been the most influential contextual variable. In addition, the Belgian regionalization process and the different stages of development of water quality policies in Belgium and the Netherlands seem to have been influential contextual developments. The Walloon unwillingness to cooperate on the clean up of the rivers Meuse and Scheldt should be related to the economic decline the Walloon region had to cope with. Finally, the case study contained some illustrations of the influence of the decision making culture of the basin states on their perceptions and strategies.

CHAPTER 6

LABORIOUS NEGOTIATIONS AND THE BREAK-THROUGH (1985-1993)

6.1 Introduction

In this chapter the description, analysis, and explanation of decision making on international Scheldt issues, which was started in Chapter 5, is continued. Chapter 6 addresses the period between 1985 and 1993. In Section 6.2 the selection of six crucial decisions, and the distinction between five decision making rounds are clarified. Subsequently, the strategic interactions in each decision making round are described. With that the first research question is answered for the period indicated. In Section 6.3 the decision making process is analyzed and explained. This section addresses the explanatory research questions three to six[1], and successively discusses the influence of the perceptions of the actors involved in the decision making process on their strategies, institutionalization and learning processes, and the influence of the context of the decision making process on the perceptions and strategies of the actors. Finally, in Section 6.4 the main conclusions of this chapter are summarized.

6.2 Reconstruction of the decision making process (1985-1993)

Selection of crucial decisions, and distinction between decision making rounds
Like in the previous chapter, the reconstruction of decision making between 1985 and 1993 mainly draws on the analysis of written material. For the reconstruction of decision making in the last rounds, however, the interviews provided interesting information on the perceptions and strategies of various actors. Appendix 12 contains the list of decisions, which were inventoried. In this series of decisions six crucial decisions, which influenced the composition of the international arena or the issues on the international agenda, could be traced, and five decision making rounds could be distinguished. Table 6.1 produces information on the crucial decisions marking the beginning and/or end of the decision making rounds, and the main actors participating in these rounds.

---

1. The second research question was answered in Chapter 4.

Table 6.1 *Decision making rounds, crucial decisions, and major actors (1985-1993)*

Round	Start situation	Major actors	End situation
V (1985-1987)	Declaration of intent of 1985	Belgian government Dutch government Belgian and Dutch Ministers of Foreign Affairs	Belgian-Dutch agreement to start negotiations on the basis of a modified declaration of intent
VI (1987-1989)	Installation of Belgian-Dutch negotiation commission (Commission Biesheuvel-Davignon)	Belgian delegation Dutch delegation Commission Biesheuvel-Davignon Belgian government Walloon government	Belgian decision to replace the chairman of the Belgian delegation,and to add representatives of the regions to the delegation
VII (1989-1991)	New Belgian delegation with regional representatives	Belgian delegation Walloon delegation Flemish delegation Brussels delegation Dutch delegation Commission Biesheuvel-Poppe	Decision of Dutch delegation to temporarily suspend negotiations, and to prepare a draft convention unilaterally
VIII (1991-1992)	Dutch decision to prepare a draft convention unilaterally	Delegations of Commission Biesheuvel-Poppe	Disagreement on Dutch and Belgian draft conventions
IX (1992-1993)	Dutch-Walloon-Flemish agreement to restart negotiations on the water conventions	Dutch government Walloon government Flemish government Brussels government Belgian government French government	Dutch decision to unlink bilateral and multilateral Scheldt and Meuse issues Decision of Scheldt and Meuse basin states and regions to start multilateral deliberations on the water quality of Scheldt and Meuse Dutch-Flemish decision to start bilateral negotiations on the water quantity of the Meuse and the 48'/43'/38' deepening programme

Round V: Modification of the declaration of intent of 1985 (1985-1987)

After the Dutch and Belgian Ministers of Foreign Affairs have signed the declaration of intent (See Section 5.2, Round IV), in Belgium strong opposition to the content of this declaration develops. The two main reasons for the Belgian objections concern the linkage between the deepening of the Western Scheldt and the Meuse water management issues, and the paragraph dealing with the construction of storage reservoirs in the Walloon region.

According to the critics, acceptance of the declaration of intent of 1985 would imply that, although the Belgian government protests against the Dutch linkage between the deepening of the Western Scheldt and the Meuse water management issues, it would accept such a linkage eventually. The line of reasoning is that indeed a straightening of the bend near Bath in the Western Scheldt, and the construction of the Baalhoek canal require a Belgian-Dutch convention, but that an improvement of the navigation channel in the Western Scheldt is a Belgian right that is formulated in the Scheldt statute. Consequently, a deepening of the navigation channel would never require a Belgian-Dutch convention.[2] With the acceptance of the declaration of intent the Belgian government would give up practically the issues of the Baalhoek canal and the bend near Bath, and start negotiations on an issue for which according to international law no convention would be needed.[3] Therefore, according to the critics *"this decision was a major tactical error"* (*"Cette décision constituait une erreur tactique fort grave"*).[4]

Beside the linkage between the 48'/43'/38' deepening programme and the Meuse water management issues, the paragraph dealing with the construction of storage reservoirs on Walloon territory appears to be controversial as well. The Dutch government is of the opinion that Belgium will only be able to guarantee a minimum flow of the river Meuse if storage reservoirs will be constructed. The Walloon region, however, is of the opinion that if no storage reservoir in the French-Walloon tributary of the river Meuse, the Houille, can be built, alternative solutions for the water quantity problems have to found. The French government had already informed the Belgian government in 1984 that France does not want to construct a storage reservoir in the Houille.[5] Mainly because of the objections of the Walloon region, a Belgian working group with representatives of national and regional governments concludes that the declaration has to be rejected.[6] The Dutch minister of Foreign Affairs gives in to the Belgian (Walloon) objections, and proposes to start negotiations on the basis of the declaration of intent of 1985, except for the passages referring to the construction of storage reservoirs on Walloon territory.[7] Thereupon, the Belgian government approves the modified declaration of intent[8], but adds that the declaration of intent does not alter the mutual rights and obligations that are formulated in the Scheldt

---

2. Strubbe (1988b, p. 241), Suykens (1995, p. 230).

3. Planchar (1993, p. 100).

4. Ibid.. According to Suykens, the critics forgot that linkages between Scheldt and Meuse issues were nothing new, because also in 1863, 1925, and in 1975 Scheldt and Meuse issues had been dealt with simultaneously (Suykens, 1995, p. 230). See also the historic overview presented in Section 4.8.

5. De Standaard, 17/18-10-1987.

6. Planchar (1993, p. 101).

7. Letter from the Dutch Minister of Foreign Affairs, van den Broek, to the Belgian Minister of Foreign Affairs, Tindemans, of 3 December 1986.

8. The Belgian government takes this decision on 16 December 1986.

statute.[9] The Belgian government also announces that it would be recommendable to involve the upstream basin state France in the negotiations at some moment in the future.[10]

Continued disagreement concerning the use of the Tessenderlo pipeline, and the erosion of the Scheldt Water Commission
The Dutch government takes the pipeline issue very seriously. Therefore, the Dutch embassy expresses its concerns about the Belgian plan to grant a permit for the discharge of waste water in the Western Scheldt via the pipeline Tessenderlo-Antwerp.[11] At the second and third meetings of the SWC the 'pipeline issue' is the main topic of discussion. At these meetings the Belgian delegation explains that the Belgian government will grant a permit for the discharge of brackish water only, if the discharges will not deteriorate the water quality of the river Scheldt at the Belgian-Dutch border.[12] In 1987 the Belgian government decides to withdraw its delegation from the SWC, because it does not want to continue formal deliberations concerning the water quality of the river Scheldt during the negotiations on the water conventions in the commission Biesheuvel-Davignon (See decision making round VI).[13] Belgium also discontinues its participation in the Technical Meuse Commission (TMC) (See Section 5.2, Round IV). Within Belgium, in particular the Walloon region does not want to continue Belgian-Dutch interactions in these commissions.[14]

Toward a policy plan for the Western Scheldt
The Dutch 'Raad van de Waterstaat' subscribes to the conclusion of the TSC-research report of 1984 that implementation of the 48'/43'/38' deepening programme in the Western Scheldt is acceptable.[15] Nevertheless, the advisory body recommends to address the problems related to the storage of polluted dredged materials, the protection of the banks along the Western Scheldt, and the protection of the tidal areas. The advisory body also refers to the issue of the (distortion of) competition between Belgian and Dutch sea ports. The participation rounds organized by the advisory body had shown that the regional and local governments in the Dutch Province Zeeland do not support the implementation of the deepening programme in

9. Letter from the Belgian Minister of Foreign Affairs, Tindemans, to the Dutch Minister of Foreign Affairs, Van den Broek, of 26 January 1987. On 30-12-1986 the Belgian Minister of Foreign Affairs had already informed the Dutch Minister of Foreign Affairs that the Scheldt Statute will continue to be fully applicable.

10. d'Argent (1997, p. 18).

11. The embassy issues a report on 5 November 1985 (Files Ministry of Transport, Public Works, and Water Management TSC-meeting 30, 7 March 1986, Ghent). On 26 June 1986, the Dutch Minister of Transport, Public Works, and Water Management opposes the use of the pipeline once more (Letter from the Dutch Minister of Transport, Public Works, and Water Management to the Belgian Minister of Public Works of 26 June 1986).

12. Files Ministry of Transport, Public Works, and Water Management, Meeting 2 of the Scheldt Water Commission (SWC), 16 January 1986. Later the Belgian Ministry of Foreign Affairs confirms the Belgian standpoint in a letter to the Dutch embassy of 26.2.1986. The third and last meeting of the SWC takes place on 23 October 1986.

13. Report of deliberations between Belgian and Dutch civil servants of 15-4-1987 (In files SWC, Dutch Ministry of Transport, Public Works, and Water Management).

14. Files TSC-meeting 33, 23 November 1989, Middelburg.

15. Advice of the Dutch advisory body, the 'Raad van de Waterstaat' concerning the 48'/43'/38' deepening programme of 30 October 1985.

the Western Scheldt unconditionally (See Section 5.2, Round IV).[16] Furthermore, no coordinated policy for the management of the Western Scheldt does exist. For these two reasons the Dutch Ministry of Transport, Public Works, and Water Management takes the initiative to start deliberations on the preparation of an integrated policy plan for the Western Scheldt.[17]

Round VI: Negotiation commission Biesheuvel-Davignon (1987-1989)

When the Belgian Council of Ministers decides to restart negotiations on the water conventions with the Netherlands on 16 December 1986, it also decides that the Belgian delegation of the negotiation commission will be chaired by the Belgian former EC commissioner E. Davignon. The Dutch government appoints the former prime minister B.W. Biesheuvel as chairman of the Dutch delegation. The record of service of both chairmen indicates the political importance attached to the issues at stake.

Only a few months later, on 3 March 1987, the Walloon Prime Minister Wathelet raises objections to the composition of the Belgian delegation. Because of the Act on the Institutional Reforms of 9 August 1980, which had given more autonomy to the Belgian regions, the regions would have to be involved in the negotiations on the water conventions.[18] According to the Walloon region there are, it's true, Walloons in the Commission Biesheuvel-Davignon, but they do not represent explicitly the Walloon region. The Walloon region takes this matter up with a committee, which was established to settle disputes concerning the distribution of competencies between the regions and the national government.[19] The Walloon Prime Minister also repeats the Walloon point of view that the negotiation commission should be extended with representatives of the upstream basin state France, because all basin states and regions would have to be involved in the negotiation on a convention for the river Meuse.[20] Finally, the Walloon region announces that it is does not agree with the linkage between the deepening of the navigation channel in the Western Scheldt and the Meuse convention.[21]

At the first meeting of the Commission Biesheuvel-Davignon on 9 March 1987, the commission decides to install two working groups: one for the river Scheldt[22], and one for

---

16. The Dutch province of Zeeland does not only formulate technical conditions to the implementation of the deepening program, such as conditions concerning the protection of the banks along the Western Scheldt against erosion, but also supports actively the policy of the Dutch national government to make a linkage between the deepening of the navigation channel in the Western Scheldt and the other 'water conventions' (Letter from the Province of Zeeland to the Dutch Minister of Transport, Public Works, and Water Management of 12 March 1986, in files of the SWC, Dutch Ministry of Transport, Public Works, and Water Management).

17. Ministerie van Verkeer en Waterstaat, Rijkswaterstaat, Directie Zeeland, Notitie nr. RFO 85.078. Proposal to formulate a policy plan for the Western Scheldt, 23-12-1985. On 29 January 1986 Dutch local, regional and national governments start the preparation of an integrated policy plan for the Western Scheldt. The discussions on the content of this policy plan are supported by an extensive research project (Scheele *et al.*, 1987, p. 105).

18. Gazet van Antwerpen, 16-10-1987.

19. Ibid.

20. Ibid.

21. Ibid.

22. The working group for the Scheldt would be chaired by the Secretary-General of the Belgian Ministry of Public Works, Paepe.

the river Meuse.[23] It is planned to install these working groups on 26 May 1987 in Brussels.[24] Because of the Walloon objections formulated above, however, the working groups would never be installed. Instead, negotiations with the Netherlands are postponed. A Belgian (Walloon) member of the commission Biesheuvel-Davignon explains the delay of the negotiations by Belgian 'intertribal disputes'.[25]

Although no formal negotiations take place, Belgium and the Netherlands continue to discuss the issue of the Tessenderlo pipeline.[26] In the discussions on this issue the Dutch compare the water quality of the Scheldt with the water quality of the Rhine[27], and emphasize that a good quality of water and sediments in the Scheldt is a joint Belgian-Dutch interest, since water and sediment pollution causes problems related to the storage of polluted dredged material. Finally, the Dutch also point to the increasing protests in the Dutch Province of Zeeland against the Scheldt pollution, and to the WVO permits the Dutch government has granted to Belgium (See also below).

On 8 August 1988 there is another important constitutional amendment in Belgium. According to this amendment most water management and infrastructure competencies are attributed to the Belgian regions Flanders, Wallonia, and the newly established Brussels Capital Region.[28] In October 1989 the Belgian national government and the governments of the three regions conclude a cooperation agreement. According to this agreement the new Belgian delegation will consist of representatives of the national government and the governments of the three regions. The chairman of the Belgian delegation, Davignon, is succeeded by the former Secretary-General of the Belgian Ministry of Public Works, Poppe. The official position of the Belgian delegation in the negotiation commission Biesheuvel-Poppe is prepared by a Ministerial Commission *ad hoc* for the Relations with the Netherlands, which consists of 20 national and regional ministers. The meetings of this commission are prepared by a working group with representatives of the national and regional administrations.[29]

Increasing protests of lower level governments and NGOs, and contacts between professionals
Because the politicians are unable to reach an agreement on pollution reduction in the Scheldt basin, the political pressure from lower level governments and NGOs, in particular in the Dutch Province of Zeeland, is increasing. In November 1988 the Province of Zeeland asks the European Commission to play a mediating role in the negotiations on the water quality of the Scheldt.[30] Environmental NGOs organize several actions. In 1988, the action group

---

23. The working group for the Meuse would be chaired by the Director-General of the Port of Liège, Planchar.

24. Planchar (1993, p. 102). Press release Ministry of Transport, Public Works, and Water Management, No. 3702, 10 March 1987.

25. Planchar in: NRC/Handelsblad, 27-10-1987.

26. See report of the informal meeting to discuss the issue of the pipeline Tessenderlo on 28 June 1988, the letter from the Dutch Minister of Transport, Public Works, and Water Management, Smit-Kroes, to her Flemish colleague, Dupré, of 22 September 1988, and the minutes of the TSC-meeting 32, 14 October 1988, Beveren.

27. For more information on the Rhine basin, see Section 4.5.1.

28. Planchar (1993, p. 103), Suykens (1995, p. 230).

29. Maes (1990, p. 8).

30. Provincie Zeeland (1993, pp. 19-20).

'Save the Scheldt' collects 20,000 signatures as a protest against the severe pollution of the river Scheldt, which are presented to the Belgian and Dutch central governments.[31] In 1989, Greenpeace presents a Scheldt Action Plan to regional and local governments and NGOs. In the same year, the Belgian Association for a Better Environment (*Bond Beter Leefmilieu*, BBL), and the Dutch Reinwater Foundation (*Stichting Reinwater*) publish a report, entitled "The Scheldt, Flemish delta as an area of ecological disaster"[32], which gives an overview of the waste water discharges in the Scheldt basin. Partly because of the increasing intensity of the protests against the absence of international action to combat Scheldt water pollution, the regional Directorate Zeeland of Rijkswaterstaat takes the initiative to start informal deliberations with Flemish water managers to discuss Scheldt water quality issues in 1987.[33] Furthermore, Belgian and Dutch experts organize the first international Scheldt symposium.[34]

Dutch decision to grant two WVO Permits to the Belgian government
On 18 May 1988, the Dutch government grants the two WVO permits the Belgian government had applied for in November 1985 (See Section 5.2, Round IV).[35] The first permit, which is valid for two years, allows Belgium to dump yearly 500,000 m³ of dredged material from the bar of Zandvliet, which is situated on Belgian territory, into the Western Scheldt.[36] The Dutch Minister of Transport, Public Works, and Water Management also decides that this will be the last permit that allows the dumping of dredged material from dredging works on Belgian territory in the Dutch Western Scheldt.[37] The second permit, which is valid for three years, allows Belgium to dump yearly 10,000,000 m³ of dredged material from the bars in the Western Scheldt into the Western Scheldt.[38] This permit contains several conditions concerning the water quality of the river Scheldt. Among other things, it states that the results of Belgian water quality policies will be taken into account in decision making on the next WVO- permit for Belgium.[39]

The conditions formulated in the WVO-Permit imply that if Belgium would not be able to improve the water quality of the river Scheldt, it would no longer receive a permit for the dumping of dredged material. In that case, maintenance costs of the navigation channel would rise dramatically, since alternative storage locations would have to be searched for.[40] In

---

31. The Dutch name of the working group is 'Redt de Schelde'. In this group several Flemish and Dutch environmental NGOs, religious and political organizations work together.

32. Bond Beter Leefmilieu, Stichting Reinwater (1989).

33. Interview F. De Bruijckere.

34. This symposium, which is called "The Scheldt, access to Antwerp", is organized in Antwerp on 1 and 2 December 1988. The Proceedings of this symposium are published in: *Water*, Vol. 7, Nr. 43/1, pp. 155-210, and: *Water*, Vol. 7, No. 43/2, pp. 211-265. At this symposium Flemish and Dutch experts discuss scientific aspects of the maintenance of the navigation channel, water and sediment pollution, and the ecology of the river.

35. Strubbe (1988a, p. 149).

36. Ministerie van Verkeer en Waterstaat, Hoofddirectie van de Waterstaat, Hoofdafdeling Bestuurlijke en Juridische zaken ,1988, Nr. RJW 51525, WVO-vergunning voor België, p. 16.

37. Ibid., p. 15.

38. Ministerie van Verkeer en Waterstaat, Hoofddirectie van de Waterstaat, Hoofdafdeling Bestuurlijke en Juridische zaken, 1988, Nr. RJW 51526, WVO-vergunning voor België, p.20.

39. Ibid., pp.19-20.

40. Strubbe (1988b, p. 241).

Belgium this is interpreted as a second infringement of the Scheldt statute within a short time span. First, the Dutch had declared that Belgium needs a permission for the implementation of the deepening program, and now it also needs a permission for the dumping of sediments.[41] For these reasons, on 8 September 1988, the Belgian government informs the Dutch government that in spite of the Dutch Surface Water Act, navigation on the Western Scheldt has to be free, and the Scheldt Statute should not be violated.[42]

For discussions and negotiations on the compliance with the conditions formulated to the WVO-permit, and the formulation of the content and conditions of a next permit, a special working group is installed, the working group for the WVO Permit of the Subcommission for the Western Scheldt of the TSC.[43]

Round VII: Negotiation commission Biesheuvel-Poppe (1989-1991)
After the chairmen of the negotiation commission have visited the Flemish and Walloon governments, negotiations are restarted on 13 October 1989.[44] In the Commission Biesheuvel-Poppe the Dutch delegation reconfirms the linkage between Scheldt infrastructure and the Meuse water quality and quantity issues.[45] The Dutch government is of the opinion that the Belgian delegation has the key for the solution of most Belgian-Dutch water management issues. The Dutch do expect that an internal Belgian agreement will be reached by a Flemish compensation of the Walloon "losses", either by a Flemish contribution to the construction of Walloon infrastructure projects, or by a financial compensation.[46] The Dutch delegation prepares several statements, which are passed over to the Belgian delegation. In short, the Dutch delegation proposes to[47]:

- assign the basic quality to all water courses in the Scheldt and Meuse basins[48];
- assign additional water quality objectives for specified river sections, such as water quality objectives for drinking water abstraction, cyprinids, salmonids, shellfish, or bathing water[49];
- complete the immission-based policy with an emission-based policy, similar to the emission-based policy developed by the Rhine basin states, i.e. the development of Action

---

41. Strubbe (1988a, p. 149).

42. Strubbe (1988b, p. 241), Files TSC-meeting 32, 14 October 1988, Beveren.

43. This working group consists of representatives of the Regional Directorate Zeeland of Rijkswaterstaat, the Antwerp Sea port Service (*Antwerpse Zeehavendienst*), which was the Flemish administration responsible for the dredging works on the bars in the navigation channel, the Flemish Agency for waste water treatment (*Vlaamse Maatschappij voor waterzuivering* VMZ), and the Flemish Administration for the Environment, Nature, Land and Water Management (*Administratie Milieu, Natuur, Land en Waterbeheer*, AMINAL). The VMZ and AMINAL were involved, because these administrations are responsible for pollution abatement in the Flemish region.

44. Files Ministry of Transport, Public Works, and Water Management of TSC-meeting 33, 23 November 1989, Middelburg.

45. TK 1988-1989, Aanhangsel van de Handelingen, 68.

46. TK 1994-1995, 24 041 No. 3.

47. Proposals made by the Dutch delegation in the Commission Biesheuvel-Poppe, numbered A to E (Files Dutch Ministry of Transport, Public Works, and Water Management).

48. In the Netherlands this basic quality is called the "algemene milieukwaliteit", whereas in Belgium the basic quality is called the "basiskwaliteit".

49. These objectives are based on the existing EC water quality Directives (See also Section 4.5.2).

Programs for Scheldt and Meuse containing reduction percentages. This would imply nothing else but an implementation of the agreements made at the North Sea Ministerial Conferences;

- inform each other on permits for waste water discharges to be granted;
- collect and exchange data on waste water discharges, which enables the basin states to monitor the implementation of the Action Programs;
- develop sanitation programs for municipal waste water discharges[50];
- develop sanitation programs for industrial waste water discharges;
- reach an agreements on thermal pollution, radioactivity, and the quality of the water beds;
- develop plans for a joint ecological management of both rivers;
- develop a joint coordinated monitoring program;
- develop an alarm system for cases of accidental pollution[51];
- coordinate national regulations concerning the prevention of accidental pollution;
- develop provisions for conflict resolution;
- install international river commissions for Scheldt and Meuse.

The Dutch delegation thinks that a Belgian-Dutch agreement on the water conventions will form a sound basis for joint negotiations with France on trilateral agreements for the Scheldt and the Meuse. According to the Dutch delegation the water quantity problems in the Meuse are caused by the Belgian diversions of water from the river Meuse basin to the Scheldt basin, because these diversions would lower the flow of the river Meuse during periods of low flows even more. Therefore, the Dutch are of the opinion that Belgium should not be allowed to divert Meuse water to the adjacent basins, unless the Dutch government is consulted.[52] This would imply that the Walloon region can no longer export Meuse water for drinking water production to the conurbations of Brussels and Antwerp. The Dutch continue their demand of a minimum flow of 28 m^3 at the Belgian-Dutch border.[53] As regards the construction of the Baalhoek Canal the Dutch delegation informs the Belgian delegation that there is much resistance against the construction of this canal in the Dutch province of Zeeland, and that the province has skipped the alignment of this canal in the provincial land-use plan.[54]

In particular the Walloon and Brussels representatives in the Belgian delegation of the negotiation commission do strongly oppose the Dutch proposals. Therefore, on 19 December 1990 the Belgian *ad hoc* Commission for the Relations with the Netherlands decides that.[55]:

- water quality objectives for the rivers Scheldt and Meuse have to be identical from the source to the mouth of these rivers;

---

50. Including agreements on sewerage percentages and waste water treatment percentages to be achieved.

51. According to the Dutch delegation, in particular in the Meuse basin such an alarm system would be indispensable, because Meuse water is used for drinking water production. The delegations should reach an agreement on (1) the obligation of industries to report cases of accidental pollution, (2) the on-line monitoring of the water quality at some crucial monitoring stations, and (3) effective procedures to communicate on cases of accidental pollution.

52. Proposals made by the Dutch delegation in the Commission Biesheuvel-Poppe, numbered A to E (Files Dutch Ministry of Transport, Public Works, and Water Management).

53. Ibid.

54. Ibid.

55. TK 1990-1991, Aanhangsel van de Handelingen, 356.

- water quality objectives have to be the same for the Meuse and the Scheldt, namely the objectives of the EC Directive 78/659 for cyprinids[56];
- negotiations on the water quality objectives should be restricted to the main courses of the river basins.[57]

According to the Walloon and Brussels Capital Regions each extra effort to ameliorate the water quality of these rivers is exclusively beneficial for the Netherlands. Therefore, if the Dutch would like to formulate water quality policies that go beyond the EC Directive 78/659 for cyprinids, they should bear the financial responsibility for the extra efforts needed to achieve these objectives.[58] In addition, the Walloon region offers the Dutch to transport Walloon drinking water to the Netherlands via a pipeline, since the Walloon drinking water industry has more capacity than it needs to satisfy the demands of the Walloon region.[59]

Belgium would like to have more information on the financial consequences, and the consequences for existing policies of the possible bilateral or multilateral agreements on the rivers Scheldt and Meuse. The *ad hoc* Commission would also like to have more information on the water quality objectives, and the use of Meuse water in the Netherlands. To what extent do the Dutch demand Meuse water for economic use, and to what extent for ecological reasons? Finally, the Commission would like to know whether the Dutch, if indeed they want to formulate international water quality policies that go beyond the EC Directive 78/659 for cyprinids, they cannot discuss such policies in a broader international context.[60]

The Dutch delegation absolutely does not agree with the Belgian standpoints, because it wants[61]:

- to reach an international agreement on the application of the other EC Directives, such as the Directives for drinking water abstraction, and shell fish water;
- to reach an agreement on the implementation of the international agreements made at the North Sea Ministerial Conferences of 1984, 1987, and 1990[62];
- to observe the recommendations formulated by the IPR (See next section);
- Belgium to start with the clean-up of its waters, just like the surrounding countries are

---

56. These objectives are less ambitious than the agreements that are made at the North Sea Ministerial Conference of 1990. Because the water quality of the river Meuse already complies with the EC-water quality objectives for cyprinids, according to the Dutch this would imply that the Walloon region would not have to take further action to clean-up the river Meuse (TK 1990-1991, Aanhangsel van de Handelingen, 356).

57. The Walloon region has problems with the broad definition of "the Meuse" in the draft Meuse convention of 1975 (See Section 5.2, Round II). According to this definition the precious ground water resources in the Walloon region, which have a high economical value, would fall under the Meuse convention. The Walloon region, however, prefers to maintain absolute sovereignty as regards the management of these ground water resources (Proposals made by the Dutch delegation in the Commission Biesheuvel-Poppe, numbered A to E, Additional notes made by and for the Dutch delegation, Files Dutch Ministry of Transport, Public Works, and Water Management).

58. Press release Dutch Ministry of Transport, Public Works, and Water Management, 9 January 1991, No. 4466.

59. Letter from the chairman of the Belgian delegation, Poppe, to the chairman of the Dutch delegation, Biesheuvel of 27 November 1990.

60. Letter from the chairman of the Belgian delegation, Poppe, to the chairman of the Dutch delegation, Biesheuvel, of 10 December 1990.

61. Press release Dutch Ministry of Transport, Public Works, and Water Management, 9 January 1991, No. 4466.

62. For more information on these conferences, see also Section 4.5.2.

doing[63];

- to improve the water quality of the Meuse not only for economic reasons, but also for ecological reasons, and to reduce the load of pollutants transported to the North Sea;
- to intensify international contacts in the Scheldt and Meuse basins.[64]

According to the Dutch, serious negotiations with Belgium on the water quality and quantity of Scheldt and Meuse have proven to be impossible. Therefore, on 8 January 1991, the Dutch delegation decides to suspend the negotiations, and to draft a convention unilaterally in order to further clarify its objectives.[65] The Dutch also postpone the technical discussions in the TSC concerning the 48'/43'/38' deepening programme and the construction of the Baalhoek Canal, until progress is made with the bottlenecks in the negotiations in the water conventions, in particular with those concerning the water quality of the Meuse.[66] According to the Dutch delegation, the only positive development is that the Netherlands and the Flemish region have almost reached an agreement on plans to restrict the use of Meuse water during periods of low flow, which may contribute to a solution of the water quantity problems the Dutch perceive in the river Meuse basin.[67]

Ever increasing pressure on the governments of the basin states
From the foregoing, it becomes clear that after the Belgian regions becoming directly involved in the negotiations, the basin states still are unable to reach an international agreement on the Scheldt and Meuse infrastructure and water management issues. Pressure on the governments of the basin states, however, to reach such an agreement, increases rapidly at the end of the eighties and the beginning of the nineties. On 1 December 1990, the Interparliamentary Council of the Benelux (IPR) approves a resolution[68], which was prepared at the first Benelux Meuse Conference.[69] In this resolution the IPR formulates several recommendations. The Meuse basin states should[70]:

- apply a river basin approach;
- install an International Meuse Commission analogous to the International Rhine

---

63. According to the Dutch as regards the development of water quality policies Belgium is 15 years behind compared with the Rhine basin states (Press release Dutch Ministry of Transport, Public Works, and Water Management, 9 January 1991, No. 4466).

64. There are hardly international contacts since Belgium has decided to withdraw the Belgian delegations from the Scheldt Water Commission and the Technical Meuse Commission.

65. TK 1990-1991, Aanhangsel van de Handelingen, 356.

66. Report TSC-meeting 34, 2 January 1991, Point 4.

67. Press release Dutch Ministry of Transport, Public Works, and Water Management, 9 January 1991, No. 4466.

68. In Dutch the IPR is the 'Raadgevende Interparlementaire Beneluxraad'; IPR (1990); Resolution of the Interparliamentary Meuse conference, Approved by the Interparliamentary Council of the Benelux on 1 December 1990.

69. The Dutch Members of Parliament Zijlstra and Lilipaly took the initiative to organize this conference on the management of the river Meuse, which took place on 12 November 1990 in Maastricht. At the conference the main results of a preparatory research report, which was made by the Institute for the Environment and Systems Analysis in Amsterdam, the "Laboratoire d' Ecologie" in Brussels, and the University of Metz, were discussed (Zijlstra and Lilipaly, 1995, p. 27). At this conference also the idea was born to organize a similar conference on the Scheldt.

70. IPR (1990); Resolution of the Interparliamentary Meuse conference, Approved by the Interparliamentary Council of the Benelux on 1 December 1990.

Commission;
- formulate a Meuse Action Programme taking into account the North Sea Action Program;
- develop an international monitoring network;
- develop an alarm system for cases of emergency;
- improve the water quality to achieve the ecological basis quality, and to safeguard drinking water abstraction from the Meuse.
- conclude negotiations on the water conventions in the short term.

Although the IPR is able to formulate a joint recommendation, the basin states do have several sharp discussions about the water conventions in this intergovernmental organization. Later, a Dutch member of the IPR would state: "*Anybody who ever has sampled the atmosphere in the IPR knows how tense the relations between colleagues could be if the water conventions were discussed.*" [71]

Because negotiations in the Commission Biesheuvel-Davignon/Poppe have been laborious from the beginning, and Belgium and the Netherlands have not yet reached an agreement on the main transboundary water management issues, the effectiveness of the Dutch linkages across Scheldt and Meuse issue-areas becomes a topic of discussion in the Netherlands. The director of the drinking water production firm in the Dutch Biesbosch, Oskam, for example, states in an interview that "*Mrs Smit-Kroes continued the linkage, as this probably is written in her Course on Negotiations, but it does not work that way [...]. You have to deal with the rivers separately, since the problems are complicated enough. You should, for example, exchange the deepening of the Western Scheldt with the sanitation of the Brussels discharges into the Senne*".[72] In part of the Dutch press the inflexibility of the chairman of the Dutch delegation, Biesheuvel, is criticized. It is stated that he holds on to the linkage, whilst the ineffectiveness of this linkage has been proven.[73]

Also some lower level governments and NGOs begin to lose their patience. Some municipalities along the Western Scheldt continue to develop spatial planning plans for the parts of the Western Scheldt situated on their territories. Because of these plans, the Flemish government will need a permit for dumping dredged material on the territory of the municipalities.[74] The competence to grant a permit to the Belgian government would enable the municipalities to make a linkage between the permit and an amelioration of the water quality of the Scheldt[75], and by that to increase their possibilities to influence decision making on international Scheldt issues. The perception of Dutch local governments may be illustrated with some statements of the mayor of the Dutch village Valkenisse, Plomp: "*when I make an inventory of the environmental measures of Brussels, I can only conclude that Belgium is a developing country*" and "*[...] my municipality cannot become the victim of the institutional chaos in a neighbouring country.*"[76] Dutch lower level governments in the province of Zeeland know very well that they are largely dependent on Belgium as regards the

---

71. TK 1995, 97th meeting, 31-8-1995, p. 5926.

72. Smit-Kroes was the Dutch minister of Transport, Public Works, and Water Management; Oskam in: Van der Veen (1989).

73. PZC, 20-11-1990.

74. Tromp and Swart (1988, p. 199).

75. The declaration of intent of 1985 had clearly shown that the Dutch national government was interested more in the water quantity and quality of the river Meuse than in the water quality of the Scheldt.

76. Camps (1990, pp. 26 and 29).

achievement of the objectives they formulate for the Western Scheldt. Therefore, they send the draft policy plan for the Western Scheldt to several Belgian government agencies, and invite the Belgian government to participate in the preparation of Western Scheldt policies. The Belgian government prefers a joint Belgian reaction by the minister of Foreign Affairs.[77] In a letter the minister states that: "[...] *it is clear that the draft of the policy plan deals with issues that, as regards their transboundary effects, are part of the negotiations between both countries. In order to avoid double work and misunderstanding, deliberations on these issues should take place in the Belgian-Dutch negotiation commission on Scheldt and Meuse issues, for that purpose by both governments installed.*"[78] In the letter it is also stated that if transboundary issues are not discussed in the Belgian-Dutch negotiation commission, they can be discussed in the existing international platforms, such as the Permanent Commission or the TSC.[79] Clearly noticeable, the Belgian government is not interested in another international forum.

Environmental NGOs continue to protest against Scheldt water pollution. An event that receives a lot of attention in the media is a protest march from the mouth of the river Scheldt, in Breskens, to the source of the river in the French place Gouy from 9 till 28 April 1990.[80] During this march several mayors of cities along the river Scheldt are visited. The participants demand, among other things, a Scheldt Action Programme and an International Scheldt Commission.

Round VIII: Dutch and Belgian draft conventions (1991-1992)

On 8 january 1991 the Dutch government had decided that it would prepare a draft convention unilaterally (See Round VII). On 15 April 1991, the chairman of the Dutch delegation, Biesheuvel, passes over a draft convention for cooperation on the management of Meuse and Scheldt to the chairman of the Belgian delegation, Poppe.[81] Among other things, the Dutch draft convention states that[82]:

- in 1995 the basic quality has to be reached;
- within a period of ten years the river basins have to be clean, with a water quality good enough for drinking water production, which does not burden the North Sea any longer;
- the parties have to coordinate the assignment of functions to the water system;
- the parties have to formulate an action programme with jointly formulated objectives;
- the dumping of polluted dredged material more downstream in the estuary has to be restricted;
- agreements have to be made on the development of an alarm system for cases of accidental pollution;
- agreements have to be made on a joint monitoring program;
- agreements have to be made on the risks of flooding;
- the 48'/43'/38' deepening programme can be implemented;

---

77. Report TSC-meeting 33, 23 November 1989, Middelburg.

78. Letter from the Belgian minister of Foreign Affairs, Eyskens, to the chairman of the Core Working Group of the Western Scheldt, De Vries-Hommes, of 12 January 1990.

79. Ibid.

80. De Volkskrant, 28-4-1990, Kuijs (1990, pp. 6-7).

81. TK 1990-1991, 21 800 XII No. 61.

82. Saeijs and Turkstra (1991, p. 208), Dutch draft convention for cooperation on the management of the rivers Scheldt and Meuse of 12 April 1991.

- the parties have to cooperate in international river basin commissions for Scheldt and Meuse with a broad mandate;
- Flanders and the Netherlands will develop a saving scheme for cases of low flow in the river Meuse.

The Dutch delegation does have several strategic and substantial arguments to conclude one integrated convention dealing with the Scheldt and Meuse issues simultaneously, and not to include the Baalhoek project in the convention. The arguments to present an integrated draft-convention are[83]:

- that the Dutch delegation would like to install one commission for each river basin with responsibilities for water quality management, water quantity management, and the management of the navigation channels;
- the relationship between the water quantity and the water quality of the Meuse, in particular during periods of low flow;
- the relation between the deepening dredging works, the quality of the environment (in particular the impacts of dredging works and the dumping of sediments), and the Scheldt regime.

The arguments to present one convention dealing with Scheldt and Meuse issues simultaneously are that[84]:

- the existing international water quality policies for the rivers Scheldt and Meuse are similar;
- Brussels, Wallonia, and Flanders do all have an interest in both Scheldt and Meuse issues;
- the water systems of the Meuse and the Scheldt are connected in various ways;
- the Dutch fresh water losses in the Scheldt-Rhine connection would have to be compensated in the Meuse basin.

Finally, the Dutch arguments not to include the Baalhoek project in the draft convention are that[85]:

- the lower level governments and NGOs in the Dutch province of Zeeland oppose the construction of this canal;
- although the alternative is less attractive from a nautical point of view, the Flemish region does have an alternative maritime access to the Waasland harbour;
- the Baalhoek Canal is no longer necessary to make concessions to Belgium, because Belgium does not want to construct storage reservoirs anyway, and the water quality policies formulated in the draft convention do not go beyond existing international policies.

The Dutch continue the linkage between Scheldt and Meuse issues. In an explanation for the Dutch Council of ministers, the Dutch delegation writes that: "*From different sides, one insisted to unlink the Scheldt and Meuse issues. The Dutch negotiation delegation, however, in accordance with the instructions given by the ministers concerned, has dealt with the issues as a package, for which parallel negotiations have to take place. Gradually, the linkage of the different issues also appeared to be very useful.*"[86] The Dutch, however, do also make

---

83. Explanatory report from the Dutch negotiation delegation for the Dutch council of ministers. Attached to the draft convention finished by the Dutch delegation on 12 April 1991.

84. Ibid.

85. Ibid.

86. Ibid.

several concessions. First, they state that if Belgium will agree with the draft convention, the Netherlands will give up its demands for the construction of storage reservoirs in the Walloon Ardennes, and the compensation of fresh water losses in the Scheldt-Rhine connection.[87] Secondly, the convention does no longer contain detailed obligations, but rather is a framework convention. Thirdly, the convention does contain obligations for the Netherlands as well.

According to the Flemish region, the Dutch draft convention forms a sound basis for further negotiations. The main Flemish objection concerns the absence of provisions for the construction of the Baalhoek Canal.[88] The Walloon region, however, has many objections to the draft conventions. Among other things, it[89]:

- prefers to conclude three separate conventions for the water quality of the Scheldt and the Meuse, the water quantity of the Meuse, and the deepening of the navigation channel in the Western Scheldt instead of one integrated convention;
- opposes the passage "cooperation on the management of the Meuse" in the title of the draft convention, because this passage would imply an abdication of sovereignty;
- does not agree with the regulations concerning the water quality of the Scheldt and the Meuse[90];
- would like to involve France in the negotiations on the water quality of Scheldt and Meuse;
- emphasizes its right to extract water from the Meuse if it wants to do so, and to feed the Albert Canal with water diverted from the Meuse in Liège;
- states once more that it does not want to construct storage reservoirs on its territory, and suggests the Netherlands once more to buy Walloon drinking water (See also Round VII).[91]

At the first plenary meeting of the negotiation commission Biesheuvel-Poppe after the presentation of the Dutch draft convention[92], the Belgian delegation is not able to give a formal reaction on the Dutch draft convention, because no internal Belgian agreement has been reached.[93] A governmental crisis in Belgium delays the negotiations on the water conventions for several months.[94] The Belgian delegation, however, continues its efforts to reach an agreement on the water conventions, and prepares three draft conventions which are presented to the Dutch delegation.[95] The three Belgian draft conventions successively deal

---

87. Suykens (1995, p. 231).

88. Ibid.

89. Ibid.

90. The Walloon politicians want to know more about the consequences of the implementation of the draft convention, and present the draft to a famous francophone lawyer, Ergec, of the Free University of Brussels. According to this expert, the convention does not add extra obligations to the obligations formulated in existing international agreements (Suykens, 1995, p. 231).

91. De Stem, 9-3-1991.

92. Meeting of plenary negotiation commission Biesheuvel-Poppe, 6 June 1991, Brussels.

93. TK 1990-1991, Aanhangsel van de Handelingen, 679, TK 1990-1991, 21 800 XII No. 61.

94. IPR (1992, p. 35).

95. The three Belgian draft conventions are presented by the chairman of the Belgian delegation, Poppe, on 21 October 1991. The draft conventions are not yet approved by the Belgian government, because no new national and regional governments have been installed yet.

with the protection of the Meuse and the Scheldt against pollution, the flow of the river Meuse, and the management of the navigation channel in the Western Scheldt.[96] The content of the draft water quality convention is more or less similar to the standpoint formulated by the Belgian *ad hoc* commission for the relations with the Netherlands on 19 December 1990 (See Round VII).[97] The water quality convention contains articles, which imply that the Dutch would (partly) bear the financial responsibility for the clean-up of the rivers Scheldt and Meuse.[98] In addition, the Belgian delegation proposes to involve France in the negotiations on the convention for the water quality of the Scheldt and the Meuse. The draft convention concerning the flow of the river Meuse contains a saving scheme for periods of low river flow. The draft convention on the management of the navigation channel in the Western Scheldt provides for the implementation of the 48'/43'/38' deepening program. This draft convention also contains a Belgian interpretation of the Scheldt Statute. Article 4 contains the following statement: "*The parties accept as point of departure of their policy that the navigation channel in the Western Scheldt should be adapted to the present and future requirements of navigation.*"[99]

Although the Dutch had proposed to conclude one integrated convention, they accept the Belgian proposal to conclude three separate conventions.[100] Furthermore, they subscribe to the Belgian viewpoint that France should be involved in the negotiations on the water quality of Scheldt and Meuse.[101] The Dutch delegation, however, does not agree with the contents of the water quality and deepening conventions drafted by the Belgian delegation. According to the Dutch delegation[102]:

- the draft convention concerning the water quality of the Scheldt and the Meuse does only contain regulations to maintain the water quality of the rivers, but no regulations to improve the water quality[103];
- the draft convention concerning the water quality of the Scheldt and the Meuse is not in accordance with the EC 'polluter pays principle', because it makes the Dutch (partly) responsible for the financing of Belgian water quality policies[104];
- the modification of the Scheldt Statute is not an issue on the Belgian-Dutch negotiation

---

96. Files Ministry of Transport, Public Works and Water Management, TSC-meeting 35, 3 December 1991, Terneuzen.

97. Letter from the chairman of the Dutch delegation, Biesheuvel, to the chairman of the Belgian delegation, Poppe, of 19 November 1991.

98. Ibid.

99. Ibid.

100. Files Dutch Ministry of Transport, Public Works, and Water Management, TSC-meeting 35, 3 December 1991, Terneuzen.

101. Ibid.

102. Letter from the chairman of the Dutch delegation, Biesheuvel, to the chairman of the Belgian delegation, Poppe, of 19 November 1991.

103. The Dutch, it's true, do not want to regulate everything in detail, but would like to state in the convention clearly that the objective of the basin states is to restore the water quality and the quality of the water beds in both rivers, to safeguard drinking water production, and to protect the North Sea (Ibid.).

104. Therefore, according to the Dutch delegation the reference in the draft water quality convention to the Rhine Action Program, which, among other principles, is based on the polluter pays principle, is out of place.

agenda.[105]

The Chairman of the Dutch delegation states: *"For the Netherlands it is impossible to conclude an "empty" convention. By that the problems of the negotiation commission would be shifted entirely to the river commissions."*[106]

New conditions to the Belgian permit for dumping dredged material

Whilst in the Commission Biesheuvel-Poppe negotiations on the water conventions continue, in the TSC Flanders and the Netherlands continue discussions on the content of a new WVO Permit, in particular on the conditions formulated in this permit.[107] One of the conditions the Dutch had formulated in the WVO-permit granted in 1988 was that the Belgian government has to publish a report with information concerning the dredging works in the Scheldt, and the water quality of the water and the water bed. According to the report published by the Flemish government, the water quality of the Scheldt at the Belgian-Dutch border (at Schaar van Ouden Doel) has not been improved significantly since 1988.[108] Therefore, the Dutch decide to formulate new, more strict conditions to the next WVO permit. The Dutch delegation of the TSC also stresses the need to create public support in the Dutch region of Zeeland, if the Flemish government would like to implement the 48'/43'/38' deepening program. According to the Dutch the extraction of large amounts of polluted sediments from the Lower Sea Scheldt may contribute to this support.[109] Although the Flemish authorities

---

105. Also in the Belgian-Dutch discussions concerning the construction of a road connection between the northern and southern banks of the Western Scheldt the Scheldt Statute plays a role. Not to hinder navigation on the Western Scheldt, Belgium prefers to construct a deeply situated tunnel. (Letter from the Belgian Director-General of water infrastructure, Demoen, to the Director-General of Rijkswaterstaat, Blom, of 9 August 1991).

106. Letter from the chairman of the Dutch delegation, Biesheuvel, to the chairman of the Belgian delegation, Poppe, of 19 November 1991.

107. The two WVO permits for Belgium granted by the Dutch Minister of Transport, Public Works, and Water Management in 1988, expire in respectively 1990 and 1991 (See also Round VI). In spite of the decision, which was taken by the Dutch Minister in 1988, not to allow the dumping of dredged material from the bar near Zandvliet (on Flemish territory) on Dutch territory from 1990, the Belgian government applies for an extension of this permit on 12 February 1990 (Letter from the Regional Directorate Zeeland of Rijkswaterstaat to the Director-General of Rijkswaterstaat, 20 June 1990. Notitie Nr. AX 90.092). The Dutch government, however, takes a decision, which is in accordance with the decision it had taken in 1988, and refuses to grant a new permit for the dumping of dredged material stemming from maintenance dredging works on the bar of Zandvliet (Files TSC-meeting 34, 4 January 1991). On 30 January 1991, the Flemish Minister of Public Works and Transport applies for another WVO-permit for the dumping of dredged material in the Western Scheldt stemming from maintenance dredging works on the bars in the Western Scheldt (Letter from the Flemish Minister of Public Works and Traffic, Sauwens, to the Dutch Minister of Transport, Public Works, and Water Management of 18 December 1990). Because of the constitutional amendment of 1988, the Flemish instead of the Belgian minister applies for this WVO-permit.

108. Ministerie van de Vlaamse Gemeenschap, Departement Leefmilieu en Infrastructuur (1991).

109. Interview F. De Bruijckere. Experts know that the fine fractions of sediment in the Scheldt river, which generally are heavily polluted, are deposited in the mixing zone of the Scheldt estuary, which is situated in the Lower Sea Scheldt. In this zone, where the fresh water of the river Scheldt is mixed with the salt water flowing into the estuary from the North Sea, there is a specific physical condition, indicated turbidity maximum. Because the port of Antwerp is situated along the Lower Sea Scheldt, intensive dredging works are necessary in the harbour docks, and their entrance channels. Until 1991, Flanders used the technique of 'sweep beam dredging' to bring back the sediments to the main stream of the river, where they are transported downstream. Eventually, the polluted sediments ended up somewhere in the Dutch Western Scheldt and the North Sea. In the working group for the WVO Permit of the TSC the Dutch launch the idea to extract polluted sediments from the water

reject the Dutch proposal in first instance, especially because they do not know where to store the polluted sediments in the densely populated conglomerate of Antwerp, they gradually develop a more positive attitude toward the ideas proposed by the Dutch.[110] On 28 November 1991 the Dutch Minister grants a new WVO Permit to the Flemish Government, which is valid until 1 January 1995.[111] In this permit the licensee is permitted to dump 10,000,000 m³ dredged material from dredging works on the bars in the navigation channel of the Western Scheldt on other locations in the Western Scheldt yearly.[112] The Permit contains several conditions. First, the licensee has to extract within a term of three years 1,300,000 ton dry mud from the Lower Sea Scheldt, at least 200,000 ton in the first year, and at least 400,000 ton in the second year.[113] Priority is given to the extraction of the most polluted sediments that can be found in the access channel to the Kalosluice.[114] Secondly, the licensee has to present a mud balance of the Lower Sea Scheldt to the Head-Engineer-Director of the Dutch Rijkswaterstaat yearly, including information on the location and estimated quantities of deposited sediments, the amount of sediments that are stored, the storage location, and the possibilities to reduce the flow of sediments to the Western Scheldt.[115] The permit also states that if the Flemish government will apply for another permit, the amelioration of the water quality of the Scheldt at the Dutch-Flemish border will be taken into account.[116] In the participation rounds Dutch local governments and environmental NGOs raise major objections to the permit. According to them the dumping of dredged material would require an Environmental Impact Assessment, and the Dutch government should not grant another permit to the Flemish government, because the Flemish region has not complied with the conditions formulated in the WVO Permit of 1988.[117]

Continued external pressure on the governments of the basin states to reach an international agreement on the management of the river Scheldt
The second Scheldt symposium, which is called "The Scheldt, Perspectives on ecological restoration"[118], is concluded with a resolution that is passed over to the Benelux Parliament. The core of this resolution is that the basin states should install an International Scheldt Commission that has to prepare an Action Programme for the rehabilitation of the river Scheldt. Furthermore, the resolution states that the navigation function of the Scheldt has to

---

system of the Lower Sea Scheldt instead of bringing these sediments back to the main course of the river. According to the Dutch, this policy strategy would be a solution for the short and middle long term. In the long run, the river Scheldt has to be cleaned up, and transport of sediments to the mouth of the river as such is a natural phenomenon.

110. Ibid.

111. Ministerie van Verkeer en Waterstaat, WVO-vergunning baggerspecie België, 28 November 1991.

112. Ibid., p. 19.

113. Ibid., Article 6.

114. Ibid.

115. Ibid., Article 5.

116. Ibid., Part I Introduction.

117. Ibid., pp. 5-17.

118. This symposium takes place on 24 October 1991 in Bergen op Zoom (The Netherlands). The proceedings of this symposium are published in: *Water*, Vol. 10, No. 60, pp. 157-212.

be guaranteed.[119]

On 13 and 14 March 1992 also the Benelux parliament formulates a resolution concerning the management of the river Scheldt, which was prepared at the Benelux Scheldt conference.[120] The resolution states that the basin states should[121]:

- clean-up the water and water beds of the river Scheldt;
- extract heavily polluted sediments from the water system;
- restore or compensate nature losses caused by the maintenance of the navigation channel, for example by the creation of flooding areas;
- reach an agreement on the deepening programme for the navigation channel of the Western Scheldt;
- install an International Scheldt Commission that has to prepare a Scheldt Action Program;
- develop international monitoring networks and an alarm system for cases of emergency;
- apply the European Directives and the commitments made at the third North Sea Ministerial Conference of 1990.

Environmental NGOs continue their efforts to raise political attention for the water pollution of Scheldt and Meuse. Most eye-catching, and very effective from an environmental point of view, are some cases of litigation against Belgian polluting industries. The Dutch Reinwater Foundation carefully examines discharges of the Flemish firm Sopar in Zelzate into the canal Ghent-Terneuzen, and together with some other environmental NGOs, institutes legal proceedings against this firm.[122] Because of a decision of the Court in The Hague the firm Sopar is forced to treat its waste water.[123] A second case of litigation concerns the cokes industry Carcoke in Brussels. Due to the attention of public and politics for this case, the responsible Brussels minister, Gosuin, forces the firm to reduce its discharges substantially.[124] Since economic prospects are bad and investments to build waste water treatment facilities are too high, the firm closes in 1993. Another case of litigation concerns the firm Cockerill Sambre in Wallonia. Belgian environmental NGOs also try to influence the water quality policies of their national and regional governments. The Belgian division of Greenpeace presents a research report on the Walloon compliance with international agreements to reduce surface water pollution, such as the agreements made at

---

119. Anonymous (1991, pp. 660-662.)

120. At the Benelux Meuse conference (See Round VII) the idea was born to organize a similar conference for the Scheldt, which takes place on 29 November 1991 in Antwerp. At this conference, the main results of a preparatory research report, which was made by the Delta Institute for Hydrobiological research (Delta instituut voor Hydrobiologisch onderzoek) are discussed (IPR, 1991a, p. 89). At the conference representatives of the Brussels Capiral Region strongly oppose some passages of the research report, because data about the Brussels Capital Region Region would be incorrect, and the region is described as the "dirty child of Europe" (IPR, 1991b, p. 11). The Brussels delegation to the conference prepares an appendix to the research report explaining the situation in the Brussels Capital Region (Ibid., pp. 11, 24, 25). Because some representatives of French local governments participate in the conference, the Benelux Scheldt conference is the first occasion where representatives of all basin states and regions are able to discuss the Scheldt water management.

121. IPR (1992, p. 25), Zijlstra and Lilipaly (1995, pp. 30-31).

122. Reinwater (1992c, p. 2). The Reinwater Foundation was founded in 1974 to protest against the salt discharges into the river Rhine by the French potassium mines.

123. Ibid.

124. Ibid.

the third North Sea Ministerial Conference.[125] The research report gives a scathing judgement of the water quality policy in the Walloon region. It states, among other things, that more than 60% of the Walloon industries does not have a discharge permit or an insufficient one, and that the Walloon region treats only about 25 % of its municipal waste water.[126]

The Pilot study OostWest
Preparatory studies for the Dutch Policy plan for the Western Scheldt (See Round V) indicated two main causes for the ecological degradation of the Western Scheldt: water and sediment pollution, and morphological disturbances in the estuary, which were caused by the creation of polders in the past, dredging works and the dumping of dredged material in the Western Scheldt.[127] According to these studies, long term solutions for the first problem are obvious and comprise a reduction of domestic and industrial waste water discharges in the whole catchment area of the Scheldt. It is recognized, however, that it will be more difficult to find solutions for the problems related to the morphological disturbances. Therefore, one of the action points formulated in the Policy plan for the Western Scheldt is to start a research project on the impacts of the creation of polders in the past, dredging works, and the dumping of dredged material on the morphology and ecology of the estuary.[128] In addition, the study has to present policy alternatives for the solution of the inventoried problems to support long term decision making on the management of the Western Scheldt. To carry out this study the working group OostWest (EastWest[129]) with representatives of the regional Directorate Zeeland, and the Tidal Waters Division (DGW[130]) of Rijkswaterstaat is installed. In December 1991 this working group issues a Pilot Study.[131] The conclusions of this study, among other things, are that[132]:

• sustainable development in the Scheldt estuary can only be realized if policy and management are based on an integrated water systems approach. Eventually the system should be defined as the entire river basin of the Scheldt (the so-called river basin approach);

• maintenance of the navigation channel in the Western Scheldt should aim at minimizing dredging activities, because these activities do have a negative impact on the ecology of the Western Scheldt;

• the area of natural flooding areas should not be diminished, on the contrary, this area should be extended with areas that presently are situated behind the dikes (In Dutch this policy alternative is called 'ontpolderen').

Later, these research results would play an important role in decision making in the issue-area of the maritime access to the port of Antwerp. First, they form an important input for the Flemish-Dutch discussions on the dredging policies in the Scheldt estuary. Secondly, they

---

125. CEDRE (1991). The research was carried out by the 'Centre d'étude du droit de l'environnement'.

126. CEDRE (1991, pp. 58-59).

127. Bestuurlijk Overleg Westerschelde (1991a, pp. 39-51).

128. Ibid., Action points N5-N9, p. 80.

129. This is an acronym for the *East*ern part of the *West*ern Scheldt.

130. DGW =Dienst Getijdewateren. Later this research institute of Rijkswaterstaat is called RIKZ (Rijks Instituut voor Kust en Zee).

131. Pieters *et al.* (1991).

132. Ibid., pp. 127-130.

are used in the Flemish-Dutch discussions on the financial compensation of nature losses that will be caused by the implementation of the 48'/43'/38' deepening program. Finally, the Pilot study introduces the policy alternative 'ontpolderen', which later would dominate Dutch discussions on a plan for nature restoration along the Western Scheldt.

The Dutch Policy plan for the Western Scheldt

On 7 March 1991 the Core working group for the Western Scheldt (BOW)[133], representing Dutch local, regional and national government agencies, finishes the Policy plan for the Western Scheldt (See Round V).[134] The main objectives of this plan is: "*The creation of a situation whereby the environment can be conserved and restored, whereby, furthermore, potential natural assets can be developed, taking into consideration the function of shipping in the region, coupled with sea port and industrial activities. This policy should also form a solid foundation for the development of fishing and recreation facilities. The importance of water-control structures should also be safeguarded.*"[135] This objective implies that the region recognizes the importance of the navigation function of the Western Scheldt. The recommendation of the Dutch advisory council 'Raad van de Waterstaat' concerning the implementation of the 48'/43'/38' deepening programme are included in the policy plan (See Round V). The water quality objectives formulated in the plan are similar to the objectives formulated in the third Policy Document on Water Management, and the agreements that were made at the second North Sea Ministerial Conference.[136] Because the participating parties know that for the achievement of these objectives they are largely dependent on Belgium[137], in the policy plan it is strongly recommended to intensify deliberations and cooperation with Belgian government agencies.[138]

On 27 June 1991 Dutch municipalities along the Western Scheldt establish the Task group for the Western Scheldt.[139] In the Task group the municipalities want to coordinate their Scheldt policies before discussing them in the Core working group for the Western Scheldt.[140] One of the objectives of the task group is to maintain the linkage between the water quality of the Scheldt and the deepening of the navigation channel in the Western Scheldt. Furthermore, the task group demands an Environmental Impact Assessment for the 48'/43'/38' deepening program, and aims at a policy plan for the whole basin similar to the Policy plan for the Western Scheldt.[141]

---

133. BOW = Bestuurlijk Overleg Westerschelde.

134. Bestuurlijk Overleg Westerschelde (1991a).

135. Bestuurlijk Overleg Westerschelde (1991b).

136. Bestuurlijk Overleg Westerschelde (1991a, p. 42).

137. See for example Bedet (1993).

138. Bestuurlijk Overleg Westerschelde (1991a, pp. 44-45).

139. In Dutch: Taakgroep Westerschelde. The Task group is a member of the Association of Dutch River Municipalities (VNR, Vereniging Nederlandse Riviergemeenten), which was founded on 24 May 1989 (VNR, 1990). A main objective of the VNR is to promote the conclusion of water conventions for Scheldt and Meuse (Ibid., p. 6).

140. Plugge (1993).

141. Ibid.

Round IX: The break-through (1992-1993)

At the beginning of March 1992 new national and regional governments are installed in Belgium. At the first meetings between the Dutch, Walloon and Flemish ministers[142], it is agreed to restart negotiations on the water conventions, and to aim at the conclusion of one 'umbrella convention' dealing with all Scheldt and Meuse issues.[143] The ministers do also want to continue the linkage between Scheldt and Meuse issues.[144] It is concluded that the Dutch cannot afford the conclusion of a convention with Flanders on the deepening of the navigation channel without the conclusion of conventions on the water quality of Scheldt and Meuse, whereas for the Belgian or Flemish government the reverse is unacceptable.[145] For the preparation of the Belgian standpoint in the negotiations with the Netherlands, a new ministerial commission with representatives of the central state and the three regions is installed.

Beside the installation of new governments in Belgium, another event in the beginning of 1992 would have a considerable influence on the course of the negotiation process. On 18 March 1992, the UN-ECE Convention on the Protection and Use of Transboundary Watercourses and International Lakes is concluded.[146] This convention, among other things, states that international cooperation on water management issues should be organized on a basin-scale, and is signed by all Scheldt and Meuse basin states and the European Union. Ever since 1990 Belgian and Dutch negotiators had facilitated and stimulated actively the negotiations on the UN-ECE Convention.[147] According to a Belgian negotiator, the lengthy deadlock in the negotiations on the water conventions was a main reason for the Belgian and Dutch efforts to stimulate the negotiations on the UN-ECE convention.[148] Two main Belgian problems always had been that France was not involved in the negotiations on the Scheldt and Meuse water management issues, and the Dutch linkage across Scheldt infrastructure and

---

142. Meetings between the Dutch minister of Transport, Public Works, and Water Management, Maij-Weggen and her new Flemish colleagues, The Minister of Public Works, Kelchtermans and the Minister responsible for the Environment, De Batselier, and her new Walloon colleague, the Minister responsible for the Environment, Lutgen, on 13 March 1992 in Brussels (IPR, 1992, p. 34). These meetings take place simultaneous to the evaluation of the Benelux Scheldt and Meuse conferences in the IPR (See Round VIII).

143. Ibid., p. 37.

144. Ibid.

145. The Ministers also use the opportunity to repeat their standpoints once more. Among other things, the Dutch Minister declares that the Dutch do not accept that international water policies in the Rhine basin are different from international Scheldt and Meuse policies: "*it cannot be the case that we use for Scheldt and Meuse a less strict regime than for the Rhine.*" ("*Het kan niet zo zijn, dat men voor de Schelde en de Maas een minder sterk regiem hanteert dan voor de Rijn.*") (Ibid., p. 36). The Walloon region states once more that it would like to involve France in the negotiations on the water quality of Scheldt and Meuse (Ibid., p. 45).

146. United Nations Economic Commission for Europe, Convention on the Protection and Use of Transboundary Watercourses and International Lakes, Geneva, E/ECE/1267, 1992. Plans to start the preparation of this convention were developed at the CSCE (Conference on Security and Cooperation in Europe)-Meeting on the Protection of the Environment (Sofia, Bulgaria, 1989).

147. Interview K. De Brabander.

148. Ibid.. According to K. De Brabander, the Belgian and Dutch delegation in the UN-ECE were able to stimulate the negotiations on the UN-ECE convention, partly because the Netherlands and Belgium successively were responsible for the coordination between the member states of the European Union. Beside Belgium and the Netherlands, France and Germany as well strongly favoured the conclusion of the UN-ECE convention.

Meuse water management issues, which created an unsolvable internal Belgian problem.[149] The Belgian and Dutch negotiators, who were involved in both the negotiations on the water conventions and the negotiations on the UN-ECE convention, expected that the conclusion of the UN-ECE convention would contribute to a solution of these problems[150], and consequently would have a positive impact on the negotiations on the water conventions.

Although the Dutch government had not yet decided to unlink the bilateral and multilateral Scheldt and Meuse issues, in 1991 it had already allowed the Dutch delegation in the commission Biesheuvel-Poppe to contact French government agencies informally. The French parties contacted were the Agence de l'Eau Rhine-Meuse (situated in the Meuse basin), the Agence de l'Eau Artois-Picardie (situated in the Scheldt basin), and the French Ministry of Foreign Affairs. One reason to contact these French agencies was that France had proven to be a good partner in international negotiations, in particular as regards the management of the river Rhine.[151] In 1992 also Germany and Luxembourg are contacted. The other basin states, i.e. France for the rivers Scheldt and Meuse, and Germany and Luxembourg for the river Meuse, react positively on the Dutch proposal to start international negotiations on the Scheldt and the Meuse.[152]

The Dutch delegation prepares a new draft convention on the management of the rivers Scheldt and Meuse, which largely draws on the draft convention the Dutch had presented in April 1991 (See round VIII).[153] However, unlike the draft convention of April 1991, the draft of April 1992 aims at the involvement of France in the negotiations on the Scheldt and the Meuse, and defines the Meuse from the source to the mouth of the river. In June 1992, the Belgian and Dutch delegations in the commission Biesheuvel-Poppe give up the idea to conclude an umbrella convention, and agree that negotiations on the water conventions should aim at three separate conventions: A multilateral Scheldt and Meuse framework convention between all basin states and the EC, and two bilateral Belgian-Dutch conventions: a first one on the distribution of the water of the Meuse, and a second one on the deepening of the navigation channel in the Western Scheldt.[154] The construction of the Baalhoek Canal is not addressed. The Flemish government would like to start a transboundary EIA-study on the Baalhoek project, but the Dutch Minister does not want to cooperate on a study until Belgium and the Netherlands would have reached an agreement on the other Scheldt and Meuse issues.[155] On 13 July 1992 the Dutch Minister Maij-Weggen, the Dutch Minister of Foreign Affairs, Van den Broek, and the Belgian Ministers of the Environment, Onkelinx, and

149. Ibid.

150. Ibid.

151. Interview R. Zijlmans.

152. TK 1991-1992, Aanhangsel van de Handelingen, 686.

153. Dutch draft convention concerning the cooperation on the management of the Scheldt and the Meuse of 14 April 1992.

154. TK 1991-1992, Aanhangsel van de Handelingen, 686.

155. Meeting between the Dutch Minister of Transport, Public Works, and Water Management, Maij-Weggen, and the Flemish Minister of Public Works and Internal Affairs, Kelchtermans of 18 June 1992 (Files Dutch Ministry of Transport, Public Works, and Water Management TSC-meeting 36, 10 December 1992, Antwerp). Although the Baalhoek projects still is on the international agenda, implementation of this project is not expected in the foreseeable future, since the construction of the canal would be extremely expensive (TK 1991-1992, Aanhangsel van de Handelingen, 853), and after the conclusion of the water conventions the Flemish region would have to allocate its budgets for the implementation of the deepening programme first.

Foreign Affairs, Claes, agree to restart negotiations on the water conventions in the short run, and aim to conclude the negotiations before the summer of 1993.[156]

The new Belgian ministerial commission preparing the negotiations with the Netherlands on the water conventions installs a commission of experts to investigate whether the Dutch draft convention on the water quality of Scheldt and Meuse goes beyond existing international agreements.[157] The result of this study is that only few parts of the Dutch draft convention go beyond international policies. The Dutch delegation thanks the Belgian delegation for the Belgian investigation and the Belgian points of attention, but concludes that apparently the Belgian delegation has hardly any objection to the Dutch draft, and therefore would like to know how negotiations should be continued now. The chairman of the Dutch delegation, Biesheuvel, adds: *"The year 1993 will be the seventh year of a negotiation commission that has never been able to negotiate really."*[158]

Although from the foregoing it becomes clear that negotiations still are laboriously, Belgium and the Netherlands are able to reach an agreement on three conventions dealing with several minor bilateral issues, which are called the "small water conventions": an Agreement for the regulation of navigation and recreation on the border Meuse, a Convention concerning a correction of the Belgian-Dutch border in the Canal Ghent-Terneuzen, and a Convention on the Protection of the banks along the Western Scheldt.[159]

Because of the continuing Belgian-Dutch disagreement, on 9 April 1992 the European Parliament formulates a resolution concerning the water quality of the river Meuse.[160] The resolution refers to the EC-guidelines 80/778/EC for drinking water, 75/440/EC for the protection of surface waters, and the agreements that were made at the North Sea Ministers Conferences. Furthermore, it recalls several accidental discharges into the Meuse threatening drinking water production in the Netherlands, and the laborious Belgian-Dutch negotiations on the water conventions. For these reasons the European Parliament demands the EC member states to comply with EC environmental guidelines, and the European Commission and the Meuse riparian states to ameliorate the water quality of the Meuse. The resolution states that only a supranational authority will be able to install an international agency for the Meuse basin. The European Parliament does not expect a rapid conclusion of Belgian-Dutch negotiations, and consequently demands the European Commission to take action. The Parliament also demands the Commission to use the European Funds for structural economic aid to stimulate the sanitation of the river Meuse basin.

---

156. Files Ministry of Transport, Public Works, and Water Management, TSC-meeting 36, 10 December 1992, Antwerp.

157. Files Ministry of Transport, Public Works, and Water Management, TSC-meeting 36, 10 December 1992, Antwerp. The study is made by a representative of the Walloon region, Smitz, a representative of the Brussels region, Laurent, and a representative of the Flemish region, Bruyneel. This working group was installed on 28 July 1992.

158. Letter from the chairman of the Dutch delegation, Biesheuvel, to Janssens of the General Directorate on Politics of the Belgian Ministry of Foreign Affairs.

159. Files Ministry of Transport, Public Works, and Water Management, TSC-meeting 36, 10 december 1992, Antwerp. These conventions are signed on 6 January 1993.

160. European Parliament, 1992-1993, Resolutie over de waterkwaliteit van de Maas d.d. 9 April 1992.

Discussions concerning compensation of the Walloon region

Because the problem caused by the Dutch linkage of Scheldt and Meuse issues was that Flanders and the Netherlands perceived that they would benefit from an agreement on the water conventions, but Wallonia did not, the Dutch had always speculated on an internal Belgian agreement, i.e. a Flemish compensation of the Walloon 'losses', which would make on agreement on the water conventions attractive to all contracting parties.[161] In the beginning of 1992 intranational Belgian discussions take place on some possibilities to compensate the Walloon region. Whilst the Flemish port of Antwerp wants to improve the maritime access to the North Sea by the implementation of the 48'/43'/38' deepening program, the Walloon port of Liège also would like to improve its navigation infrastructure. The Walloon region is the demanding party for two main infrastructure projects. First, it would like to modernize the Albert canal, which is the liaison between the port of Liège and the port of Antwerp.[162] Secondly, it would like to construct a diversion canal from Oelegem to Zandvliet thus providing for a good connection between the Albert canal and the Scheldt-Rhine connection.[163] Since these infrastructure works were not in the interest of Flanders, but on the contrary would cause spatial planning problems near Antwerp, Flemish politicians are unwilling to cooperate on the implementation of these projects. Therefore, some Walloon politicians suggest to link these two infrastructure issues to the Belgian-Dutch negotiations on the water conventions. If Flanders would help Wallonia with the construction of the two infrastructure projects, Wallonia would be willing to discuss the international Meuse issues, thus increasing the chance on an international agreement on the deepening program, which would be beneficial to the Flemish region.

Not only the possibility of a Flemish compensation of the Walloon region, but also a Dutch compensation of the Walloon region was discussed. Wallonia was interested in a new sluice near Ternaaien, a straightening or deepening of the Meuse near Eijsden, and an improvement of the Meuse route.[164] These projects would improve the connection between the port of Liège, the Rhine and Rotterdam. For the implementation of these infrastructure projects, however, the Walloon region is dependent on the Netherlands. According to one respondent, a Dutch compensation of Walloon losses was discussed during the negotiations in the Commission Biesheuvel-Poppe. Discussions on such as compensation, however, were discontinued because of the succession of the Minister of Transport in the Walloon region. The succeeding Walloon minister did not stem from the region of Liège, and probably therefore showed less interest in the infrastructure projects improving navigation possibilities to the port of Liège. This might be explained by the fact that Walloon politicians are more dependent on their regions.[165]

Neither of the linkages discussed above are made, because a compensation of the Walloon region was no longer necessary after the Dutch decision to unlink the bilateral Flemish-Dutch and the multilateral Scheldt and Meuse water quality issues (See second section below).

---

161. TK 1994-1995, 24 041 No. 3.

162. Le Soir, 8-4-1992.

163. Planchar (1993, pp. 144-145).

164. COIB (1994).

165. Interview C. De Villeneuve.

Plea for a sanitation fund

During the past decade it has become clear that Belgium, in particular the Walloon and Brussels Capital Regions, oppose the draft water quality conventions prepared by the Dutch, mainly because they expect the implementation of these conventions to be extremely costly. According to these parties, extra investments in waste water treatment facilities would be beneficial for the Netherlands only. The draft convention on the water quality of Scheldt and Meuse presented by the Belgian delegation on 21 October 1991 also contained articles which would make the Netherlands financially responsible for the clean-up of the rivers Scheldt and Meuse (See Rounds VIII). Because of the Walloon and Brussels objections to Dutch draft conventions, and because after almost 25 years the basin states still have not been able to reach an agreement on the water conventions, and to clean-up the rivers Meuse and Scheldt, some Dutch members of the Benelux Parliament propose to establish a sanitation fund for the clean-up of the river Meuse.[166] These MP's do expect that if the Dutch government will not install such a fund, the clean-up of the river Meuse will take many more years. In the Meuse case, the establishment of a sanitation fund would imply that the Dutch government finances partly the construction of waste water treatment infrastructure in, for example, the Walloon region. The Dutch MP's refer to the conclusion of the Rhine chloride agreements, which entailed a financial compensation of the French state by the governments of the other Rhine basin states, among which the Dutch government.[167] In addition to the Benelux parliament, the three southern provinces of the Netherlands, Northern-Brabant, Limburg, and Zeeland, discuss the idea of a sanitation fund at a joint symposium.[168]

The Dutch Minister of Transport, Public Works, and Water Management, Maij-Weggen, however, is of the opinion that the financial compensation in the Rhine river negotiations was the first and last time that the Dutch government used this strategy, and opposes the establishment of a sanitation fund to clean-up the rivers Meuse and/or Scheldt. The main reason for this is that the installation of such a fund would not be in accordance with the international environmental legal 'the polluter pays principle'.

The Dutch decision to unlink bilateral and multilateral issues

Since the conclusion of the UN-ECE Convention the Dutch and Flemish governments face a problem. Ever since the Dutch decision to link two Belgian infrastructure projects to the water quality and quantity of the Meuse, and the water quality of the Scheldt in 1967, the Dutch had continued to link the Scheldt infrastructure issues to the Meuse water management issues. The involvement of the upstream basin state France, a party that has no interest in the deepening of the navigation channel in the Western Scheldt, in the negotiations on the water quality of Scheldt and Meuse, however, would make it hardly possible to continue the linkage between the deepening of the navigation channel in the Western Scheldt and the multilateral Scheldt and Meuse water management issues. It, therefore, would also make it impossible to make a Flemish-Dutch package deal including a Flemish-Dutch agreement on the water quality of the Scheldt. In Flanders such as package deal was seen as an alternative, in case the Walloon region would continue to refuse cooperation on the clean-up of the rivers Scheldt and Meuse. After the conclusion of the UN-ECE Convention, however, a bilateral Flemish-Dutch convention on the water quality of the Scheldt would have been an

---

166. PZC, 12-5-1993, Zijlstra, K. in NRC, 9-6-1993.

167. Dieperink (1997, pp. 170-171, 212).

168. Provincie Noord-Brabant (1993).

anachronism. As regards the involvement of France in the negotiations, a Dutch negotiator states: "*Obviously, we did not only have to deal with the UN-ECE convention, but also with the linkage between the water conventions. This played a role until the Dutch delegation itself did not believe in the linkages any more. At a given moment that turning point had come.*" "*But as long as we continued the linkage it was in fact impossible to involve France, because that would have implied a French involvement in the deepening issue.*"[169] As regards the Dutch strategy another Dutch negotiator states: "*Although the French reactions were positive, the Dutch hesitated to take the decision.*" (to involve France in the negotiations).[170] The same negotiator states: "*In the Netherlands some people were in favour of a linkage between the Meuse and the Scheldt, between bilateral and multilateral issues.*"[171], and "*Some people, among which myself, had been in favour of the involvement of France in the negotiations for a very long time.*"[172] One of the respondents stated that historical events may explain partly the Dutch hesitance to unlink the Scheldt infrastructure and Meuse issues. In 1927, the Dutch Minister of Foreign Affairs was sent home by the First Chamber of the Dutch Parliament, because he would have served the Dutch interest insufficiently. The Dutch government absolutely wanted to prevent a similar event.[173]

Although the UN-ECE Convention was concluded in March 1992, and Dutch civil servants have already contacted representatives of the other basin states in the beginning of 1992, the Dutch government does not yet decide to unlink the bilateral and multilateral Scheldt and Meuse issues in 1992. In the beginning of 1993 it becomes clear that the Belgian federalization process is not yet completed, and that treaty-making competencies will be attributed to the Belgian regions. This would enable France and the Netherlands to negotiate directly with the Belgian regions, and in particular would enable the Netherlands and the Flemish region to conclude Flemish-Dutch conventions concerning the bilateral Scheldt and Meuse issues. Beside the conclusion of the UN-ECE Convention, also these intranational changes in Belgium make it necessary to evaluate the Dutch strategy. After the succession of the Dutch Minister of Foreign Affairs, Van den Broek, by Minister Kooijmans, which is accompanied by a new policy of the Ministry of Foreign Affairs, in June 1993 the Dutch government decides to unlink the bilateral and multilateral Scheldt and Meuse issues. This strategic change implies that the Dutch approval of the deepening programme is no longer linked to the amelioration of the water quality in the rivers Scheldt and Meuse. The Dutch minister of Transport, Public Works and Water Management states: "*I have gradually come to a change of course in some direct deliberations with my Flemish colleagues.*"[174] According to the Dutch minister the federalization has shaped a situation in which "*Wallonia will be less than ever prepared to ameliorate the water quality of the Meuse in exchange for a Dutch cooperation on a Flemish interest.*"[175] From 1993 the Dutch do no longer speculate on an internal Belgian arrangement for compensation of Walloon 'losses', but accept that the new Belgian regions have to be approached separately. Some Dutch negotiators think positively

---

169. Interview C. De Villeneuve.

170. Interview R. Zijlmans.

171. Ibid.

172. Ibid.

173. Interview anonymous. See also the historical overview presented in Section 4.8.

174. TK 1992-1993, 23 075 No.1

175. TK 1993-1994, 23 536 No.1.

about the conclusion of the third stage of the Belgian state reforms, and the attribution of treaty-making competencies to the regions. As one negotiator states: "*Every time we were confronted with a Belgian State unable to act internationally. Therefore, we are pleased with the Sint Michiels agreements*[176], *although relations are complicated. Sometimes a complex situation is better than nothing.*"[177] Whilst on the one hand for some Dutch negotiators the completion of the Belgian federalization process is a relief, on the other hand for some Belgian negotiators the Dutch decision to unlink the 48'/43'/38' deepening program, and the multilateral Meuse issues is a relief: "*Meuse and deepening of the Scheldt were no longer connected by the disastrous Dutch linkage.*"[178] Flanders has been waiting long for a new Dutch policy. The reaction of the Flemish Prime Minister Van den Brande on the Dutch decision to unlink the bilateral and multilateral issues is: "*We have formulated the new situation in Belgium in a legal text now. That clearly made more impression. It is typical for the business-like Dutch that they want to see it formally on paper first.*"[179] Witness the questions put by Dutch Members of Parliament, the Dutch decision to unlink bilateral and multilateral issues comes as a surprise.[180] The Dutch parliament adopts a motion declaring that in case the water conventions with Wallonia and Flanders will be unlinked, the water quality of the Meuse has to be safeguarded.[181] Some MPs have severe critics on the Dutch Scheldt and Meuse policy. According to them the Dutch strategic change has come much too late, because all Flemish and Dutch political parties, except for the Dutch liberals (the VVD), have proposed to unlink the Scheldt and Meuse issues for many years.[182] According to these MP's, the Dutch strategy to speculate on a Flemish compensation of Walloon losses has failed miserably, and because of this failure Belgian-Dutch relations are damaged unnecessarily.[183] In addition, the Dutch Minister of Transport, Public Works, and Water Management, Maij-Weggen, would lack diplomatic skills, because she has stepped on Walloon toes several times.[184]

Shortly after the Dutch decision to unlink the bilateral and multilateral issues, the commission Biesheuvel-Poppe is dissolved, and new delegations are composed.[185] These delegations no longer consist of a fixed number of persons. Their composition depends on the subjects that are discussed. From 1993 negotiations on the water conventions are split up into multilateral negotiations between the Scheldt and Meuse riparian states, and bilateral Flemish-Dutch negotiations. The multilateral negotiations aim at the conclusion of conventions on the water quality of Scheldt and Meuse. Parallel to the multilateral negotiations Flanders and the Netherlands start negotiations on the following bilateral

---

176. The Sint Michiels agreements are the agreements concluding the third stage of the Belgian state reforms.

177. Interview C. de Villeneuve.

178. Suykens (1995, p. 232).

179. Het Financieele Dagblad, 22-6-1993.

180. TK 1992-1993, 23 075 No. 1, Annex.

181. TK 1992-1993, 23 075 No. 2.

182. See, for example, Zijlstra, K. in NRC, 9-6-1993.

183. Ibid.

184. Ibid.

185. TK 1992-1993, Aanhangsel van de Handelingen, 621.

issues[186]:
*   the deepening of the navigation channel in the Western Scheldt according to the 48'/43'/38' deepening program;
*   the flow of the river Meuse.[187]

In addition, they agree to start discussions on:
*   the maritime access to the Waasland harbour[188];
*   a possible extra deepening programme in the Western Scheldt[189];
*   the bilateral aspects of the water quality of Scheldt and Meuse.[190]

Flanders and the Netherlands agree on a linkage between the 48'/43'/38' deepening programme and the flow of the river Meuse.[191]

Dutch evaluation of the relationship with its southern neighbour
Because of the Belgian state reforms, in particular the attribution of treaty-making competencies to the regions, the Dutch government has to reflect on its relationship with Belgium. In 1993, a governmental report on the new administrative relations with Belgium is issued.[192] In addition, the Dutch Ministry of Transport, Public Works and Water Management prepares a report on the bilateral relations with Belgium.[193] As regards water management the latter report states that: "*Different but not necessarily conflicting interests exist: Belgium can influence the discharge of sediments, water quality and quantity. The Netherlands can on the other hand to some extent control the navigation channels*"[194] As regards the international contacts of the Ministry the document states: "*In general it can be stated that more or less permanent contacts of in particular the regional directorates are intense and reasonably frequent. At higher levels there are considerably less contacts, whilst at the central level sometimes a vacuum exists. The cause of the vacuum partly is the administrative impasse in Belgium. In cases where complex files with consequences for other issues or international legal consequences were at discussion, demarcation disputes between Belgian governments became visible. [...] Perhaps more important are differences in decision-making culture, which in the Netherlands is more aiming at deliberations and consensus-building than in Belgium, where decision-making was to a larger extent based on rigid power balances and political motives.*" [195]

---

186. TK 1993-1994, 23 075 No. 3.

187. The Flemish region is able to conclude a bilateral agreement with the Netherlands on the water quantity of the river Meuse, because the Flemish region diverts Meuse water via the Albert Canal and the Zuid-Willems Canal.

188. The Waasland harbour is the name of the Antwerp harbour on the left bank of the river Scheldt. These discussions in fact address the Baalhoek alignment of 1975.

189. The so-called second stage or 50' program.

190. The bilateral aspects of the water quality of the Scheldt have been discussed for years in the working group for the WVO Permit of the Subcommission for the Western Scheldt of the TSC.

191. TK 1993-1994, 23 530 No. 6.

192. TK 23 536 No. 1.

193. COIB (1994).

194. Ibid., p. 4.

195. Ibid., p. 5.

International research and information exchange
In the beginning of the nineties international research and information exchange is intensified, and transboundary networks of professionals are developed gradually. Two important research projects are the project OostWest and the International Scheldt Group (ISG).

On 10 December 1992, two Dutch experts, Pieters and Turkstra, present the results of the Pilot study OostWest to the Technical Scheldt Commission (See Round VIII)[196], which decides to formalize the working group OostWest as a working group of the Subcommission for the Western Scheldt. From that moment OostWest formally is a joint Dutch-Belgian research project. Belgian participants in this research projects are the Institute for Nature conservation (IN)[197], and the Antwerp Sea port Service.[198] Before the meeting of 10 December, the Flemish had given a reserved reaction on the results of the Pilot Study. "*It should be emphasized that the presented solutions are part of the research stage. Under any circumstances they are policies or policy proposals. It is too early for that, also since Belgium did not yet participate in the discussion.*"[199] The Flemish keep their contribution to the research project 'low profile', and state that only after the conclusion of the Convention on the deepening of the navigation channel in the Western Scheldt Flemish authorities are able to participate fully in this research project.[200] A main topic of discussion in the project OostWest are the ideas of 'ontpolderen'. According to ecological experts, the area of natural flooding areas along the Scheldt estuaries should be extended with areas that presently are situated behind the dikes. Some Flemish are afraid that the Dutch would formulate new conditions to the implementation of the deepening program.[201] By that time nobody could know that a few years later one of the policy alternatives that was presented in the pilot study, namely 'ontpolderen', would become much more problematic for the Dutch government than for the Flemish one.

Beside the project OostWest another international research project is started. The water quality managers in the Scheldt basin have been waiting for a Scheldt convention, and the installation of an International Scheldt Commission for many years. In the beginning of 1992, there still is no prospect on an international agreement on the water conventions. External pressure on the authorities responsible for water quality management, however, has increased tremendously. For these reasons Flemish and Dutch water managers take the initiative to establish an informal working group to carry out preparatory work for a formal International Scheldt Commission, which would have to be installed in the future. The working group is called the International Scheldt Group (ISG)[202] The main objective of the ISG is to describe

196. Report TSC-meeting 36, 10.12.1992, Antwerp.

197. IN =Instituut voor Natuurbehoud.

198. Antwerpse Zeehavendienst.

199. Claessens *et al.* (1991, p. 188).

200. Interview H. Smit.

201. Interview anonymous.

202. The idea to establish the ISG is born at an international conference on European water policy in Flanders, where the Rhine, Scheldt and Meuse were used as case studies (European water congress, Antwerp-Belgium, 19-20 September 1991). Representatives of the Flemish Environment Agency (VMM) and the regional Directorate Zeeland of Rijkswaterstaat (RWS-Zld) who met at this conference thought that frequent meetings between water managers of the Scheldt basin states would be useful. The first name of the working group is ISG-DWS (International Study Group/Description of the Water quality of the Scheldt basin). Later this name is changed into ISG (International Scheldt Group). Organizations participating in the ISG are the Agence de l'Eau

the water quality (emissions and immissions), hydrology, ecology and morphology of the Scheldt basin with the use of existing data and documentation. Also an inventory of the administrative organization of water management in the basin states and regions is made. Another objective is to enhance the cooperation between the Scheldt basin states.[203] Because of the ongoing laborious negotiations between the Scheldt basin states, the installation of a basin-wide working group is a difficult task, which requires careful diplomacy. The initiators ask prof. C. Heip of the research institute CEMO to become chairman of the group.[204] The CEMO contacts water management agencies in the other basin states and regions, France, Wallonia and Brussels. In first instance the Walloon and Brussels water management agencies do not react. Later the chairman of the ISG succeeds in convincing these parties of the importance of the informal working group. The minutes of the third meeting of the ISG, however, state "*Although the Walloon colleagues see no problem in joining ISG/DWS, until this moment they have not attended one single meeting.*"[205] The French emphasize the need for comparable efforts of all basin states. The leader of the French delegation states in his opening speech of the meeting in Douai (France) "*Only if the different participants make comparable efforts to establish the aims of international agreement, the water quality of the river Scheldt will improve considerably.*"[206] At the fourth meeting representatives of all basin states and regions are present.[207] The fifth meeting of the ISG is a workshop of two days, and is organized by the Walloon region.[208] At this meeting the Director-General of the DGRNE of the Walloon Ministry states that "[...] *participation of the Walloon region is no problem as long as ISG/DWS acts as an informal study group, and as long as ISG/DWS has no political objectives.*"[209]

In the ISG, the Dutch RIKZ takes the responsibility for the design of a Scheldt GIS.[210] According to the chairman of the ISG one of the problems with the Dutch was that the RIKZ

Artois-Picardie (France)/ Ministère de la Région Wallone, DGRNE (Wallonia)/ Ministère de la Région de Bruxelles-capitale, ARNE (Brussels)/ VMM and IN (Flanders)/ RWS-Zld, RWS-DGW, and the CEMO (The Netherlands). The Project Secretariat is run by the regional directorate Zeeland of Rijkswaterstaat. In the first months of the project the Water division of the Belgian federal Institute of Hygiene and Epidemiology participates. In 1993 the activities of this division are taken over by the VMM and the Ministère de la Région Wallonne/ DGRNE.

203. ISG/DWS/7.

204. Main reason to choose prof. C. Heip is that this person works at an independent research institute. Secondly, he is an expert on the Scheldt water management problems. Thirdly, prof. Heip is a Belgian, who works in the Netherlands, and therefore is aware of cultural differences between these two basin states. Finally, he masters the Dutch and the French language.

205. Minutes of the third meeting of ISG/DWS on 3 December 1992, Douai (ISG/DWS/29). The first meeting took place on 29 April 1992 in Yerseke (ISG/DWS/6), and the second meeting on 10 September 1992 in Aalst (ISG/DWS/20).

206. Minutes of the third plenary meeting on 3 December 1992, Douai, ISG/DWS/29, Opening by mr. Grandmougin of the Agence de l'Eau Artois-Picardie.

207. Minutes of the fourth meeting of the ISG-DWS, 26 March 1993, Aalst (ISG/DWS/33). At this meeting a Walloon representative remarks that in Wallonia the emphasis lies on the management of the river Meuse, but that the region nevertheless is interested in exchange of experiences with the other participants. A representative of the cabinet of Minister Gosuin states that in the Brussels Capital Region a reorganization is going on.

208. Minutes of the workshop ISG-DWS, 2-3 June 1993, Namur (ISG/DWS/40).

209. Ibid.

210. ISG/DWS/9.

wanted to do too much itself. *"That has caused quite some distrust, the GIS system. I have clearly experienced that."*[211] A main discussion point during the meetings of the ISG is the provision of emission data. In Wallonia, data on emissions of industries sometimes are confidential, and consequently cannot be exchanged easily.[212]

Developments in the Dutch province of Zeeland
On 1 September 1992 the Dutch Minister of Transport, Public Works, and Water Management declares that according to international (EC) law an EIA is not obligatory for the implementation of the deepening program.[213] Furthermore, she argues that there are no other good reasons to start a new study on the impacts of the deepening program, because such a study was published by the TSC in 1984, and because the 'Raad van de Waterstaat' had organized a participation round, and all parties had been given the opportunity to express their concerns. The Task group for the Western Scheldt clearly states that it does not agree with the point of view of the Dutch minister, mainly because of the (preliminary) results of the Oost West-project, which, unlike the research report published in 1984, indicates significant negative impacts of the implementation of the deepening programme on the ecology of the Scheldt estuary. Furthermore, the Task group emphasizes once more the importance of a linkage between the implementation of the deepening programme and the water quality of the river Scheldt.[214]

On 22 October 1992 the parties who developed the Policy plan for the Western Scheldt sign a covenant, and by that promise to comply with the objectives formulated in that policy plan (See Round VIII).[215]

6.3   Analysis and explanation of the decision making rounds V to IX

In the following sections, the course of the interactions in the five decision making rounds distinguished in the period between 1985 and 1993 will be analyzed and explained. In Section 6.3.1 the influence of the perceptions of the national governments, lower level governments, and NGOs on their strategies will be analyzed, and with that research question 3 will be answered for the period between 1985 and 1993. Subsequently, in Section 6.3.2 the strategic interactions and institutionalization processes will be discussed, and with that research question 4 will be answered for the period indicated. In Section 6.3.3 the strategic and cognitive learning processes will be analyzed, and there research question 5 will be answered for the same period. To conclude, in Section 6.3.4 the influence of the context, which was described and analyzed in Chapter 4, on the perceptions and strategies of the actors involved in the decision making process will be analyzed, and research question 6 will be answered

---

211. Interview C. Heip.

212. Minutes ISG-workshop Namur, ISG/DWS/40, 7 June 1993, and minutes of a meeting of representatives of the DGRNE, the University of Liège, RIKZ and RWS-Zld on the exchange of data between the Scheldt-GIS and the LIFE 2 project (see Chapter 7), on 6 July 1994.

213. Letter from the Dutch Minister of Transport, Public Works, and Water Management to the Second Chamber of the Dutch parliament of 1 September 1992.

214. Letter from the Task Group for the Western Scheldt to the Dutch Minister of Transport, Public Works, and Water Management of 15 October 1992.

215. Bestuurlijk Overleg Westerschelde. Covenant of 22 October 1992.

for the second case study period distinguished in this thesis.

6.3.1 Influence of perceptions of actors involved in the decision making rounds V to IX on their strategies

Perceptions and strategies of the Scheldt basin states and regions
Table 6.2 produces information on the main strategies of the Scheldt basin states and regions in each decision making round. In the following, the strategies of these actors are related to their perceptions of (inter)dependence.

The Dutch government, the gradually learning party
Table 6.3 produces information on the linkages that were made in the period between 1985 and 1993. Like the linkages that were made in the period between 1967 and 1985, these linkages also are in accordance with the theory that in cases where the decision rule is unanimity, and there are heterogenous preference intensities across issue-areas, the chance on the occurrence of tactical issue-linkage is high. In 1985, the Belgian and Dutch Ministers of Foreign Affairs reached an agreement on linkages across several Scheldt and Meuse issue-areas. Although since 1975 the Walloon region had opposed continuously to the building of storage reservoirs on Walloon territory, the Belgian and Dutch Ministers of Foreign Affairs still mentioned these storage reservoirs in their declaration of intent. Only after the Walloon Region had made clear once more that it did not want to build these reservoirs, and for that reason did not accept the declaration of intent, the Dutch minister proposed to delete the passages dealing with the building of storage reservoirs on Walloon territory. This decision may be interpreted as the result of a strategic learning process. The Dutch government had learned that negotiations with Belgium are impossible, if it would insist on the building of storage reservoirs, and therefore adapted its objective and strategy. The Dutch government, however, did continue the linkage across Scheldt and Meuse issue-areas it had reached agreement on with the Belgian Minister in 1985, because it speculated on a Flemish compensation of Walloon 'losses', either financially or by cooperation on infrastructure projects.

From 1986 to 1989 the Dutch government was hardly able to negotiate with its southern neighbour, because the Belgian federalization process had created a situation in which the Belgian national and regional governments were unable to formulate a joint Belgian standpoint for the negotiations with the Netherlands. Because of the laborious negotiations in the commission Biesheuvel-Davignon/Poppe, and the absence of a platform where Scheldt quality issues could be discussed, the Dutch used the Technical Scheldt Commission, where normally only Scheldt infrastructure issues were addressed, to start discussions concerning the water quality of the Scheldt. In these discussions, the Dutch Ministry of Transport, Public Works, and Water Management started to use the resource it had since a few years, namely the competence to grant a WVO-Permit to Belgium, to exert influence on Belgium by making a new linkage. From 1986 the Belgian government would receive a permit to dump dredged material in the Western Scheldt only if it would clean-up the river Scheldt. Later, the Flemish region had to extract large amounts of polluted sediments from the Lower Sea Scheldt.

Table 6.2 *Main strategies of Scheldt basin states and regions in decision making rounds V to IX*

Round	Actor	Type of strategy	Strategic actions
V (1985-1987)	Belgian government (Walloon region)	Reactive aimed at maintaining autonomy	Does not accept proposal to build one or two storage reservoirs in Wallonia
	Dutch government	Reactive interactive strategy: Negotiating (Strategic learning)	Proposes to modify declaration of intent
	Belgian government	Reactive interactive	Proposes to start negotiations on the basis of the modified declaration of intent
		Reactive aimed at maintaining autonomy	Emphasizes agreements formulated in the Scheldt Statute
VI (1987-1989)	Walloon region Belgian government	Reactive aimed at maintaining autonomy	Decide to postpone all deliberations with the Dutch government
	Dutch government	Reactive interactive strategy, issue-linkage	Links Belgian WVO Permit to improvement of the water quality of the Scheldt
	Belgian national and regional governments	Offensive interactive	Propose to restart negotiations on the water conventions
VII (1989-1991)	Dutch delegation	Offensive interactive	Presents several documents dealing with Scheldt and Meuse issues in negotiation commission
	Walloon delegation Brussels delegation Belgian government	Reactive interactive	Do not accept Dutch proposals, and propose to comply with EC-guideline 78/659 for cyprinids
	Dutch government	Reactive aimed at maintaining autonomy	Does not accept Belgian proposal, and decides to suspend negotiations

Round	Actor	Type of strategy	Strategic actions
VIII (1991-1992)	Dutch delegation	Offensive interactive	Presents proposal draft convention
		Negotiating (Strategic learning)	Gives up demand for storage reservoirs in Wallonia
		Reactive interactive, issue-linkage	Links WVO Permit for Flemish region to extraction of polluted sediments from the Lower Sea Scheldt
	Walloon region	Reactive	Does not accept Dutch proposal for draft convention
	Belgian delegation	Reactive	Presents proposal three draft conventions
	Dutch delegation	Reactive	Does not accept Belgian proposal for draft conventions
IX (1992-1993)	Dutch government	Offensive interactive	Approaches other Scheldt and Meuse basin states to participate in multilateral deliberations
		Reactive interactive (Strategic learning)	Decides to unlink the 48'/43'/38' deepening program, and the Scheldt and Meuse water quality issues
	French government	Reactive interactive	Is willing to participate in deliberations concerning Scheldt and Meuse
	Walloon government Brussels government	Reactive interactive (Strategic learning)	Are prepared to start deliberations on water quality of Scheldt and Meuse
	Flemish government	Reactive interactive, issue-linkage	Links water quantity Meuse to deepening program
	Dutch government	Reactive interactive, issue-linkage	Links deepening programme to water quantity Meuse

Table 6.3 *Issue-linkages in the decision making process on international Scheldt issues (1985-1993)*

Year	Actor proposing issue for negotiation	Proposed issue	Linking actor	Linked issues
1985	Belgian government	Baalhoek Canal 48'/43'/38' deepening programme	Dutch government	Water quality Meuse Water quantity Meuse
	Dutch government	Water quality Meuse Water quantity Meuse	Belgian government	Baalhoek canal 48'/43'/38' deepening programme
1988	Belgian government	WVO Permit	Dutch government	Water quality Scheldt
1991	Flemish government	WVO Permit	Dutch government	Extraction of polluted sediment from the Lower Sea Scheldt
1993	Dutch government	Decision to unlink bilateral and multilateral Scheldt and Meuse issues		
1993	Flemish government Dutch government	Deepening programme Water quantity Meuse	Dutch government Flemish government	Water quantity Meuse Deepening programme

When representatives of the Belgian regions were added to the Belgian delegation, and negotiations were restarted in 1989, the Dutch government faced Walloon and Brussels Capital Regions that did not want to reach an agreement with the Netherlands on the implementation of existing international water quality policies, in particular the agreements made at the North Sea Ministerial Conferences. According to the Dutch delegation, acceptance of the Belgian proposals would have lead to the conclusion of 'empty agreements' that would do more harm than that they would contribute to an improvement of the water quality. Therefore, the Dutch government decided to suspend the negotiations, and to prepare a draft convention unilaterally.

Dutch draft conventions The Dutch draft convention of April 1991 is fundamentally different from the draft water quality conventions of 1975, because the new draft emphasizes the need for a *joint* effort to clean-up the rivers Scheldt and Meuse. It does not only contain water quality objectives at the Belgian-Dutch border, but contains obligations for Belgium *and the Netherlands*. Furthermore, unlike the conventions of 1975, the Dutch draft of 1991 is an integrated convention addressing the main river functions simultaneously. Some Dutch reasons for this were the relationship between the water quantity and the water quality of the river Meuse during periods of low river flow, and the relationships between dredging and dumping of dredged material in the Scheldt estuary, the dissemination of pollutants, and the morphology and ecology of the Scheldt estuary. The Dutch draft of 1991 is clearly inspired by new management concepts, such as integrated water management and the river basin approach. In addition, the draft of 1991 does no longer contain detailed policies, but has the character of a framework convention, which contains the main principles of cooperation between the basin states. A main reason for this was that at that time international policies had already been developed in other international arenas, such as the North Sea Ministerial Conferences, and the EC. Finally, the Dutch presented an alternative way to deal with the water quantity problems in the Meuse, which would make the Dutch less dependent on the Walloon region: a Flemish-Dutch saving scheme. This strategic change may be interpreted as a strategic learning process, because the Dutch had learned that the Walloon region would definitely not accept the construction of storage reservoirs on its territory. This strategic change, however, may also be interpreted as a cognitive learning process, because several research reports had indicated that the construction of storage reservoirs in the Meuse would have negative ecological consequences. When after the presentation of their draft convention in 1991, the Dutch still face resistance of the Walloon and Brussels Capital Regions, probably the lowest point in the Belgian-Dutch relationship is reached.

Because of the laborious negotiations and the lengthy deadlocks, in the Netherlands a discussion started on the strategies to be used in the negotiations with Belgium. Some members of the Dutch delegation, Dutch MPs, and several others opposed the linkage between the deepening of the Western Scheldt and the Meuse water management issues, because this linkage would have been the main reason for the deadlocks in the negotiation process. Others, however, among which the Dutch Minister of Foreign Affairs, continued to speculate on a Flemish compensation of the Walloon 'losses'.

The Dutch also tried to compensate the Walloon region by cooperation on the construction or improvement of infrastructure projects in the access routes to the Walloon port of Liège. Because of the political situation and political changes in the Walloon region, however, this strategy turned out to be unsuccessful. Beside a compensation by cooperation on the construction of infrastructure projects, also a financial compensation of the upstream basin states and regions, in particular the Walloon region, became a topic of discussion. Some Dutch MP's proposed to establish a sanitation fund. The advocates of this fund pointed to the economic rationality of a sanitation fund.[216] In addition, these advocates were afraid that if the Dutch government would not establish such a fund, the sanitation of the river Meuse would take many more years. The adversaries, among which the Dutch Minister of Transport, Public Works, and Water Management, declared that the establishment of such a fund would not be in accordance with an important principle of international environmental law, namely the polluter pays principle.[217]

Whilst on the political level negotiations went laboriously, in the beginning of the nineties the Dutch Ministry of Transport, Public Works, and Water Management took the initiative for several projects aiming at international research and information exchange, such as the

---

216. The idea of a sanitation fund is strongly related to discussions on the efficiency of environmental investments. Possibly, an investment of the last 100 guilders available for waste water treatment in the Netherlands would generate more benefits if they would be invested in the Walloon part of the basin than in the Dutch part.

217. One may also argue that the establishment of a sanitation fund does not stimulate economic rationality, because individual polluters do not bear the full costs of their polluting activities, and therefore are no longer able to make economically efficient decisions. This argument, however, does not count for the so-called 'altlasten', i.e. the historical pollution.

ISG and the research project OostWest. These projects enabled the Dutch to discuss the scientific or cognitive aspects of the issues at stake with the other basin states.

Strategic changes Through the years 1992 and 1993, the Dutch government changed its strategy fundamentally. Until 1992 the Dutch government had never tried to involve the upstream basin state France in the negotiations on the Scheldt and Meuse, mainly because the involvement of France in the negotiations on the water conventions would have made it impossible to continue the linkage between the bilateral Scheldt infrastructure and the multilateral Scheldt and Meuse issues. In 1992, however, the Dutch government changed its strategy, contacted French government agencies, and drafted the first multilateral convention for cooperation on the management of the rivers Scheldt and Meuse. There are several supplementary explanations for this important strategic change. First, when the Dutch delegation contacted the French government agencies, the conclusion of the UN-ECE Convention on the Protection and Use of Transboundary Watercourses and International Lakes was shortly expected. According to this convention, which was concluded in 1992, all basin states of a transboundary river basin should prepare river basin policies jointly. Secondly, there gradually had grown a discrepancy between principles of the Dutch national water policy, such as the water systems and river basin approach, and the Dutch strategy in the negotiations on Scheldt and Meuse. Thirdly, the Dutch knew France as a good partner in the Rhine river negotiations. Finally, by the involvement of France in the negotiations on Scheldt and Meuse, the Dutch would meet the wishes of the Walloon region.

The second crucial strategic change took place in 1993. In the beginning of this year, the Dutch government decided to unlink the bilateral and multilateral Scheldt and Meuse issues. Again, there were several supplementary reasons for this strategic change. First, as mentioned before, the involvement of France in the negotiations on Scheldt and Meuse would make it hardly possible to maintain the linkage between the bilateral Flemish-Dutch deepening issue, and the multilateral Scheldt and Meuse issues. Secondly, because of the conclusion of the Belgian federalization process, and the attribution of treaty-making competencies to the Belgian regions, the Dutch government did no longer speculate on a Flemish compensation of Walloon 'losses'. Instead, the Dutch government re-evaluated the relationship with its southern neighbour, and realized that the Belgian regions had to be approached separately. Thirdly, the strategic change may also be interpreted as a strategic learning process, because the linkage between Scheldt infrastructure and Meuse water management issues did not seem to have been very effective. The case study showed that Dutch MP's and others had questioned the effectiveness of the linkage since long. Others, however, wanted to continue the linkage, even after the French government had reacted positively on the Dutch proposal to start multilateral negotiations on the management of the rivers Scheldt and Meuse. That the Dutch government decided to unlink the bilateral and multilateral Scheldt and Meuse issues only in 1993, indicates a lack of learning capacity. Although the linkage across Scheldt infrastructure and Meuse water management issues the Dutch had maintained ever since 1967 had proven to be largely ineffective, the Dutch government did not change its strategy until 1993.[218] There is, it's true, some evidence that the linkage of the deepening programme to the water quality of Scheldt has stimulated the development of Flemish water policies, because the Flemish region wanted to build up a reservoir of goodwill. The linkage across the Scheldt and Meuse issue-areas, however, seems

---

218. The linkage was ineffective *from the point of view of the Netherlands*, i.e. the linkage does not seem to have contributed much to the achievement of the *Dutch objectives* of (international) water policy.

to have had negative impacts only. First, this linkage caused an almost continuous deadlock in the negotiations, and consequently no international institutional arrangements could be developed. These deadlocks, among other things, hindered cooperation between the basin states in the ISG, the research project OostWest, and transboundary cooperation on the development of a policy plan for the Western Scheldt. The final reason why the Dutch linkage indeed may be conceived of as disastrous, was that the linkage caused a deep distrust between the basin states and regions, and reinforced the bad reputation the Dutch already had in the eyes of many Belgians. Because cultures do only change in the long run, one may expect that decision making on international Scheldt and Meuse issues will be hindered by this culture of distrust for several years at least. Although, in first instance, there certainly were good reasons for the Dutch to speculate on a Flemish compensation of Walloon 'losses', it would have been wise to change the Dutch strategy after it turned out that the Belgian parties did not want or were not able to arrange such a compensation for many years.

The Belgian national government, loser of power
Belgian decision making on the strategies to be used in the negotiations with the Netherlands clearly illustrates that international decision making may be conceived of as a two-level game. The Belgian national government continuously faced the problem of intranational disagreement. The main problem was that the region of Flanders perceived a big interest in the conclusion of the water conventions, whilst the region of Wallonia did not. Several attempts were made to reach a Flemish-Walloon consensus on the water conventions, but all failed. In 1985, the Belgian national government had expressed its willingness to give the regions financial support for the development of waste water treatment infrastructure. The national government, however, did not arrange a Flemish compensation of Walloon losses, and did not want to involve Walloon-Flemish infrastructure issues in the negotiations on the water conventions. Consequently, the national government was hardly able to reach a Walloon-Flemish consensus on the water conventions. Eventually, the standpoints presented by the Belgian delegation mostly were rather similar to the standpoints of the Walloon (and later also the Brussels Capital) Region. Because of the ongoing Belgian federalization process many competencies were attributed to the regions, and the power of the Belgian national government decreased continuously. In 1989 representatives of the Belgian regions Flanders, Wallonia, and Brussels were added to the Belgian delegation. This however, did not change the problematic situation that on the one hand the Belgium delegation had to formulate one joint Belgian standpoint, whilst on the other hand the Belgian regions had absolutely different perceptions of the international Scheldt and Meuse issues at stake. Because of the internal disputes in Belgium, from 1985 to 1993 the Belgian government was hardly able to participate in the international negotiations on the Scheldt and the Meuse in a normal way. From 1993, when also most treaty-making competencies were attributed to the Belgian regions, the Belgian federal government would only play a minor role in the negotiations on the water conventions.

The Walloon region, maintainer of autonomy
The Walloon region did not perceive an interest in any of the international Scheldt or Meuse issues at stake, and therefore mainly used reactive strategies aimed at maintaining autonomy. There are several supplementary explanations for this. First, because Walloon water quality policies were poorly developed, the conclusion of international water quality agreements would have implied that the region would have to implement and impose costly water quality

policies. The economy in the Walloon region, however, was declining[219], and the political priorities were with economic development rather than with environmental policies. Secondly, the region itself did not perceive urgent water quality problems, because it had abundant clean (ground) water in the Ardennes to produce drinking water. Thirdly, the region did not accept the linkage across Scheldt and Meuse issue-areas, and did not accept that the Netherlands, which had taken the initiative to conclude a convention on the water quality of the Scheldt and the Meuse, did not involve the upstream basin state France in the negotiations. Fourth, because the region was very well aware of the (economic) value of the Meuse water, the region wanted to maintain its autonomy as regards the water quantity management in the river Meuse. In particular the Belgian-Dutch proposal to construct storage reservoirs in the Walloon Ardennes was a thorn in the flesh of the Walloon region.

The Walloon region tried to maintain its autonomy in several ways. First, it opposed the paragraph dealing with the construction of storage reservoirs on Walloon territory in the declaration of intent of 1985. Secondly, it postponed negotiations on the commission Biesheuvel-Davignon for formal reasons. Thirdly, the region discontinued international interactions in the SWC and the TMC. Fourth, it opposed Dutch draft conventions that would go beyond existing international policies. The region even did not want to reach an agreement on the implementation of the North Sea Ministerial declarations. Fifth, the region aimed at a rather limited scope of the Scheldt and Meuse water management conventions, because it wanted to reach an agreement on the water quality of the main course of the river only. Finally, the region also reacted cautiously on the Flemish-Dutch initiative to establish an informal study group, the ISG, and emphasized that the activities of the ISG should be restricted to research and information exchange only.

The region, however, also searched for possibilities to contribute to a solution of the problems the Dutch perceived in the Meuse basin. First, the region suggested the Netherlands to buy Walloon drinking water, which could be transported to the Netherlands. Secondly, the region declared that if the Netherlands would like to formulate water quality policies that go beyond the water quality policies of the Walloon region, the Dutch government could finance the implementation of necessary policies itself. In the beginning of the nineties, the region also used reactive interactive strategies, and tried to link some Flemish-Walloon infrastructure issues to the Meuse river negotiations.

After the attribution of treaty-making competencies to the Belgian regions, the involvement of France in the Scheldt and Meuse negotiations, and the Dutch decision to unlink the bilateral and multilateral Scheldt and Meuse issues, the Walloon region changed its strategy into a more interactive strategy, and declared that it was willing to participate in multilateral negotiations on the Scheldt and the Meuse.

The Brussels Capital Region, a good partner of Wallonia
Unlike the Walloon and Flemish regions, which had received many water management competencies in 1980, the Brussels Capital Region only received these competencies after the Belgian constitutional amendment of 1988. Until 1988, hardly water quality policies had been developed in Brussels, and the city did not have waste water treatment plants at all. From 1988 a new administration and water quality policies had to be developed. After the

---

219. See also Section 4.7.2. In Belgian history the Walloon region always had been the prosperous part of Belgium, whereas the Flemish part was relatively poor and underdeveloped. In the case study period, however, the Flemish economy started to flourish, whereas the Walloon economy was declining.

Brussels Capital Region became involved in the negotiations on the water conventions in 1989, the region mostly had similar points of view as the Walloon region, and also mainly used reactive strategies aimed at maintaining autonomy. In the negotiations the Walloon and Brussels Capital Regions formed a rather stable coalition. Like the Walloon region also the Brussels Capital Region, which is situated along the Senne, a tributary of the river Scheldt, wanted to formulate international water quality objectives for the main course of the river Meuse only. After the attribution of treaty-making competencies to the Belgian regions in 1993, the Brussels Capital Region started direct negotiations with the Netherlands, France, and the other Belgian regions.

The Flemish Region, victim of Dutch linkages
The Flemish region faced an almost unsolvable problem. The Dutch government did not want to cooperate on the infrastructure projects that were necessary to maintain or improve the maritime access to the port of Antwerp unless it would be compensated with agreements on the water quality and quantity of the Meuse, and the water quality of the Scheldt. For the last issues, however, the Flemish region perceived a dependence on the other basin states and regions, in particular on the regions of Wallonia and Brussels. As was shown above, until 1993 these regions mainly used reactive strategies aimed at maintaining autonomy.

Although the Flemish region protested against the Dutch linkages between Scheldt infrastructure and Meuse issues, especially because the Scheldt Statute would oblige the Dutch to cooperate on the maintenance and improvement of the navigation channel to Antwerp, the Flemish region tried to meet the Dutch wishes. The region clearly tried to build up a reservoir of goodwill that it would need for the implementation of the deepening program. First, it developed investment plans for the construction of waste water treatment infrastructure in the Flemish region. Secondly, some Flemish parties opposed the use of the pipeline Tessenderlo, because the use of this pipeline would annul the Antwerp efforts to ameliorate the water quality of the Scheldt. Thirdly, the Flemish region agreed with the water quality conditions formulated in the WVO permits. Fourth, as regards the Meuse water quantity issue, the Flemish region was prepared to conclude a Flemish-Dutch convention on the distribution of Meuse water in periods of a low river flow, and a saving scheme. Fifth, the Flemish region approved the installation of a working group of the TSC, which would study, among other things, the ecological impacts of the implementation of the 48'/43'/38' deepening program. Finally, in the early nineties the Flemish region also was prepared to conclude a bilateral Flemish-Dutch convention concerning the water quality of the river Scheldt, if the Walloon region would continue its resistance against such a convention. The Flemish attempts to meet the Dutch wishes may be interpreted as a strategic learning process. The region knew that it was dependent on the Netherlands as regards the implementation of the deepening program, and therefore tried to build up a reservoir of goodwill.

Whereas the Flemish region tried to meet the Dutch wishes, it did not meet the Walloon wishes. Although the region had a big interest in the conclusion of the water conventions, and the deadlock in the negotiation could have been solved by a Flemish compensation of the Walloon 'losses', the Flemish region did not offer financial compensation to the Walloon region, neither did it make extra linkages with Walloon-Flemish infrastructure issues.

The case study clearly showed that also in this period of the (recent) history of the Belgian-Dutch relations concerning the management of the river Scheldt, the interpretation of the Scheldt Statute of 1839 was a major issue.[220] The Flemish and/or Belgian government tried to use the Scheldt Statute to reduce their dependence on the Netherlands. First, they used the Scheldt Statute to oppose the Dutch linkage of the 48'/43'/38' deepening programme to the other water conventions. Secondly, they used the Statute to oppose the (formulation of conditions in the) WVO Permit for the dumping of dredged material in the Western Scheldt. Finally, the Belgian delegation tried to include the Belgian interpretation of the Scheldt Statute in a Belgian draft convention on the implementation of the 48'/43'/38' deepening programme of 1991.

France, the new party in the international arena
The French Ministry of Foreign Affairs, and the water agencies Rhin-Meuse (situated in the Meuse basin) and Artois-Picardie (situated in the Scheldt basin) did react positively on the Dutch invitation to participate in the negotiations on the water quality of Scheldt and Meuse, and with that used reactive interactive strategies. Since 1992, the Agence de l'Eau Artois-Picardie participated in the informal study group, the ISG. In this informal study group, the French emphasized the need for comparable efforts by all basin states to clean-up the river Scheldt.

Perceptions and strategies of lower level governments and NGOs
Lower level governments in the Dutch Province of Zeeland gradually began to loose their patience. After more than 20 years of intermittent negotiations, Belgium and the Netherlands were still unable to reach an agreement on the water conventions. Therefore, the Province of Zeeland demanded the European Commission, which has an intervention position, to play the role of mediator, and to discontinue the impasse in the negotiations. Most actors in the Province of Zeeland opposed the Dutch policy to link the deepening programme to the Meuse water management issues, like in the declaration of intent of 1985, and emphasized more often the need to link the deepening programme to the water quality of the Scheldt. Therefore, many actors in the Province of Zeeland also opposed the Dutch decision to unlink the bilateral and multilateral Scheldt issues in 1993. Partly because the Dutch national government seemed to be more interested in the water quality and quantity of the Meuse than in the water quality of the Scheldt, some Dutch municipalities along the Western Scheldt tried to reduce their dependence on the national governments of the basin states. They took decisions, which made it obligatory to apply for a permit based on the Dutch Act on Land-use Planning for dredging and dumping activities on their territories. By that the municipalities did get an indispensable resource for the implementation of the deepening program. In addition, the municipalities founded the Task group for the Western Scheldt, and by that formed an important coalition.

The region also used the strategy of network activation, and asked several Belgian government agencies to participate in the BOW. This strategy, however, failed because of the ongoing negotiations at the political level.

Environmental NGOs took several actions to get political attention for the environmental problems in the Scheldt basin, such as a protest march, the collection of signatures, the issuing of research reports, and the drafting of Action Programs. Very effective from an environmental point of view were several cases of litigation against Belgian polluting firms, some of them leading to a closure of the firm. Dutch lower level governments and NGOs

---

220. For more information on the Statute of the Scheldt, see the historical overview presented in Section 4.8.

also opposed the decision of the Dutch government to grant another WVO Permit to the Belgian government, mainly because Belgium had not complied with the condition formulated in the preceding permit, and the water quality of the river Scheldt had not improved.

6.3.2 Strategic interactions and institutionalization processes

Development or erosion of international institutional arrangements and policies
The reconstruction of the decision making process clearly shows that in the period between 1985 and 1993, just like in the period between 1975 and 1985, Belgian-Dutch interactions on the water conventions can be characterized better by conflict than by cooperation. The linkages that were made by the Belgian and Dutch Ministers of Foreign Affairs in 1985 failed for exactly the same reason as the linkages the Dutch had made in 1967. They did not create a situation in which all parties perceived that they would be better off with an agreement than without one. The Walloon region continued its resistance against a package deal from which it would loose. Although the Dutch government slightly changed its objectives and strategies, it maintained the linkage between the deepening of the Western Scheldt and the Meuse water management issues. Although the Flemish protested against the Dutch linkage, they continued to aim at a package deal with the Netherlands, and did not compensate the Walloon region. The Dutch continued to speculate on a Flemish compensation of the Walloon 'losses', the Flemish region wanted the Dutch to unlink the Scheldt infrastructure and Meuse issues, and the Walloon region waited until one of the other parties would compensate the Walloon region, the Dutch would decide to unlink the Scheldt infrastructure and the Meuse issues, and/or until they would involve France in the negotiations. Summarizing, neither of the parties was willing to change its strategy, but all did want or expect the other parties to change their strategies. Because of the rigid mutual strategies the negotiations got in an impasse, and hardly institutional arrangements did develop. On the contrary, existing ones, the SWC and the TMC, did erode. In this context, the statement made by the chairman of the Dutch delegation in 1992 that 1993 was going to be the seventh year of a negotiation commission that had never been able to negotiate really, should be interpreted as a cry of despair (See Round IX).

In the beginning of the nineties, transboundary contacts at the administrative level were intensified. In the ISG representatives of water management agencies of all basin states and regions started to discuss the scientific aspects of the Scheldt water quality issues, and in the working group OostWest Flemish and Dutch water managers started joint research on the morphology and ecology of the Scheldt estuary. The extremely political character of the issues at stake, and the continuous disagreement on the political level, however, hindered communication between experts on these issues.

Interesting institutionalization processes took place within the Technical Scheldt Commission. With the discussions concerning the water quality of the Scheldt in the Working Group of the WVO permit, and the research on the consequences of dredging and dumping on the morphology and ecology of the Scheldt estuary in the working group OostWest, the scope of activities of the TSC was extended gradually with water quality and ecological issues. The installation of these working groups clearly indicates a progressive development of international institutional arrangements. In the TSC also concrete international policies were developed, such as policies concerning the maintenance dredging works in the Western Scheldt, the dumping of dredged material, and water quality policies. There are two supplementary explanations for the development of these Flemish-Dutch institutional arrangements and policies. The first explanation is the occurrence of cognitive learning processes (See Section 6.3.3). Experts had recognized the intellectual coherence between the dredging works, the dumping of dredged material, and the morphology and ecology of the Western Scheldt. Consequently, in discussions on the management of the navigation channel of the Western Scheldt these issues were addressed simultaneously. The second explanation is a more strategic one. Because of its dependence on the Netherlands as regards the implementation of the deepening program, the Flemish region tried to meet the demands of the Netherlands, and to build up a reservoir of goodwill.

Belgian-Dutch climate of distrust
The continuing disagreement between the parties reinforced mutual bad reputations, and increased Belgian-Dutch distrust. The tenacity of the Dutch in maintaining the linkage across Scheldt and Meuse issue-areas gave them a bad reputation in the eyes of many Belgians. The reputation of the Dutch can be characterized by the statement of Suykens concerning the Dutch decision to unlink Scheldt infrastructure and Meuse issues: "*Meuse and deepening of the Scheldt were no longer connected by the disastrous Dutch linkage.*" (See Section 6.2, Round IX). According to many Belgians, the Dutch had used a disastrous linkage strategy that had frustrated the negotiations, and the linkages that were made to the 48'/43'/38' deepening programme and the WVO permit that Belgium needed for the maintenance of the navigation channel were perceived as major violations of the Scheldt Statute.

On the other hand, the inability of the Belgian state to reach a Walloon-Flemish consensus, and to react on Dutch proposals, and the continuous Belgian unwillingness to cooperate on the clean-up of the rivers Scheldt and Meuse, and to conclude an agreement on the distribution of the Meuse water, gave the state of Belgium a bad reputation in the eyes of the Dutch. Because of the continuous discharge of untreated waste water a Dutch research institute describes the city of Brussels as "*the dirty child of Europe*" (Section 6.2, Round VIII), and a Dutch mayor complains about the institutional chaos in Belgium and states that Belgium is a developing country as regards environmental management (Section 6.2, Round VII). Also the expression of the Dutch MP about the tense atmosphere in the Benelux Parliament clearly illustrates that in the beginning of the nineties the Belgian-Dutch negotiation climate was bad at best (Round VII).

6.3.3 Learning processes

Some strategic changes of the basin states and regions, which took place in the period between 1985 and 1993, may be interpreted as the result of a strategic learning process. These examples were given in Section 6.3.1, where the dynamics of the perceptions and strategies of the individual basin states were discussed.

In the previous section, it was concluded that in the period between 1985 and 1993 transboundary research networks developed gradually. Especially the research project OostWest shed new light on the morphological dynamics and ecology of the Scheldt estuary. It indicated the relationship between infrastructure and maintenance dredging works in the Scheldt estuary, the dumping of dredged material, and the morphology and ecology of the estuary. The Pilot Study also introduced the new policy alternative, 'ontpolderen'. Although experts started to discuss these issues simultaneously, in the period between 1985 and 1993, this new knowledge did not yet have a significant impact on the decision making process. In Chapter 7 it will be shown that the results of the research project OostWest did play an important role in decision making on nature compensation and restoration in the Scheldt estuary in the period between 1993 an 1997.

6.3.4 Influence of the context on the perceptions and strategies of the Scheldt basin states and regions

The in depth study of the "action rationality" of the Scheldt basin states and regions and the interactions between them, made it possible to assess the influence of contextual factors and developments on the decision making process. In the following, these factors and developments will be discussed.

Underlying hydrological structure of the issues at stake
Like in the previous period, it is most likely that the underlying hydrological structure of the issues at stake has had a large impact on the decision making process. The Dutch continued to use their relative power advantage in the issue-area of the maritime access to the port of Antwerp to influence decision making on international Meuse issues, whilst the Belgian parties continued to use their relative power advantage as regards the water management of Scheldt and Meuse. Also after the Dutch decision to unlink bilateral and multilateral Scheldt and Meuse issues power asymmetries caused by the underlying hydrological structure of the issues at stake still had an important influence on the decision making process, because Flanders and the Netherlands agreed on a linkage between the issue of the deepening of the navigation channel in the Western Scheldt (relative Dutch power advantage) and the water quantity of the river Meuse (relative Flemish power advantage).

International context
Because of the continued linkages across Scheldt and Meuse issue-areas (See Table 6.3) decision making on international Meuse issues was the main international context of decision making on international Scheldt issues.

Although to a lesser extent, also the developments in the Rhine basin had an impact on the decision making process. The Dutch delegation used (the organization of) decision making on international Rhine issues as an example, and proposed to organize decision making in the Scheldt and Meuse basins in a similar way. It proposed to install river basin commissions, which would have to develop Action Programs. The Dutch Minister also argued that it is unacceptable that there is a much less strict regime for the rivers Meuse and Scheldt than for the river Rhine. The Walloon region used the example of Rhine river cooperation in another way. It referred to the financial compensation of France, because of the French efforts to reduce salt discharges into the Rhine basin, and suggested that the Dutch government should contribute financially to the clean-up of the rivers Meuse and Scheldt in the Walloon region.

In addition to the developments in the adjacent Meuse and Rhine basins, several other international developments seem to have had a considerable influence on the decision making process. Undoubtedly, the conclusion of the UN-ECE Convention on the Protection and Use of Transboundary Watercourses and International Lakes in 1992 was the most important development. The conclusion of this convention made it necessary to involve France in the negotiations on Scheldt and Meuse, and therefore made it hardly possible for the Dutch to continue the linkage between the deepening of the navigation channel in the Western Scheldt and the water quality of Scheldt and Meuse. Together with the attribution of treaty-making competencies to the Belgian regions, the conclusion of the UN-ECE Convention formed a main reason for the Dutch strategic change in 1993, which would change the composition of the international arenas, and the course of the decision making process fundamentally.

Also EC and North Sea policies played an important role in the decision making process.

The Dutch used these policies more often to legitimate their proposals to clean-up the rivers Scheldt and Meuse. The Belgian government, and in particular the Walloon and Brussels Capital Regions, used EC-Directives to legitimate their proposals as well. They were willing to comply with a rather dated EC Directive, and for a long time did not want to discuss the implementation of the North Sea Ministerial Agreements with the Dutch. The Benelux formulated several recommendations concerning the management of the rivers Scheldt and Meuse. The influence of these recommendations on the international negotiations, however, could not be assessed completely.

As mentioned in Section 6.3.1, the Dutch Province of Zeeland demanded the European Commission (EC) to intervene in the international negotiations, and to play a mediating role. Although the EC would never play this role, this seems to have been a wise suggestion of the Province of Zeeland. Because of the distrust that had grown between the parties, and the lengthy deadlocks in the negotiation process, a professional mediator could have played a useful role. To put it differently, it would have been worth trying, because the situation could hardly have become worse. On the one hand, the mediator could have advised the Netherlands to unlink the deepening of the navigation channel in the Western Scheldt and the Meuse water management issues, whilst on the other hand he could have recommended strongly the Belgian regions to clean-up their waters, and to cooperate on the development of international water quality policies, and on a convention on the distribution of Meuse water during periods of low flow. Obviously, a mediator would only have been able to make a positive contribution to the decision making process, if this mediator would have been accepted by all parties.

Cultural context
Also in this case study period there is some evidence for the influence of cultural characteristics of the basin states and regions on their perceptions and strategies. A very clear example is the discussion on a possible Dutch compensation of the Walloon region by cooperation on the construction of infrastructure projects to improve navigation possibilities to the port of Liège. Because succeeding Walloon Ministers came from different regions, they had different preference intensities concerning these infrastructure issues. This may be explained by the Walloon clientilism (See Section 4.6.4). Another example concerns the Walloon statements about the activities of the ISG, which seem to confirm that Walloon (Belgian) civil servants do have rather limited mandates compared with the Dutch ones. The Dutch initiative to use their GIS in the study group ISG, and the problems related to this may be an illustration of the typical Dutch overestimation of their ideas. Finally, in the ISG different traditions as regards the public access to information became apparent.

Intranational context
Undoubtedly, also in the period between 1985 and 1993 the Belgian federalization process was the most influential intranational development. In this period the unitary state of Belgium was transformed into a full federal state. The constitutional amendment of 1988 entailed the attribution of most water management competencies to the Belgian regions. Because of this major step in the Belgian federalization process, Belgium postponed negotiations with the Netherlands from 1987 until 1989. After this major state reform the Belgian regions Wallonia, Flanders, and the Brussels Capital Region became directly involved in the negotiations on the water conventions in 1989. Discussions on further state reforms continued, and in 1993, with the conclusion of the Sint Michiels agreements, Belgium became a full federal state. In that year the regions did receive treaty-making competencies,

which enabled them to negotiate directly with other states. The case study has clearly shown that beside the conclusion of the UN-ECE Convention, the completion of the Belgian state reforms was a second important contextual development that stimulated the Dutch to unlink the bilateral and multilateral Scheldt and Meuse issues.

Beside the state reforms, also the different stages of development of water quality policies in the basin states and regions continued to influence the decision making process. These differences, which were described and analyzed in Section 4.7, may explain why the Walloon and Brussels Capital Regions did not want to formulate water quality objectives that would go beyond the objectives of the EC-Directive 78/659 for cyprinids.

6.4 Conclusions

In this chapter decision making on international Scheldt issues between 1985 and 1993 was described, analyzed, and explained.

The analysis of written material and some interview reports enabled the author to generate a rather long list of relevant decisions taken by the participants in the decision making process between 1985 and 1993. In this series six crucial decisions, and five decision making rounds could be distinguished. The crucial decisions marking the beginning and/or end of the decision making rounds were the Belgian-Dutch declaration of intent of 1985, the Belgian-Dutch agreement to start negotiations on the basis of a modified declaration of intent, the Belgian decision to involve the three Belgian regions directly in the negotiations with the Netherlands in 1989, the Dutch decision to prepare a draft convention unilaterally in 1991, the agreement between national and regional ministers to aim at an 'umbrella convention' in 1992, and finally the Dutch decision to unlink bilateral and multilateral Scheldt and Meuse issues in 1993.

The strategies of the main actors involved in the decision making process were related to their perceptions. This analysis shed light on the action rationality of the individual actors. It showed that also in this period the basin states continued to have different perceptions of the issues at stake. Furthermore, the linkages that were made in the negotiation process seem to confirm the theory that in cases where decision making is based on unanimity and there are heterogenous preference intensities across issue-areas, the probability that movement away from the status quo involves issue-linkage is high. The case study also showed that the basin states and regions used rather rigid strategies, which caused lengthy impasses in the decision making process, erosion of existing institutional arrangements, and distrust. Incapability of strategic learning seems to have hindered both the Netherlands and the Flemish region to achieve their objectives.

Because lower level governments and NGOs were not directly involved in the negotiations on the water conventions, they tried to influence these negotiations indirectly. The actors in the province of Zeeland in particular emphasized the need to link the deepening programme to the water quality of the Scheldt. In this period, the municipalities along the Western Scheldt also took land-use decisions, which would reduce their dependence on the national governments as regards the implementation of the deepening program.

The strategic analysis also made it possible to trace the (contextual) factors or developments that have influenced the perceptions and strategies of the actors involved in the decision making process, and consequently the course and outcomes of this process. Like in the preceding period, the underlying hydrological structure of the international Scheldt issues was an important contextual factor. The developments in the Meuse basin and the conclusion of the UN-ECE Convention were the most influential developments in the international context. The North Sea Ministerial declarations and the EC policies, however, also influenced the international negotiations on the Scheldt and Meuse issues. Beside the different stages of development of water quality policies in the basin states and regions, the ongoing Belgian federalization process, and in particular the attribution of treaty-making competencies to the Belgian regions were the most influential intranational developments. Finally, there is some evidence for the influence of the cultural characteristics of the Scheldt basin states on the decision making process.

CHAPTER 7

CONCLUSION AND IMPLEMENTATION OF THE WATER CONVENTIONS (1993-1997)

7.1   Introduction

This chapter contains the description, analysis, and explanation of decision making in the last case study period, which extends from 1993 to 1997. The structure of the chapter is very similar to the structure of the two preceding chapters. In Section 7.2 the decision making process is described. First, the selection of crucial decisions, and the distinction between five decision making rounds is clarified. Subsequently, the strategic interactions between the Scheldt basin states and regions in these decision making rounds are described. By that the first research question is answered for the period between 1993 and 1997. In Section 7.3 the strategic interactions are analyzed and explained, and the explanatory research questions three to six are answered for the period from 1993 to 1997.[1] Section 7.3 successively addresses the influence of the perceptions of the actors involved in the decision making process on their strategies, institutionalization and learning processes, and the influence of contextual variables on the perceptions and strategies of the actors. Finally, in Section 7.4 the main conclusions of the last part of the case study are summarized.

7.2   Reconstruction of the decision making process (1993-1997)

Selection of crucial decisions, and distinction between decision making rounds
The reconstruction of the decision making process draws on the analysis of written data, among which a large number of newspaper articles, and on the analysis of reports of interviews with participants in the decision making process. These analyses made it possible to inventory a series of decisions taken by the actors involved in the decision making process, which can be found in Appendix 12. In this series of decisions seven crucial decisions, which influenced the composition of the international arena or the international issues discussed,

---

1. Research question 2 was answered in Chapter 4, which contains a description and analysis of the context of decision making on international Scheldt issues.

could be traced.[2] Table 7.1 produces information on the crucial decisions marking the beginning and/or end of the decision making rounds, and the main actors participating in these rounds. In the next sections, the strategic interactions in each decision making round will be described.

Table 7.1 *Decision making rounds, crucial decisions, and major actors (1993-1997)*

Round	Start situation	Major actors	End situation
Xa (1993-1994)	Flemish-Dutch agreement to deal with the water quantity of the Meuse and the 48′/43′/38′ deepening programme simultaneously (Flemish-Dutch agreement on a parallel discussion of the deepening programme and the alignment of the HSL)	Flemish-Dutch negotiation commission (Delegations of) Flemish government Dutch government	Agreement between delegations on draft Conventions on the flow of the river Meuse and the deepening of the navigation channel in the Western Scheldt The Dutch government refuses to sign the convention on the deepening programme as long as no basic agreement is reached on the alignment of the HSL
Xb (1993-1994)	Decision of Scheldt and Meuse basin states and regions to start deliberations on the water quality of Scheldt and Meuse	Multilateral negotiation commission (Delegations of) Dutch government French government Brussels government Walloon government Flemish government	Agreement between delegations on draft Conventions on the protection of the Scheldt and the Meuse The Flemish government refuses to sign the Scheldt or Meuse conventions as long as the Dutch government does not sign the Convention on the deepening of the navigation channel in the Western Scheldt Other parties sign the multilateral conventions

2. Two of these crucial decisions, the Flemish-Dutch agreement to deal with the deepening programme and the water quantity of the Meuse simultaneously, and the Flemish-Dutch agreement on a parallel discussion of the deepening programme and the alignment of a new high speed train, the HSL (Hoge Snelheids Lijn), were taken shortly after another. Therefore, no separate decision making round between these two decisions is distinguished in the case description.

Round	Start situation	Major actors	End situation
XI (1994-1995)	Flemish-Dutch disagreement concerning HSL and Deepening programme	Flemish government Dutch government	Flemish and Dutch governments sign Conventions on: 1. the deepening programme 2. the water quantity of the Meuse 3. the revision of the Scheldt regulations Flemish government signs Conventions on: 4. the protection of the Meuse 5. the protection of the Scheldt
XIIa (1995-1997)	Provisional installation of the International Commission for the Protection of the Scheldt against pollution (ICPS)	Delegations provisionally installed ICPS	France ratifies the multilateral Scheldt and Meuse conventions
XIIb (1995-1997)	Conclusion of the Convention on the deepening of the navigation channel in the Western Scheldt	Flemish government Dutch government Delegations TSC Participants BOW NGOs Public	Dutch Parliament approves special legislation on the implementation of the deepening programme Advice of the Commission of wise persons on nature compensation

Round Xa: Bilateral negotiations (1993-1994)

After the Dutch government had decided to unlink the bilateral and multilateral Scheldt and Meuse issues, the Flemish and Dutch governments agreed to start bilateral negotiations on the distribution of the water of the Meuse in periods of a low river discharge, and the 48'/43'/38' deepening programme in the Western Scheldt. In addition, they agreed to discuss the second maritime access to the Waasland harbour, a possible further deepening of the Western Scheldt, and the bilateral aspects of the water quality of the Scheldt and the Meuse (See Section 6.2, Round IX). In May 1993 informal bilateral discussions are started.[3] The Flemish-Dutch negotiations, however, would become more complicated. On 16 June 1993 the Flemish Prime Minister, Van den Brande, proposes parallel negotiations on the 48'/43'/38' deepening programme and the alignment of a new high speed train (HSL) from

---

3. The first informal meeting takes place on 13 May 1993, and the second one on 7 September 1993.

Antwerp to Rotterdam, and the Dutch Prime Minister, Lubbers, agrees with this proposal.[4] The Flemish government has by far most interest in the implementation of the deepening programme, whereas the Dutch government is very interested in rapid decision making in Flanders on the alignment of the HSL between Antwerp and the Belgian-Dutch border.[5] The Dutch and Flemish governments prepare a ministerial declaration, which should express the Flemish and Dutch willingness to solve a number of bilateral issues. In the meantime, the Flemish and Dutch delegations have almost reached an agreement on draft Conventions on the deepening of the navigation channel in the Western Scheldt, and the flow of the river Meuse. Negotiations on the alignment of the HSL, however, are rather laborious, and the Flemish and Dutch delegations have not yet reached an agreement.[6] Because the Flemish and Dutch Prime Ministers have reached an agreement on parallel negotiations on the deepening of the navigation channel and the alignment of the HSL, the Dutch declare that they will not sign a Convention on the deepening of the navigation channel in the Western Scheldt as long as no agreement on the alignment of the HSL is reached.[7] At that time, in Flanders the Flemish proposal to link decision making on the deepening programme and the alignment of the HSL is seen as a tactical mistake.[8] One of the problems the Flemish region faces is that the Belgian federal state has important competencies as regards decision making on rail infrastructure.[9]

The Antwerp alderman Devroe reacts furious on the statements made by the Dutch Minister, Maij-Weggen, that the Dutch will not implement the deepening programme as long as no agreement on the alignment of the HSL is reached, and states that according to international law (The Scheldt Statute) the implementation of the deepening programme is a Flemish right.[10] The Antwerp port association (AGHA[11]) declares that if no agreement on the deepening programme will be reached, the Belgian state should appeal to the International Court of Justice.[12] Because of the Dutch refusal to sign the 48'/43'/38' deepening programme, the Flemish parties postpone discussions in the Rhine-Scheldt-Delta platform (RSD[13]) on possibilities of cooperation between the ports in the Rhine Scheldt delta.[14]

---

4. TK 1993-1994, 23 075 No. 3. The alignment discussed at the meeting is part of the HSL alignment from Paris to Amsterdam (Letter from the Dutch Minister of Transport, Public Works and Water Management, Jorritsma-Lebbink, to the chairman of the Commission for Infrastructure and Water Management of the Dutch Parliament of 20 January 1995).

5. Trouw, 17-6-1993; De Volkskrant, 17-6-1993; Het Financieele Dagblad, 22-6-1993.

6. TK 1993-1994, 23 530 No. 6.

7. Trouw, 28-10-1993.

8. Ibid.

9. PZC, 26-2-1994.

10. PZC, 28-10-1993.

11. AGHA =Antwerpse Gemeenschap voor de Haven.

12. PZC, 5-2-1994.

13. RSD =Rijn Schelde Delta.

14. PZC, 14-6-1993; PZC, 24-11-1993.

Although competition between the two main ports still exists, in the eighties and nineties international developments in navigation shed another light on the relationship between Antwerp and Rotterdam. Gradually, the need for more cooperation between the ports is recognized (See for example Lemstra, 1990). Transport experts expect that in the long run shipowners will no longer choose between specific main ports, but between regions to deliver their goods. Therefore, in the long run, shipowners will not choose between Antwerp or

Although for completely different reasons, in the Netherlands several parties oppose the linkage between the deepening programme and the HSL as well. The three southern provinces of the Netherlands, the Provinces of Zeeland, Northern-Brabant, and Limburg, which are all situated in the Scheldt or Meuse basins, the Task group for the Western Scheldt, and several environmental NGOs oppose the Dutch decision to unlink the negotiations on the deepening programme and the water quality of the Scheldt and the Meuse[15], and the decision to make a linkage between the deepening programme and the HSL. According to these critics the new Dutch linkage clearly illustrates that the environmental problems in Scheldt and Meuse are made subordinate to the economic interests of the Dutch 'Randstad'.[16] Some municipalities prepare decisions based on the Dutch Act on Land-Use Planning, which will oblige the Flemish government to apply for a permit, if it would like to carry out dredging works or to dump dredged material on municipal territory. The Municipal Executive of Oostburg states that: *"such a decision will enable us to keep a finger on the pulse"*[17]

In the beginning of March 1994, the Dutch and Flemish delegations reach an agreement on the distribution of the costs of the deepening programme.[18] The Dutch government, however, refuses to sign the convention on the deepening programme as long as no agreement is reached on the alignment of the HSL.

Flemish-Dutch research in the TSC
During the negotiations in the commissions Biesheuvel-Davignon and Biesheuvel-Poppe, the construction of the Baalhoek canal had a rather low priority, and the Dutch Province of Zeeland had already skipped the alignment of the canal in its land-use plans.[19] Since 1993, however, the construction of the Baalhoek canal got a place on the international agenda again. The Flemish region is interested in a policy analysis of the second maritime access to the Waasland harbour, and of the 50' or second stage of the deepening programme. Flanders and the Netherlands agree that decision making on the second stage of the deepening programme will require an EIA.

In addition to the planning of research on the impacts of these two infrastructure projects, Flanders and the Netherlands continue research on the geomorphology and ecology of the Scheldt estuary in the working group OostWest of the TSC. Since the presentation of the

---

Rotterdam, but, for example, between the Rhine Scheldt Delta or a region in Asia. In that case, a strengthening of the competitive position of Antwerp, for example by the construction of the Baalhoek canal, would be beneficial to the development of Rotterdam and vice versa. To explore possibilities of cooperation in the Rhine Scheldt Delta, governments in this region established the platform Rhine-Scheldt-Delta (RSD). Table 4.2 (Section 4.2.2) lists the ports situated in the RSD. For more information on the activities of the RSD, see O-RSD (1994).

15. See, for example, the letter from the Secretary of the 'Stichting het Zeeuwse Landschap', Prof. dr. Nienhuis, to the Head-Engineer-Director of the regional Directorate Zeeland of Rijkswaterstaat, Saeijs, of 27 October 1993.

16. PZC, 6-10-1993, 20-10-1993, 25-10-1993.

17. Letter from the Municipal Executive of Oostburg to the regional Directorate Zeeland of Rijkswaterstaat of 5 January 1994.

18. PZC, 26-3-1994. The main issues addressed in the formal negotiations on the 48'/43'/38' deepening programme, which started on 5 November 1993, are the compensation of nature losses and the distribution of the costs.

19. PZC, 18-6-1993.

pilot study in the TSC, OostWest has become a joint Flemish-Dutch research project. Among other things, the project team carries out research on the impacts of the implementation of the 48'/43'/38' deepening programme on the morphology and ecology of the Scheldt estuary, on the long term developments in the Scheldt estuary, and on the impacts of the policy alternative 'ontpolderen' (See also Section 6.2, Round VIII).[20] A main conclusion of the research is that 'ontpolderen' positively influences the ecology of the Western Scheldt, because it increases the tidal volume of the estuary, the amount of water flowing over de bars in the navigation channel, and consequently increases erosion in the navigation channel. This would make it possible to reduce the intensity of the maintenance dredging works.[21] The regional Directorate Zeeland of Rijkswaterstaat presents the research results to a group of Flemish and Dutch experts, the so-called review team[22], which passes a positive judgement on most research results. The experts, however, also conclude that some of the proposals "[...] *are not undisputed. The project group wants to diminish the dredging works in the estuary by increasing the tidal volume. This can be done by 'ontpoldering', but this solution will encounter a lot of resistance. Public support is lacking and interested parties will rightly ask whether no alternative solutions exist.*"[23] A more general, but nevertheless important conclusion of the review team is that the dredging activities in the estuary indeed do have a negative impact on the ecology of the Scheldt estuary.[24]

Rijkswaterstaat intends to use the recommendations made by the review team, and decides to continue research on the impacts of 'ontpolderen' along the Scheldt estuary, and other possibilities to mitigate the negative side effects of dredging and dumping in the Scheldt estuary.[25]

Round Xb: Multilateral negotiations and the Flemish decision to re-establish the linkage between bilateral and multilateral issues (1993-1994)

In 1993 it was decided that simultaneously with the Flemish-Dutch negotiations (See Round Xa), the Scheldt and Meuse basin states and regions would start multilateral negotiations on the water quality of the Scheldt and the Meuse (See Section 6.2, Round IX). The formal

---

20. V&W, RWS-Zld (1994a).

21. Ibid, p. 7-8.

22. Ibid.

23. RWS-Zld (1994, p. 5).

24. Therefore, a policy analysis of the construction of the second maritime access to the Waasland harbour should take into account that the construction of this canal might make it possible to decrease the intensity of the dredging works in the eastern part of the Western Scheldt (Ibid.). Ships with a large draught would no longer have to navigate via the eastern part of the Western Scheldt, but could use the Baalhoek canal. In addition, the construction of the canal would increase the tidal volume, and with that erosion in the navigation channel as well. The experts do also suggest that a better division of tasks between the ports along the coast, such as the port of Zeebrugge, and the ports situated more inland, such as the port of Antwerp, could contribute to a long term solution of the ecological problems caused by the continuous dredging works in the Scheldt estuary (Ibid.). Furthermore, the review team recommends to carry out research on alternative possibilities to reduce the intensity of the dredging works, for example by optimizing dredging strategies. Finally, they recommend to reserve money for better communication and public information, and the involvement of interest groups in decision making on Scheldt policies (Ibid., p. 8).

25. V&W, RWS-Zld (1994b, p. 5).

negotiations on these issues are preceded by two informal meetings.[26] Participants in these meetings, for which the Dutch took the initiative, are all Scheldt and Meuse basin states and regions, i.e. France, Germany, Luxemburg, Belgium, the Flemish, Walloon, and Brussels Capital regions, and the Netherlands. In these informal deliberations France and the Netherlands demand the Belgian parties to provide more information on the distribution of competencies in Belgium, in particular on the responsibilities for the compliance with the conventions to be discussed.[27] After the two informal meetings all parties are prepared to start formal multilateral negotiations on conventions for the Scheldt and the Meuse, and the installation of international river commissions for these rivers.[28] Furthermore, they agree that the negotiations will be based on draft conventions to be prepared by each party, and that the UN-ECE Convention for the Protection and Use of Transboundary Watercourses and International Lakes[29], and other relevant international conventions and declarations will serve as point of departure.[30] Finally, it is agreed that France will take the initiative to start the formal negotiations on the multilateral Scheldt and Meuse conventions.

The new party France to some extent plays the role of mediator between the Belgian regions and the Netherlands.[31] According to the Dutch, the involvement of France in the negotiations on the Scheldt and Meuse was very important, because the Walloon region would be more inclined to listen to the French than to the Dutch.[32] The French interest in the conclusion of multilateral conventions for the Scheldt and the Meuse may be explained partly by the conclusion of the UN-ECE convention (which, however, has not yet been signed by France). In addition, the French are of the opinion that these conventions may contribute to the public support in France for costly measures to clean-up the rivers, since these measures would only be accepted if similar measures are taken in all basin states and regions.[33]

In their bilateral discussions, the Flemish and Dutch delegations had agreed to formulate joint standpoints in the multilateral negotiations, and therefore form a rather homogeneous coalition. The main objective of the Dutch and Flemish delegations is to implement integrated river basin management, and to act in accordance with the UN-ECE Convention. They aim at the installation of international river basin commissions for the Scheldt and the Meuse with competencies to develop or coordinate policies on the water quality and the water quantity of both the groundwaters and surface waters in these basins. The Walloon and Brussels Capital Regions, however, prefer to restrict international cooperation to the quality of the surface water only. The Walloon preference may be explained partly by the distribution of

---

26. These meetings take place on 9 and 10 June 1993, and on 20 September 1993 in The Hague (TK 1993-1994, 23 075 No.3).

27. TK 1993-1994, 23 536 No. 1, p. 21.

28. Ibid.

29. United Nations Economic Commission for Europe, Convention on the Protection and Use of Transboundary Watercourses and International Lakes, Geneva, E/ECE/1267, 1992.

30. TK 1993-1994, 23 075 No. 3.

31. Suykens (1995, p. 232), Interview C. De Villeneuve.

32. See, for example, the Director of the Dutch drinking water production firm 'Brabantse Biesbosch', Oskam, in: NRC, 6-11-1993. According to the Director-General of Rijkswaterstaat, Blom, it is better to leave the initiative to France. He thinks that the more noise the Dutch will make, the less they will get." (NRC, 28-1-1994).

33. TK 1994-1995, 24 041 No. 5.

competencies for water quality and water quantity management among two ministers.[34]

Beside the functional scope of the Scheldt and Meuse conventions, also the geographical scope of these conventions is a major topic of discussion. France, Flanders, the Netherlands, Luxemburg and Germany would like the international river commissions to develop policies for the tributaries of the rivers Scheldt and Meuse as well. The Walloon and Brussels Capital regions, however, would like to restrict international cooperation to the water quality of the main course of the rivers only.[35]

Apart from the discussions on the functional and geographical scope of the conventions, also the participation of NGOs in the international river commissions to be installed is a topic of discussion. The Dutch and Flemish delegations would like to involve NGOs in the international river commissions to be installed. The Walloon region, however, opposes the involvement of NGOs.[36] According to the Dutch minister Jorritsma: "[...] *this is a sensitive issue. There are countries that are used to great openness and involvement of environmental NGOs and other interest groups. The Netherlands belong to them. I have experienced the same discussion in the Rhine ministers conference. There also are countries that are extremely reserved on that issue.*"[37]

End of March 1994, after four formal meetings[38], the delegations reach an agreement on the texts of the Meuse and Scheldt conventions. The delegations have already prepared a ceremonial meeting on 26 April in France for the signing of these conventions, when on 21 April 1994 the Flemish Prime Minister, van den Brande, informs the Dutch Prime Minister, Lubbers, that the Flemish government has approved the two multilateral Conventions on the protection of the Scheldt and the Meuse, and the Convention on the flow of the river Meuse, but will not sign these conventions until the Dutch government will sign the Convention on the deepening of the navigation channel in the Western Scheldt.[39] In addition, the Flemish government postpones the approval of the Convention on the revision of the Scheldt regulations, which provides for new arrangements concerning the safety of navigation on the

---

34. TK 97-5929; TK 1995-1996, 24 041 No. 33a.

35. See for example NRC, 28-1-1994. Another topic of discussion concerns the definition of the rivers Scheldt and Meuse and their basins in the delta of the rivers Rhine, Scheldt, and Meuse. In this delta, which is situated in the Netherlands, the hydrological borders between the three river basins are not unambiguous. The Flemish delegation proposes to include, among others, the Hollands Diep, the Haringvliet, the Nieuwe Waterweg, and some other water courses near the city of Rotterdam in the definition of the Meuse basin, mainly as it wants to prevent a distortion of competition between Antwerp and Rotterdam. After the Dutch delegation has explained that the international Rhine conventions do already apply to the Nieuwe Waterweg and the other watercourses near Rotterdam, it is decided that the Meuse convention will not apply to these water courses. The delegations, however, decide that, in spite of the fact that about 2/3 of the water in the Hollandsdiep and the Haringvliet stems from the Rhine basin, these water courses will be part of the river Meuse as defined in the Meuse convention. Apart from the border between the Rhine and Meuse basins, also the border between the Meuse and Scheldt basins is not unambiguous, because some parts of the Dutch Province of Northern-Brabant receive Meuse water, but drain into the Western Scheldt (Information provided by R. Zijlmans).

36. TK 97-5932.

37. TK 97-5932.

38. The first formal meeting took place in December 1993 in Paris, the second one in February 1994 in Charleville-Mézières, the third one in March 1994 in Namur, and the final meeting took place at the end of March 1994 in Liège.

39. PZC, 21-4-1994; NRC, 21-4-1994.

Western Scheldt.[40] The Flemish decision comes rather unexpected for the Dutch government, because it implies that the Flemish region re-establishes the linkage between bilateral and multilateral Scheldt and Meuse issues, which it had always opposed. The Dutch Minister of Transport, Public Works, and Water Management, Maij-Weggen, however, does not seem to be really impressed by the Flemish decision. First, she emphasizes that the parallel discussions of the alignment of the HSL and the deepening programme was a Flemish proposal.[41] Secondly, she states that for the Dutch it is most important that the Walloon region has signed the conventions, because the Dutch use Meuse water for drinking water production.[42]

On 26 April 1994, France, the Walloon and Brussels Capital Regions, and the Netherlands sign the multilateral Conventions on the protection of the Meuse and the Scheldt in Charleville-Mézières (France).[43] Germany, Luxemburg and the Federal state of Belgium are observers to these agreements.[44] It takes another round of Flemish-Dutch deliberations until the Dutch and Flemish governments reach an agreement, and the Flemish region is willing to sign the multilateral conventions.

The POM-initiative and the Walloon-Dutch relationship
In 1993 seventeen Dutch municipalities situated along the river Meuse, among which the city of Rotterdam, and two drinking water production firms in the Netherlands start the Project Research Meuse (POM[45]). This project is very similar to the Project Research Rhine (POR[46]), which had been started in the Rhine basin.[47] In both projects, lower level governments and NGOs affected by river pollution contact the main upstream polluters, and try to reach an agreement on a reduction of waste discharges, which has the form of a covenant. The Dutch parties, however, carry a big stick, and state that if the polluters do not want to cooperate on pollution reduction, they will institute legal proceedings against them.[48] The Dutch Minister of Transport, Public Works, and Water Management, Maij-Weggen, announces that the Dutch national government will not institute legal proceedings against Walloon firms, because such legal actions could obstruct the negotiations with the Walloon region. Nevertheless, she expresses her sympathy with the POM-initiative.[49] In the Walloon region, however, the POM-initiative creates bad blood, and is not accepted.[50] A Dutch negotiator

---

40. According to the Royal Commissioner of the Province of Zeeland, van Gelder, the Flemish ministers cancelled the planned meeting for the signing of this convention, because they expected severe critics of the Antwerp region if they would sign this convention, as long as the Convention on the deepening of the navigation channel in the Western Scheldt would not have been signed by the Dutch government (PZC, 26-3-1994).

41. NRC, 27-4-1994.

42. Ibid.

43. Trb. 1994 No. 149 and No. 150.

44. Joint communiqué of the Netherlands, France, the Walloon region, and the Brussels Capital region of 26 April 1994; TK 1993-1994, 23 075 No. 4.

45. POM = Project Onderzoek Maas.

46. POR = Project Onderzoek Rijn.

47. For more information on the POR initiative in the Rhine basin, see Dieperink (1997, pp. 265-267).

48. NRC, 1-12-1993a.

49. NRC, 1-12-1993b.

50. The Director of the port of Liège, Planchar, in: NRC, 30-11-1993.

stated: "*When you use the word POM in Wallonia, feelings start running high. The project is perceived as a violation of the Walloon sovereignty.*"[51]

Beside the POM-initiative, also the Walloon-Dutch discussions on some cases of accidental pollution in the Meuse basin characterize the Walloon-Dutch relationship. After some cases of accidental pollution in the river Meuse, the Dutch Minister Maij-Weggen complains once more about the poor water quality of the river Meuse. At that moment the Walloon minister of Environmental Affairs, Lutgen, has enough of the negative publicity made by the Dutch, which is perceived as a smear campaign against Wallonia, and reacts uncommonly hard. He writes a letter to the Dutch Prime Minister, Lubbers, in which he complains about the Dutch way of acting. Furthermore, he states in interviews that[52]:

- the Dutch Minister Maij-Weggen gives incorrect information on the water quality of the river Meuse, because the river Meuse is not heavily polluted;
- the Dutch seem to have forgotten that the Dutch discharge of phosphates in the river Meuse is more than the discharge of France and Belgium together;
- the drinking water production firm in Antwerp uses Meuse water (extracted from the Albertkanaal) as well, but, unlike the Dutch drinking water firms, never does complain about the water quality of the Meuse;
- the Walloon government takes the decisions on the content and conditions of permits for waste water discharges in the Walloon region, and the Dutch government should stay out of Walloon policy making (this is a reaction on the Dutch POM-initiative);
- the Walloon region intends to comply with EC-policies, and makes efforts to clean-up the river Meuse for the benefit of the citizens of Rotterdam, who need the Meuse water to produce drinking water. Minister Lutgen adds that he does not even take into account that the Dutch have severely polluted their groundwater resources with nitrates, and that the Rhine water is too polluted for drinking water production.[53]

Continued external pressure on the governments of the Scheldt and Meuse basin states, and regional and local initiatives
On 12 June 1993 the Interparliamentary Council of the Benelux (IPR) formulates another recommendation concerning the Meuse and the Scheldt, which states that the IPR is pleased with the Dutch decision to unlink the issues of the deepening of the navigation channel in the Western Scheldt and the improvement of the water quality of the Scheldt and the Meuse, and that at long last the Dutch government has complied with the frequently expressed request of the IPR to unlink these issues.[54] Furthermore, the recommendation states that the deepening programme should be implemented as soon as possible, and that the Dutch decision to unlink the bilateral and multilateral issues has made it possible to implement the IPR Scheldt and Meuse resolutions of 22 November 1990 and 29 November 1991 (See Section 6.2, Rounds VII and VIII). The IPR forcibly recommends to start negotiations between the governments of all Scheldt and Meuse basin states on the installation of Scheldt

---

51. Interview C. de Villeneuve.

52. NRC, 30-11-1993 and 3-12-1993.

53. The Dutch drinking water production firm 'Brabantse Biesbosch' indeed does prefer the use of Meuse water for drinking water production, mainly because Rhine water does contain too high salt concentrations to use this water for drinking water production (See also NRC, 2-12-1993).

54. IPR recommendation concerning the Meuse and the Scheldt of 12 June 1993.

and Meuse commissions, and the formulation of Scheldt and Meuse Action Programmes.[55] A final recommendation is that the governments should prepare joint proposals for the solution of the financial problems related to an improvement of the water quality, according to the principles of the UN-ECE Convention.[56] This part of the recommendation suggests that the solution of the financial problems related to the clean-up of the rivers would be a joint responsibility of the basin states, and seems to refer to the idea of a sanitation fund, which was discussed before (See Section 6.2, Round IX).[57]

The three Dutch southern provinces, Limburg, Northern-Brabant, and Zeeland do explicitly plead for the establishment of a sanitation fund.[58] In addition, the three cooperating provinces would like to be involved in the negotiations on the Scheldt and the Meuse.[59] The Dutch Minister of Transport, Public Works, and Water Management replies that she will do her best in the international negotiations to create possibilities to involve the provinces in the international river commissions to be installed, but that there is no good reason to involve the provinces in the international negotiations preceding the installation of these commissions.[60] Furthermore, she opposes the idea of a sanitation fund for the same reason as discussed before (See Section 6.2, Round IX, and Section 6.3.1).[61]

The Task group for the Western Scheldt tries to contact municipalities and water boards in Flanders and Wallonia.[62] One of the objectives is to support Belgian municipalities developing environmental policies. The bigger municipalities, such as Antwerp and Brussels, however, do not react to the Dutch proposal to cooperate internationally.[63]

Beside the municipalities also environmental NGOs establish transboundary contacts. In 1993 environmental NGOs of all basin states and regions organized a study week, and in September 1994 the project 'Scheldt without frontiers' is started formally.[64] In this project the environmental NGOs intend to coordinate their actions for a clean-up of the river Scheldt, and to restore nature in its basin.[65]

---

55. Ibid.

56. Ibid.

57. One may argue that this recommendation is not internally consistent, because the 'polluter pays principle' is a main principle of the UN-ECE Convention, and the establishment of a sanitation fund is not in accordance with that principle.

58. Letter from the Provincial Executives of Zeeland, Northern-Brabant, and Limburg to the Dutch Minister of Transport, Public Works, and Water Management of 22 June 1993. Representatives of the provinces had discussed the idea of a sanitation fund at the symposium "Water zonder grenzen, Naar een pan-Europees waterbeleid." (Provincie Noord-Brabant, 1993).

59. Ibid.

60. Letter from the Dutch Minister of Transport, Public Works, and Water Management to the Provincial Executives of Zeeland, Northern-Brabant and Limburg of 8 October 1993.

61. Ibid.

62. PZC, 12-6-1993.

63. PZC, 4-8-1993.

64. Grenzeloze Schelde, Escaut sans Frontières.

65. During the action 'Save the Scheldt' (See Section 6.2, Round VI), Flemish and Dutch environmental NGOs had already organized joint actions. In the project 'Scheldt without frontiers' the first structural contacts with Walloon, Brussels and French environmental NGOs are established. The cooperating NGOs, among other things, organize projects aiming at nature education, produce a joint report describing the water quality in the Scheldt basin and the organization of the water quality management in the Scheldt basin states and regions

International research and information exchange
Whilst the multilateral negotiations on the Scheldt and Meuse conventions are started, the delegations in the ISG are finishing a progress report describing the water quality of the Scheldt, and the organization of water management in the Scheldt basin states and regions (See Section 6.2, Round IX).[66] According to the chairman of the ISG, the organizers of the ISG-meetings faced difficulties to convene the participants, when the formal multilateral negotiations on the water conventions had been started. *"At the end of the ISG one experienced the influence of the negotiations. Then it became more difficult. Delegations of the ISG were not allowed by their Ministers to visit the meetings of the ISG anymore. The politicians put pressure on their own civil servants. That was the case in all countries, in the Netherlands, in France, in Brussels, everywhere. And then we had to phone continuously in order to ask them to come."*[67] Although the exchange of information and joint research in the ISG is not a smooth process, it is the first time that water managers of all basin states meet on a regular basis. The former chairman of the ISG stated: *"The same work could have been done later, but we did not know that the Scheldt convention would be concluded. Anyway, the ISG has improved the personal relations between the people."*[68] The ISG also contributes to the establishment or improvement of the bilateral contacts between the Scheldt basin states and regions. According to some participants of the ISG, Flemish and Walloon water managers met for the first time at the ISG-meeting in Namur. *"Flemish and Walloon civil servants, from the same country, for the first time discussed with each other in the ISG."*[69] In addition, the Dutch-French contacts are intensified.[70] The ISG-project is concluded with a progress report describing the water quality in the Scheldt basin, and the organization of water management in the basin states.[71]

The ISG turns out to be a fruitful platform for joint initiatives. First, two organizations participating in the ISG, AMINAL-IN and RWS-Zld, take the initiative to organize a two-day ecological workshop in Antwerp (ISG-DES).[72] Secondly, at the end of 1993 Flemish and Dutch water managers express the intention to organize a basin-wide, multilateral third

---

(Rooy, de *et al.*, 1993), and an Action Programme. For more in formation on this project, see De Rooy and Wijffels (1993).

66. Draft texts of this progress report are discussed at the sixth meeting of the ISG, on 17 December 1993 in Middelburg.

67. Interview C. Heip.

68. Ibid.

69. Ibid.. Interview L. Santbergen.

70. RWS-Zld, RWS-DGW and the Agence de l'Eau Artois-Picardie organized several bilateral meetings. A first meeting already took place in 1991. Other bilateral French-Dutch meetings took place in 1992, on 15 July 1993, and on 17 June 1994.

71. ISG (1994a,b). In addition, an informative pamphlet is issued (ISG, 1994c).

72. International Study Group/ Description of the Ecology of the Scheldt; ISG/DWS/10; This workshop is organized on 21 and 22 June 1994 in Antwerp; The Walloon region is not able to participate in this workshop, mainly because it has a lack of personnel capacity (Minutes of meeting between representatives of DGRNE, University of Liège, RIKZ and RWS-Zld on the exchange of data between the LIFE 2 project and the Scheldt GIS on 6 July 1994.

Scheldt symposium.[73] Thirdly, some ISG-members want to apply for an EC-LIFE-subsidy for the development of a basin-wide Decision Support System (DSS) for the water quality management in the Scheldt basin.[74] Eventually, because experts do have different preferences concerning the models to be used, two instead of one DSSs are developed: a first one for the estuarine part of the basin and the coastal zone (LIFE I)[75], and a second one for the river Scheldt till the Flemish-Dutch border (LIFE II).[76] Both models should become a tool for decision making on international water quality policies, for example in the International Commission for the Protection of the Scheldt. One of the main objectives of both projects is to enhance cooperation between the basin states.[77]

---

73. VMM, IN, RWS-Zld, and RIKZ; Proposal for the organization of a third Scheldt symposium of 9 November 1993. At the sixth meeting of the ISG this proposal is discussed (Minutes of the sixth meeting of the ISG, ISG/50). At this meeting it appears that France and the Brussels Capital region do have two main objections to the proposal. First, they are of the opinion that there is too much emphasis on the water *policy* in the proposal, whilst at the symposium only the scientific aspects of the management of the Scheldt basin should be discussed (Ibid.; At this meeting no representatives of the Walloon region are present). Secondly, they do not want to organize a Scheldt symposium under the flag of the ISG, because the ISG has no formal basis. Therefore, the organization of the third Scheldt symposium is postponed until the Convention on the protection of the Scheldt is signed.

74. In first instance, the VMM would take the initiative to set up a working group that should prepare the application for an EC-LIFE subsidy for the development of a Scheldt-DSS. However, because the Dutch have more experience with the application for EC-subsidies, the VMM asks Rijkswaterstaat to prepare the application (Interview M. Bruyneel). Thereupon, the Dutch RIKZ and the research institute Delft Hydraulics develop a plan to extend the Dutch water quality model SAWES to the Scheldt basin. After the Dutch parties have finished their project proposal, a meeting with representatives of all basin states and regions is organized to discuss this proposal. At this meeting it turns out that the Walloon government agency, the DGRNE, is interested in the development of a DSS for the water quality of the Scheldt basin as well, but prefers to use the PEGASE model, which is developed by the University of Liège (Interview M. Bruyneel), and is used by the DGRNE for some time. The Dutch, however, do not want to work with this model for technical reasons, mainly because the model would be too detailed, and does not take into account parameters, such as heavy metals (Interview F. de Bruijckere). Because the DGRNE refuses to work with the SAWES model, and the RIKZ refuses to work with the PEGASE model, the development of one joint basin-wide DSS turns out to be impossible. The VMM does not want to pass the opportunity to intensify its contacts with the Walloon region (Ibid.), and states that it is willing to cooperate with Wallonia on the further development of the PEGASE model, and with the RIKZ on the further development of the SAWES model. The Dutch, however, decide to demand the Belgian Federal Ministry of Public Health and the Environment to cooperate on the development of a DSS for the Scheldt (Interview M. Bruyneel).

75. Participants in this project are the Belgian Federal Ministry of Public Health and the Environment, the Free University of Brussels, RWS-Zld, RIKZ, and the Dutch research institute Delft Hydraulics. Main contractor of this project is RIKZ.

76. Participants in this project are the Agence de l'Eau Artois-Picardie, the DGRNE, the VMM, and the universities of Liège, Namur and Brussels. Main contractor of this project is the VMM.

77. Agence de l'Eau Artois-Picardie (France), Ministère de la Région wallone (Belgium), Vlaamse Milieu Maatschappij (Belgium) (1993), Proposal LIFE-programme, Development of a computer decision support framework for the assessment of reduction of specific waste water discharge in the river Scheldt basin; Beleidsondersteunend systeem van het waterkwaliteitsbeheer in het estuarium van de Schelde en de kustzone. Voortgangsverslag per 31 december 1994.

Round XI: A Flemish-Dutch agreement and the conclusion of the water conventions (1994-1995)

Toward a Flemish-Dutch agreement

The Flemish choice to re-establish the linkage between bilateral and multilateral issues of 21 April 1994, and the Dutch linkage between the deepening programme and the alignment of the HSL (See Round Xb) cause another impasse in the negotiations on the water conventions. On 1 July 1994 the Flemish and Dutch delegations reach an agreement on the final text of the Convention on the deepening of the navigation channel in the Western Scheldt.[78] The Dutch government, however, does not want to sign this convention immediately, because of the linkage mentioned above. Nevertheless, it allows the Flemish region to start investigations on the wrecks in the Western Scheldt that have to be recovered before the deepening programme can be implemented.[79]

Both Flemish and Dutch actors heap criticism on the Dutch decision not to sign the Convention on the deepening of the navigation channel in the Western Scheldt until Flanders and the Netherlands would have reached an agreement on the alignment of the HSL. The AGHA suggests once more that Belgium should start litigation at the International Court of Justice.[80] The Royal Commissioner of the Province of Northern-Brabant, Houben, states that a further delay of the Dutch approval is at the expense of the atmosphere between the Netherlands and Flanders, and impedes the interregional cooperation, for example between the Flemish region and the three southern Dutch provinces.[81] The steering group of the RSD,[82] the Chambers of Commerce of Antwerp and Rotterdam[83], and the chairman of the Dutch national port Council, Brokx[84], all do oppose the Dutch refusal to approve the deepening programme, since by that the Dutch government would hinder the development of cooperation between the ports in the Rhine-Scheldt-Delta, in particular between the ports of Antwerp and Rotterdam.

The Dutch national government, however, perceives a strong interest in decision making on the alignment of the HSL, and therefore, in spite of the severe criticism on its policy, does not renounce immediately the Flemish-Dutch agreement to deal with the HSL and the deepening issues simultaneously. Only about six months after the meeting in Charleville-Mézières, at a first meeting between the Dutch Prime Minister, Kok, and Minister of Foreign Affairs, Van Mierlo, and the Flemish Prime Minister, van den Brande, Flanders and the Netherlands reach an agreement on the procedures that will be followed in decision making on the alignment of the HSL.[85] From that moment, the Dutch government is willing to approve the Convention on the deepening of the navigation channel in the Western Scheldt,

---

78. Regional Directorate Zeeland of Rijkswaterstaat, Memo RVO-94.081 of 18 November 1994.

79. PZC, 26-4-1994.

80. PZC, 5-7-1994.

81. PZC, 27-8-1994.

82. PZC, 1-6-1994.

83. Cobouw, 3-10-1994.

84. PZC, 30-9-1994.

85. This meeting takes place on 25 October 1994. Letter from the Dutch Minister of Transport, Public Works and Water Management, Jorritsma, to the chairman of the Commission for Transport and Water Management of the Dutch parliament of 20 January 1995.

and Flanders and the Netherlands are able to reach an agreement on the water conventions. On 1 December 1994, delegations of the Flemish and Dutch governments declare that they will sign the[86]:
1. Convention on the deepening of the navigation channel in the Western Scheldt;
2. Convention on the flow of the river Meuse;
3. Convention on the revision of the Scheldt regulations.
In addition, the Flemish region will sign the[87]:
4. Convention on the protection of the Scheldt;
5. Convention on the protection of the Meuse.[88]
Furthermore, the parties agree on a detailed planning of the joint decision making on the alignment of the HSL Antwerp-Rotterdam.[89]

Preparation of a new WVO permit for the Flemish region, and objections to the implementation of the 48'/43'/38' deepening programme
Since 1985, the Belgian respectively Flemish governments had to apply for a WVO Permit for dumping dredged material in the Western Scheldt. The second WVO permit, which was granted to the Flemish region on 28 November 1991, expires on 1 January 1995 (See Section 6.2, Round VIII). On 29 April 1994, the Flemish government applies for another permit for dumping dredged material with a duration of six years.[90] Because in 1994 the negotiations on the Convention on the deepening of the navigation channel in the Western Scheldt are expected to be concluded successfully, the Dutch Ministry of Transport, Public Works and Water Management anticipates on the conclusion of this convention. In the working group for the WVO Permit of the TSC, Flemish and Dutch experts jointly prepare a permit for the period from 1 January 1995 to 31 December 2000, which takes into account the possibility of the implementation of the 48'/43'/38' deepening programme. Implementation of this programme would make it necessary to increase the intensity of the dredging works, and the dumping of dredged material in the Western Scheldt. In the short run, this increase is caused by non-recurring infrastructure dredging works. In the long run, however, the amount of dredged material to be dumped in the Western Scheldt will increase permanently, because of the increased intensity of the maintenance dredging works. Like the preceding WVO permits, the new permit also contains conditions aiming at an improvement of the water quality in the Western Scheldt. According to the new permit, in the period between 1995 and 1997 the licensee would have to extract 300,000 tons of dry mud from the Lower Sea Scheldt

---

86. TK 1994-1995, 22 026, 23 075 No. 10; Dutch-Flemish communiqué of 1 December 1994 concerning the water conventions and the HSL Antwerp-Rotterdam.

87. Ibid.

88. France, the Walloon and Brussels Capital regions, and the Netherlands have already signed these last two conventions (See Round Xb).

89. Beside the agreements on the conclusion of the water conventions and the alignment of the HSL, the Flemish and Dutch government declare that they will sign a Convention on cultural cooperation between the Netherlands and Flanders, and a joint declaration concerning (Dutch-Flemish communiqué of 1 December 1994 concerning the water conventions and the HSL Antwerp-Rotterdam):
• the Flemish-Dutch cooperation on environmental protection, land-use planning, public works, transport, economy and employment, welfare, technology, and scientific research;
• transboundary cooperation on the provincial and municipal level;
• Flemish-Dutch cooperation in the Benelux and the EC.

90. Draft WVO Permit for the Flemish region of 1994.

yearly. Like in the preceding permit, priority would be given to the extraction of polluted sediments from the access channel to the Kalosluice. Furthermore, the draft permit states that in 1997, an assessment will be made whether the amount of sediment to be extracted yearly should be adjusted. Compared with the previous permits, the new permit contains several new conditions as well. Among other things, it states that the Flemish government has to provide information to the Head-Engineer-Director of the regional Directorate Zeeland of Rijkswaterstaat on siltation rates and changes of depth in the entrance channels, current amounts of mud in the Lower Sea Scheldt, and the ratio of marine to fluvial mud in the Lower Sea Scheldt, including the docks, in 1998 and 2000.[91]

In the Netherlands, several lower level governments and NGOs oppose the content of the draft WVO permit for the Flemish region. The Provincial Executive of Zeeland criticizes the anticipation on the implementation of the 48'/43'/38' deepening programme, because it is of the opinion that decision making on the deepening programme requires an EIA-study.[92] After the Dutch delegation has announced that the Flemish region will contribute 44,000,000 guilders to projects for nature compensation, the Province drops its objections to the WVO-permit.[93] The municipality of Flushing proposes to formulate additional conditions in the WVO permit, and to stimulate the Flemish region to reduce the transboundary load of pollutants that flow into the Flemish region from Wallonia, Brussels, and France.[94] The Zeeland Environment Federation (ZMF) opposes the draft WVO permit for various reasons.[95] First, it alleges the Flemish region not to have complied with the conditions formulated in the preceding permit, i.e. the water quality of the river Scheldt has not improved significantly. Secondly, according to the ZMF the Flemish region would not live up to the agreements that have been made at the North Sea Ministerial Conferences. Thirdly, the ZMF also criticizes the anticipation on the deepening programme, because an EIA would have to be finished first.

Perhaps more surprising, also in the Flemish region some parties oppose the implementation of the deepening programme. Flemish environmental NGOs doubt whether the deepening programme is cost-effective, and argue that only a redistribution of tasks between the ports of Antwerp and Zeebrugge is a real solution of the problems of the maritime access to the port of Antwerp.[96] The Flemish environmental NGO BBL makes public an actualized secret cost-benefit analysis of the Flemish government of 1986, which casts a doubt on the cost-effectiveness of the deepening programme.[97] The Flemish

---

91. Draft WVO Permit 1994, See also Verlaan *et al.* (1997, p. 256).

92. PZC, 21-9-1994; In 1992, the Dutch government had decided that according to international (EC) law an EIA is not obligatory for the implementation of the deepening programme. Furthermore, it was argued that there are no other good reasons to start a study on the impacts of the deepening programme, because such as study had already been issued by the TSC in 1984, and the 'Raad van de Waterstaat' had formulated an advice concerning the deepening programme (See Section 6.2, Round IX).

93. PZC, 21-12-1994.

94. Letter from the municipality of Flushing to the Regional Directorate Zeeland of Rijkswaterstaat of 29 June 1994.

95. On 7 September 1994, representatives of the Ministry of Transport, Public Works, and Water Management, the licensee, and the ZMF discuss the content and conditions of the draft WVO permit (Report meeting on 8 September 1994).

96. PZC, 7-9-1994.

97. De Volkskrant, 29-11-1994; Haagse Courant, 29-11-1994.

government, however, reacts immediately, and states that the report of the BBL is dated, does contain many weaknesses, and does not change the Flemish intention to sign the water conventions as soon as possible.[98] In spite of the criticism, on 4 November 1994 the Dutch Minister of Transport, Public Works, and Water Management issues the new WVO permit for the Flemish region, which is valid until 31 December 2000.

Signing of the water conventions
Following the agreements made on 1 December 1994, on 11 January 1995, the Flemish Region and the Netherlands sign the Convention on the revision of the Scheldt regulations in Middelburg.[99] A few days later, on 17 January 1995, they sign the Convention on the deepening of the navigation channel in the Western Scheldt, and the Convention on the flow of the river Meuse, and the Flemish Region signs the Conventions on the protection of the Scheldt and the Meuse in Antwerp.[100] At this memorable day, 27 years after Belgium and the Netherlands have reached an agreement to start negotiations on, among other issues, the water quality of the Scheldt and the Meuse and the water quantity of the Meuse, and 10 years after they have agreed to start negotiations on, among other issues, the 48'/43'/38' deepening programme, the negotiations on the water conventions are concluded. These conventions are the first important international conventions signed by the Flemish region. Later the Dutch Minister would say in the Dutch parliament: "*To be honest, I do not want to hear the word 'linkage' anymore. My Belgian colleagues got enough of it as well. Where two are fighting, two are guilty. In the past it mutually happened each time and subsequently lead nowhere.*"[101], and: "*Suspicion concerning all these topics that existed the last 20 to 25 years, and that in fact had begun when the Netherlands and Belgium became independent states, should be stopped at last.*"[102]
In the next sections, the contents of the water conventions are discussed.[103]

Convention on the protection of the Scheldt[104]
Contracting parties to the Convention on the protection of the Scheldt are the Netherlands, France, the Flemish region, the Brussels Capital region, and the Walloon region. Whilst the three Belgian regions are contracting parties, the State of Belgium is not. The Convention

---

98. Ministerie van de Vlaamse Gemeenschap, Administratie Waterinfrastruktuur en Zeewezen, Reactie op de persmededeling van de Bond Beter Leefmilieu, 29 November 1994.

99. TK 1994-1995, 23 075 No. 5.

100. TK 1994-1995, 24 041 No. 3. At this day environmental NGOs declare that the Conventions on the protection of the Scheldt and the Meuse do not guarantee a clean-up of these rivers, because they do not contain clear policy objectives, and terms (PZC, 17-1-1995). Furthermore, they demand the right to participate in the commissions, or to become observer at least (Ibid.). Finally, the environmental NGOs repeat their request for an EIA-study on the 48'/43'/38' deepening programme, and express the need of 'ontpolderen' to compensate nature losses caused by the implementation of this programme (Gazet van Antwerpen, 18-1-1995).

101. TK 97-5930.

102. TK 97-5939.

103. Readers who are interested in the decision making process, but are less interested in a detailed description of the contents of the water conventions may skip the next sections, and continue to read from Round XIIa.

104. For a discussion on the legal aspects of the Conventions on the protection of the Scheldt and the Meuse, see Maes (1996).

on the protection of the Scheldt is a so-called framework convention. It contains the main principles of international cooperation on the protection of the Scheldt, and provides for the installation of an international basin commission. Unlike the Convention on the improvement of the navigation channel in the Scheldt near Bath the Belgian and Dutch delegations had reached an agreement on in 1975 (See Section 5.2, Round II), the Convention on the protection of the Scheldt does not contain water quality objectives.

The Scheldt is defined from the source to the mouth in the sea, including the Sea Scheldt and the Western Scheldt.[105] The Scheldt basin is defined as the area draining into the Scheldt or her tributaries.[106] The Convention contains the following objective: "*In the spirit of the UN-ECE Convention on the protection and use of transboundary water courses and international lakes, the Contracting Parties cooperate, taking into account their common interests and the specific interests of each of them, in a spirit of good neighbourship, to maintain and improve the quality of the Scheldt.*"[107] To reach this objective the parties take measures in the part of the Scheldt basin situated on their territories.[108] The contracting parties install an International Commission on the Protection of the Scheldt against pollution (ICPS).[109]

Principles of international cooperation included in the Convention are the precautionary principle, the prevention principle, the polluter pays principle, and the principle of pollution abatement at the source.[110]

The Convention also states that the contracting parties will act in a similar way in the whole basin to prevent distortion of competition.[111] In addition, each contracting party will make an effort to realize integrated management of the Scheldt[112], and the contracting parties will start deliberations to ensure the conditions for a sustainable development of the Scheldt and its basin.[113] Furthermore, they will protect, and where possible improve, the quality of the aquatic ecosystem of the Scheldt.[114] One article explicitly deals with the quality of the water bed, and states that the contracting parties inform each other on their policies concerning the management of the sediments of the Scheldt, and will coordinate these policies according to their needs.[115] In addition, this article states that the dumping and downstream displacement of polluted dredged material will be restricted as much as possible.[116] The articles 5 to 8 of the Convention deal with the installation, organization, and tasks of the ICPS. The main task of the ICPS is to prepare a Scheldt Action Programme

---

105. Verdrag inzake de bescherming van de Schelde, Artikel 1 sub a.

106. Ibid., Artikel 1 sub c.

107. Ibid., Artikel 2.1.

108. Ibid., Artikel 3.1.

109. Ibid., Artikel 2.2.

110. Ibid., Artikel 3.2.

111. Ibid., Artikel 3.3.

112. Ibid., Artikel 3.4.

113. Ibid., Artikel 3.5.

114. Ibid., Artikel 3.6.

115. Ibid., Artikel 4.1.

116. Ibid., Artikel 4.2.

(SAP) with policy objectives and policy measures to reach these objectives.[117] These policies should be directed at point sources and non-point sources of pollution, to improve the water quality, and to preserve and improve the ecosystem.[118] Other tasks of the commission are to[119]:

- collect and assess data to detect the main sources of pollution;
- coordinate monitoring programmes;
- inventory and stimulate the exchange of data on the main sources of pollution;
- assess the effectiveness of the action programmes;
- exchange information on water policies of the contracting parties;
- exchange information on projects for which an impact assessment is obligatory, and which have a significant transboundary impact on the quality of the Scheldt;
- exchange information on best available technologies;
- stimulate international cooperation on research;
- be a framework for discussions concerning actions to be taken in transboundary tributaries and canals of the river system of the Scheldt;
- advise contracting parties on the cooperation provided for in the Convention;
- coordinate the different national and regional warning and alarm systems for cases of accidental pollution;
- cooperate with other international commissions which have similar tasks for adjacent water systems;
- issue a public annual report on its activities;
- deal with other issues within the area of application of the Convention, as requested by the contracting parties in mutual agreement.

The ICPS consists of delegations of the contracting parties.[120] Each contracting party appoints at most eight delegates, among which the leader of the delegation.[121] The ICPS has a rotating chair for a period of two years.[122] The ICPS meets once a year or more often at the request of at least two delegations.[123] Meetings can be held at the ministerial level.[124] The Commission installs working groups to assist the performance of its tasks according to its needs.[125] The decision making rule in the ICPS is unanimity.[126] French and Dutch are the official working languages of the ICPS.[127] The permanent secretariat to assist the commission has its seat in Antwerp.[128] Observer to the ICPS can be any state that is not a contracting

---

117. Ibid., Artikel 5 sub d.

118. Ibid.

119. Ibid., Artikel 5.

120. Ibid., Artikel 6.1.

121. Ibid.

122. Ibid., Artikel 6.2.

123. Ibid., Artikel 6.3.

124. Ibid.

125. Ibid.

126. Ibid., Artikel 6.4; If one of the delegations withholds its vote, this does not stand in the way of unanimity.

127. Ibid., Artikel 6.5.

128. Ibid., Artikel 6.6.

party, and whose territory is situated in the basin of the Scheldt.[129] This logically can only be the Federal state of Belgium. The European Community can become an observer as well.[130] Observers are allowed to participate at the meetings of the ICPS, but do not have voting right.[131] Each contracting party bears the costs made by its delegation in the ICPS and its working groups.[132] The division of the other costs made by the ICPS, among which the costs of the Secretariat, is more or less based on the surface area and the number of inhabitants of the basin in the basin states.[133]

Convention on the deepening of the navigation channel in the Western Scheldt
Contracting parties to the Convention on the deepening of the navigation channel in the Western Scheldt are the Flemish region and the Netherlands. The Convention provides for the implementation of the 48'/43'/38' deepening programme for the navigation channel in the Western Scheldt.[134] Because the Contracting Parties regulate 'de commun accord' aspects of the navigation on the river Scheldt, some Fleming interpret the Convention as an implementation of the Scheldt Statute.[135] The preamble of the convention states that other conventions concerning the Scheldt remain in full force, which implies that the convention does not alter the Scheldt statute. In addition, it states that the maintenance and further development of the navigation function of the Western Scheldt should take place in harmony with the other functions of the Western Scheldt, among which the nature function. According to a Belgian negotiator, the convention marks a temporary end to the Belgian-Dutch disagreement on the interpretation of the Scheldt statute.[136] The implementation of the 48'/43'/38' deepening programme includes the following works[137]:
1. the salvage of wrecks and other obstacles in the navigation channel and at the anchorages;
2. the local protection of the banks along the Western Scheldt;
3. works for nature restoration;
4. the local deepening of the navigation channel, and possibly a local deepening or displacement of the anchorages.
The technical details of these works are described in the Appendices A to C of the Convention. Article 3 of the Convention regulates the division of tasks. The Dutch government is responsible for the works listed (1) to (3)[138], whilst the Flemish government

---

129. Ibid., Artikel 7.1 sub a.

130. Ibid., Artikel 7.1 sub b.

131. Ibid., Artikel 7.3.

132. Ibid., Artikel 8.1.

133. TK 1994-1995, 24 041 No. 3; Verdrag inzake de bescherming van de Schelde, Artikel 8.2: The Netherlands: 10%; France: 30%; the Walloon Region: 10%; the Flemish Region: 40%; the Brussels Capital Region: 10%.

134. Verdrag tussen het Vlaams Gewest en het Koninkrijk der Nederlanden inzake de verruiming van de vaarweg in de Westerschelde, Artikel 2.1 (Trb. 1995 Nr. 51).

135. See, for example, Strubbe (1995, p. 235). See also the historic overview presented in Section 4.8.

136. Ibid.

137. Verdrag tussen het Vlaams Gewest en het Koninkrijk der Nederlanden inzake de verruiming van de vaarweg in de Westerschelde, Artikel 2.

138. Ibid., Artikel 3.1.

is responsible for the implementation of the infrastructure dredging works, listed (4).[139] The contracting party, which is responsible for the implementation of the works, has to apply for the necessary permits as well.[140] Article 5 regulates the division of costs. The Dutch have to pay for the salvage of wrecks and the local protection of the banks up to fl 54,000,000,-, the costs of nature restoration as far as these costs exceed fl 44,000,000, -, and the costs of monitoring the impacts of the implementation of the deepening programme.[141] The Flemish region has to pay all other costs. Reasons for the Dutch contribution to the costs of the deepening of the navigation channel are that the Dutch have a small economic interest in the implementation of the deepening programme, since these works do also improve the maritime access to the Dutch harbours of Flushing and Breskens, and that the Netherlands receive an income from VAT. Finally, the Dutch pay to be neighbourly.[142]

Beside the implementation of the 48'/43'/38' deepening programme, the convention provides for a new statute for the TSC.[143] The TSC monitors the preparation, implementation and maintenance of the deepening dredging works provided for in the Convention, and the impacts of these works on the water system of the Western Scheldt.[144] Article 8 of the Convention states that contracting parties will jointly study other projects related to the development of the navigation function of the Western Scheldt, among which the second maritime access to the Waasland harbour (Baalhoek Canal), and a possible further deepening of the navigation channel in the Western Scheldt, the so-called second stage of the deepening programme.[145] It also states that these studies will take into account the relevant Flemish and Dutch Acts and regulations, including those concerning EIA.[146] Finally, this Article makes it possible to implement the second stage of the deepening programme without the conclusion of a new convention, if Flanders and the Netherlands would reach an agreement on the implementation of this programme.[147] Unlike the Convention on the protection of the Scheldt, the Convention on the deepening of the navigation channel in the Western Scheldt contains provisions for dispute settlement.[148]

Other conventions
Since they are less relevant for the management of the river Scheldt, the other conventions are only shortly discussed. The Convention on the protection of the Meuse is almost equal to the Convention on the protection of the Scheldt. Most striking of the Meuse Convention is that the Brussels Capital Region is contracting party, although it is not situated in the

---

139. Ibid., Artikel 3.2.

140. Ibid., Artikel 3.3.

141. Ibid., Artikel 5.

142. TK 1994-1995, 24 041 No. 5.

143. Verdrag tussen het Vlaams Gewest en het Koninkrijk der Nederlanden inzake de verruiming van de vaarweg in de Westerschelde, Artikel 2; Till then the formal basis of the TSC was a Protocol of deliberations between ministers of Belgium, Luxemburg, and the Netherlands in Luxemburg on 29, 30, and 31 January 1948.

144. Verdrag tussen het Vlaams Gewest en het Koninkrijk der Nederlanden inzake de verruiming van de vaarweg in de Westerschelde, Artikel 2.1.

145. Ibid., Artikel 8.1.

146. Ibid., Artikel 8.2.

147. Ibid., Article 8.3.

148. Ibid., Artikel 9.

Meuse basin. The reason for this is that the Brussels Capital Region uses Meuse water for its drinking water production.[149] The region, however, has a special status. It only has voting right as regards decision making on issues related to the use of Meuse water for drinking water production or its financial contribution to the International Commission for the Protection of the Meuse (ICPM).[150] The Meuse basin states Luxemburg and Germany are no contracting parties, but observers to the convention.[151] Consequently, the contracting parties to the Scheldt and Meuse Conventions are the same.[152]

Contracting parties to the Convention on the flow of the river Meuse are the Flemish region and the Netherlands.[153] The Convention provides for an alternative to the 50 m³/s guarantee in Liège, and to the construction of storage reservoirs in the Walloon region (See Section 5.2, Round II). It regulates the Flemish and Dutch diversions of water from the river Meuse, the reduction of water losses in periods of low river flow, and the cooperation on research and development of the Border Meuse (*Grensmaas*), the part of the river Meuse that forms the border between Flanders and the Netherlands. Furthermore, the preamble and Article 8 state that the Convention provides for the compensation of water losses in the Kreekrak sluices (See Section 5.2, Round I). Point of departure of the regulation for water distribution in cases of low flow is that each party has a right on an equal amount of water from the river Meuse.[154] For the implementation of the measures the parties install a working group for the regulation of the flow of the river Meuse.[155] Later, the Walloon region would become observer in this working group.

Contracting parties to the Convention on the revision of the Scheldt regulations are Belgium, the Flemish region, and the Netherlands.[156] The Convention provides for a modification of the Scheldt regulations of 1839 and 1842, and contains new regulations concerning pilotage, and the joint supervision of the pilotage.

Round XIIa: Preparation of the first Scheldt Action Programme (1995-1997)

Continued negotiations in the provisional International Commission for the Protection of the Scheldt against pollution
Since the negotiations on the water conventions have been concluded and, among other conventions, the Conventions on the protection of the Scheldt and the Meuse have been signed, as regards the water quality of Scheldt and Meuse the eyes are focused on the International Commissions for the Protection of the Scheldt and Meuse against pollution

---

149. TK 1994-1995, 24 041 No. 3.

150. Verdrag inzake de bescherming van de Maas, Artikel 4.

151. TK 1994-1995, 24 041 No. 3.

152. For more information on the Convention on the protection of the Meuse, see Bouman (1996), Gosseries (1995), Tombeur (1995), and Smitz (1995).

153. Verdrag inzake de afvoer van het water van de Maas (Trb. 1995 Nr. 50).

154. Ibid., Artikel 3.1.

155. Ibid., Artikel 5; For more information on this convention see for example Bouman (1996) and Merckx (1995).

156. Verdrag tussen het Koninkrijk der Nederlanden, het Koninkrijk België en het Vlaams Gewest tot herziening van het Reglement ter uitvoering van artikel IX van het Tractaat van 19 april 1839 en van hoofdstuk II, afdelingen 1 en 2, van het Tractaat van 5 november 1842, zoals gewijzigd, voor wat betreft het loodswezen en het gemeenschappelijk toezicht daarop (Scheldereglement) (Trb. 1995, Nr. 48).

(ICPS and ICPM).[157] Anticipating ratification of the Convention on the protection of the Scheldt, on 11 May 1995 the contracting parties install a provisional ICPS in Antwerp.[158] Referring to the laborious negotiations on the water conventions and the provisional installation of river commissions for the Scheldt and the Meuse, the chairmen of the Flemish and Dutch delegations in the ICPS state: "*It was a lot of talking, but there were few results. Fortunately, however, time now is ready for the opposite: no words, but action!*"[159] Like the negotiations on the multilateral Conventions for the protection of the Scheldt and the Meuse, the two river Commissions are strongly related as well. The same states and regions are contracting parties to the Scheldt and the Meuse conventions, and many representatives of these basin states and regions are members of both Commissions. The Dutch government meets the wishes of Dutch lower level governments, and composes a broad delegation for the ICPS with representatives of the Province of Zeeland and the Task group for the Western Scheldt.[160] At the first meeting of the plenary commission, the Administrator-General of the VMM, Van Sevencoten, is nominated as chairman until the end of 1997.[161] The Frenchman Lefébure is appointed as Secretary-General of the ICPS.[162] Finally, at the first meeting of the provisional commission three working groups are installed[163]:

- a working group on the water quality (working group 1)[164];
- a working group on emissions (working group 2)[165];
- a temporary working group for transboundary cooperation and Joint Environmental Projects (JEPs) (working group 3).[166]

The main task of the working groups is to collect the necessary information for the first Scheldt Action Programme (SAP). Working group 1 has to make a description of the water quality of the Scheldt, and a coordinated inventory of physical, chemical and ecological measurements.[167] Working group 2 is responsible for an inventory of point sources and non point sources of pollution in the basin. This group also has to draft proposals for coordinated

---

157. Shortly after the conventions have been signed, the Dutch minister of Transport, Public Works and Water Management, Jorritsma, states that the completion of the Action Programmes for the Scheldt and the Meuse within a year is feasible (De Stem, 18-1-1995) Later, she makes a more realistic prognosis, and declares that the Action Programmes for the Scheldt and the Meuse will be finished mid 1997 (TK 1995, 97th meeting, 31-8-1995, p. 5931).

158. Press release of the ICPS, Antwerp, 11 May 1995.

159. Hoogland and Bruyneel (1995, p. 129).

160. PZC, 22-12-1994; De Stem, 11-1-1995.

161. Press release of the ICPS, Antwerp, 11 May 1995.

162. Until 1 January 1996 he would also act as provisional Secretary-General (SG) of the International Commission for the Protection of the Meuse (ICPM). At this date the Dutchman Zijlmans becomes the SG of the ICPM.

163. Press release of the ICPS, Antwerp, 11 May 1995.

164. Working group 1 is chaired by the Dutchman Saeijs (Rijkswaterstaat-Directie Zeeland).

165. Working group 2 is chaired by the Frenchman Grandmougin (Agence de l'Eau Artois-Picardie).

166. Working group 3 is chaired by the Walloon De Kerckhove (Ministère de la Région wallone, DGRNE) The JEPs were part of a programme of the EC stimulating cross-border cooperation on environmental issues. Because of a delay of the implementation of this programme, however, in the case study period no JEPs for the Scheldt basin would be developed.

167. Press release of the ICPS, Antwerp, 5 December 1995.

warning and alarm procedures for cases of accidental pollution.[168] Working group 3 has to make a proposal for the relation between existing transboundary platforms for cooperation and the ICPS.[169] In addition, it has to draft proposals for JEPs that could be financed by the EC.[170]

The parties involved in preparing the first SAP do have different levels of ambition. The French chairman of working group 2, Grandmougin, for example, states in an interview that the Scheldt can at the most become a clean sewer[171], whereas the Dutch Rijkswaterstaat is of the opinion that a restoration of the ecosystem of the estuary and the river Scheldt is possible.[172] According to several respondents, the economic problems in northwestern France influence the French perception of the international pollution issue.[173] "*The French business community states that it is willing to treat its waste water, but doubts whether this is useful, when downstream parties continue their waste discharges. Therefore, France pleads for similar policies in all basin states.*"[174] As in the French part of the basin, in the Walloon region the economic situation continues to be rather troublesome as well. Consequently, political priorities are with economic development rather than with environmental policies.[175] A Dutch negotiator stated: "*The most important problem in the Scheldt basin is a socio-economic one. The upstream parts of the Scheldt basin do not belong to the most prosperous parts of Europe.*"[176]

At the first meetings of the provisional ICPS it turns out that the contracting parties do interpret the Convention on the protection of the Scheldt differently.[177] The discussions on the functional and geographical scope, which took place in the multilateral negotiation commission (See Round Xb), are continued in the provisional river commission. The Flemish and Dutch delegations prefer to start cooperation on the water quality and quantity of both the surface water and the ground water of the Scheldt basin. The Walloon and Brussels delegations, however, repeat the standpoints they had formulated in the multilateral

---

168. Ibid.

169. Ibid.

170. Ibid.

171. Smit *et al.* (1995, p. 1).

172. Hendriksen (1995, p. 3).

173. Interviews C. de Villeneuve and R. Zijlmans. The French department of Nord-Pas de Calais, which is situated in the Scheldt basin, has about the highest unemployment rate of France.

174. Interview C. de Villeneuve.
Because of the organization of French water management, in particular the organization of the *Comités de Bassin*, the target groups, among which the industries, do have much influence on the formulation of water quality policies. Several respondents refer to the differences between the position of the French water agency Rhin-Meuse in the negotiations on a Meuse Action Programme, and the position of the water agency Artois-Picardie in the negotiations on the SAP. The water agency Rhin-Meuse would favour more ambitious international policies than Artois-Picardie, which may be explained by the relatively bad economic situation in the French part of the Scheldt basin, and the experiences of the water agency Rhin-Meuse in the International Commission for the Protection of the Rhine.

175. Interview M. Bruyneel.

176. Interview R. Zijlmans, L. Santbergen in (Anonymous, 1997a, p. 2).

177. See, for example, L. Santbergen in (Anonymous, 1997a, p. 2). The first plenary meetings of the ICPS take place on 11-5-1995 in Antwerp, on 5-12-1995 in Antwerp, on 16-4-1996 in Ghent, and on 18-6-1997 in Antwerp.

negotiations preceding the conclusion of the water conventions, and want to restrict the international cooperation to the water quality of the surface waters only. In addition to the functional scope, also the geographical scope remains a topic of discussion. France, the Netherlands and Flanders prefer a broad interpretation of the Scheldt convention, and would like to discuss the water quality in the whole basin, i.e. the river Scheldt and her tributaries. The Walloon and Brussels delegations want to restrict the formulation of joint water quality objectives to the main course of the river. France is prepared to discuss water quantity issues as well, if the geographical scope of the convention is extended to the region between the Scheldt basin and the Sea, thus accompanying the IJzer basin. The reason for this is the hydrological interlinkage by canals feeding Dunkerque with Scheldt waters. This still is open to discussion, but would be in accordance with the proposal for a EU Framework Directive.[178] The Dutch prefer to extend the geographical scope with the Grevelingen, The Eastern Scheldt, and the Veerse Meer. Although this Dutch proposal is not accepted, it might be accepted in the future, since it is in accordance with the EU Framework Directive as well.[179]

At a bilateral Walloon-Dutch meeting, which takes place on 4 March 1996, the parties reach an agreement on the interpretation of the Scheldt and Meuse conventions. The compromise is that the water quality and ecological objectives will be formulated for the main course of the river only, whilst the policy measures will be taken in the tributaries of the rivers as well. Information on the policies for the subbasins will be presented to the Commission, but the basin states remain responsible for the implementation of these policies themselves.[180]

Since in the ICPS no water quantity issues are addressed, in 1997 some basin states and regions intend to apply for an EC-Interreg-subsidy for a joint project in which the water quantity issues in the Scheldt basin should be addressed. Furthermore, the members of the ICPS expect that after the EC Framework Directive for Water is issued[181], the Scheldt and Meuse basin states and regions have to reopen negotiations on the Scheldt and Meuse conventions because these conventions do not address water quantity issues.

Apart from the discussions on the functional and geographical scope of the ICPS, i.e. on the interpretation of the Convention on the protection of the Scheldt, the parties discuss the list of parameters to be monitored. Each party has different traditions, and consequently different preferences concerning the parameters to be monitored. Although in the working

---

178. Commission proposal for a council directive establishing a framework for community action in the field of water policy. 26.02.1997.

179. Because hydrological criteria do not give a definite answer, the Dutch classification of the basins in the Dutch delta is based on pragmatic considerations. The Dutch preferred to assign all salt and brackish waters in the delta, among which the Eastern Scheldt, to the Scheldt basin, and all fresh waters to the Meuse basin. Two advantages of this would be that experts on the management of salt water systems would have to be represented in the ICPS and its working groups only, and that the province of Northern-Brabant would not have to be represented in both the river commissions for the Scheldt and the Meuse, but in the ICPM only (Information provided by R. Zijlmans).

180. These Walloon-Dutch contacts are continued with a meeting between the Dutch Ministers Jorritsma and Lutgen, which takes place on 2 October 1996. At this meeting, the ministers decide to organize regular meetings at high administrative level.

181. Commission proposal for a council directive establishing a framework for community action in the field of water policy. 26.02.1997.

group for the water quality, the Dutch delegation proposes a long list of parameters[182], the parties agree that, in first instance, the inventory of emissions will be limited to the five parameters that are recognized by all parties.[183] In these discussions the measurement of radioactivity is addressed as well. The Belgian regions do not want to take into account parameters of radio activity, because the federal state of Belgium, which is not a contracting party to the Scheldt and Meuse conventions, has the competencies in this field.[184]

Another topic of discussion is the relationship between the ICPS and the existing transboundary organizations or projects. The Dutch delegation opposes plans to integrate some Flemish-Dutch transboundary basin committees in the ICPS[185], because the scope of the transboundary basin committees, which includes water quality and quantity of both groundwater and surface water, is broader than the scope of the ICPS, whose activities are limited to the water quality and ecology of the main stream of the river Scheldt.

In June 1997 the first official report of the Scheldt basin states is issued.[186] This report contains a description of the water quality of the Scheldt in 1994, and will enable the contracting parties to monitor the impacts of the policies to be developed in the ICPS. On 20 November 1997 the contracting party France ratifies the Scheldt and Meuse conventions at last.[187] When this chapter was completed[188], it is expected that the ICPS will be installed formally on 9 March 1998.[189] The first SAP will contain policies for the short term (5 years), the medium term (15 years), and the long term (one generation). In the short term, the ICPS focuses on the coordination of the water quality policies already planned by the individual basin states and regions.[190]

Improving relations between the basin states and regions
In 1996 the judgements of representatives of the basin states and regions involved in decision making in the ICPS concerning the progress made in this commission show much similarity. They judge positively the decision making *process*. According to the respondents mutual relations are gradually improving, and the negotiation climate is rather good. After the meetings between the Walloon and Dutch delegations in the river commissions, and the bilateral Walloon-Dutch ministerial meeting, the Dutch-Walloon relationship is improving as well. In the ICPS and the ICPM the first stable contacts between water managers of Wallonia and the Netherlands are established. According to a Dutch negotiator: "*up to and including the negotiations on the Convention on the protection of the Scheldt, discussions with Belgium, in particular the Walloon region, took place on the subpolitical level of the ministerial cabinets, with the exception of the two informal meetings preceding the formal negotiations. We were never able to establish contacts with the administration, where the work was done.*

---

182. Interview J.-M. Wauthier.

183. Press release of the ICPS, Ghent, 16 April 1996.

184. Interview anonymous.

185. For more information on these commissions, see Appendix 6 describing the administrative organization of water management in the Scheldt basin.

186. ICBS/CIPE (1997).

187. Lefébure (1997, p. 8).

188. This chapter is completed in December 1997.

189. Lefébure (1997, p. 8).

190. Ibid.

This has changed after the provisional installation of the ICPS."[191] A Walloon member of the ICPS states that the Walloon region has gradually changed its policy, and is more willing to cooperate internationally now. He does, however, recommend a slow and smooth process, because otherwise the delegations inevitably have to cope with conflicts.[192] Another respondent states that: "*the Dutch position has changed gradually. In the past, the Dutch requested much and did not receive anything. By now they are more realistic, and possibly will get something.*"[193] Finally, the former chairman of the ISG states: "*I have to say that within a few years the relations have improved considerably. The time that I was working in the ISG and now are as different as night and day.*"[194] Although the relations between the Scheldt basin states and regions are gradually improving, they do not organize a multilateral third Scheldt symposium, as they had planned (See Round Xb). Even though the water conventions are concluded, the third Scheldt symposium, like the first and second ones, is a bilateral Flemish-Dutch meeting.[195]

Expectations concerning the content of the first SAP do not run high
Whilst the relations between the parties are improving, as regards the content of the first SAP the expectations of neither of the parties do run high. One of the members of the commission states. "*One experiences clearly that the parties hold on to their national policies. The first SAP will be based on the action programmes of the individual basin states.*"[196] A Walloon member of the ICPS states: "*I do not think that the ICPS will be able to speed up the construction of waste water treatment plants in the Scheldt basin, because the commission does not provide the necessary financial resources to construct these plants. The exchange of information in the ICPS, however, may, for example, enable the parties to build better*

---

191. Interview C. de Villeneuve.

192. Interview J.-M. Wauthier.

193. Interview anonymous.

194. Interview C. Heip.

195. On 17 February 1995, after the Flemish region has signed the multilateral Conventions on the Protection of the Scheldt and the Meuse, French, Dutch and Flemish water management agencies continue the discussions on the organization of a third Scheldt symposium (Minutes of a preparatory meeting of 17 February 1995). At this meeting representatives of the Agence de l'Eau Artois-Picardie propose to organize a Scheldt symposium in France, especially because they would like to use the symposium for the presentation of their new five year programme (Ibid.). In the beginning of 1995, when all problems related to the organization of a multilateral Scheldt symposium seem to be solved, the city of Antwerp launches the idea to organize a symposium on the water management of the Scheldt estuary. The organizing committee attempts to organize one joint symposium, but the city of Antwerp wants to organize the symposium in Antwerp and the Agence de l'Eau Artois-Picardie does want to organize the symposium in France. Therefore, it is decided to organize two Scheldt symposia, a first one for the estuary and a second one for the upstream part of the Scheldt basin. The first part of the third Scheldt symposium is organized by the city of Antwerp on 6 and 7 December 1995, and addresses the water management of the Scheldt estuary (Proceedings third international Scheldt Conference, Integrated Water management of the Scheldt estuary. Antwerp). The second part of the third symposium, which would be organized by the Agence l'Eau Artois-Picardie, is planned in 1996 (Final announcement of the third Scheldt symposium), but has not yet been organized (in December 1997). At the first part of the third Scheldt symposium the Dutch Royal Commissioner in the Province of Zeeland, van Gelder, proposes to organize the fourth Scheldt symposium in Flushing, at the mouth of the river Scheldt, in 1998.

196. Interview anonymous.

waste water treatment plants, or to build them on a better place."[197] A Flemish member states: "*I do not think that the treatment of waste water will be quickened by the ICPS, but that the ICPS is installed, exactly because the parties were ready and willing to treat their waste water. In the Flemish, but also in the Walloon region, rather large budgets are allocated for water quality policies. The already planned policies are that costly that future Flemish ministers will face problems to finance these policies.*"[198]

Although in December 1997, no basin-wide water quality policies are formulated yet, and the expectations concerning the content of the first SAP do not run high, there are some indications of an improving water quality in the Scheldt basin. The large scale investment programmes, in particular in the Flemish region, seem to bear fruit.[199]

Round XIIb: Implementation of the Convention on the deepening of the navigation channel in the Western Scheldt (1995-1997)

After the conclusion of the Convention on the deepening of the navigation channel in the Western Scheldt, the Flemish and Dutch governments face the challenge to implement the 48'/43'/38' deepening programme. In addition, the Dutch government has to develop a plan to compensate the nature losses caused by the implementation of this programme on Dutch territory. The implementation of the deepening programme and decision making on nature compensation are interrelated processes, both intellectually and strategically. Nevertheless, it is possible to describe these processes separately. In the following sections, successively the implementation of the 48'/43'/38' deepening programme and decision making on nature compensation will be described.

Implementation of the 48'/43'/38' deepening programme

The implementation of the 48'/43'/38' deepening programme is a lengthy and complicated process, in which several decision making rounds may be distinguished. Table 7.2 produces information on these decision making rounds, which will be described below.

Round 1: Complicated legal procedures, and an agreement between the Dutch Ministry and environmental NGOs (1/1995-11/1995)

For the implementation of the 48'/43'/38' deepening programme, which entails infrastructure dredging works on several bars in the Western Scheldt[200], the Flemish government needs several permits. Because the water conventions are concluded, and the Dutch national government wants to be a good neighbour, it helps the Flemish government to prepare the application for these permits. Local and regional governments in the province of Zeeland and environmental NGOs, however, who have never been directly involved in decision making on the deepening programme, do not want to cooperate on the implementation of this programme unconditionally. Because of the many legal procedures, these actors do have ample possibilities to delay the implementation.

---

197. Interview J.-M. Wauthier.

198. Interview H. Maeckelberghe.

199. PZC, 1-4-1994; PZC, 4-10-1994; Anonymous (1995b, p. 3); Interview with C. Heip in: Anema (1997).

200. For information on the technical details of the implementation of the 48'/43'/38' deepening programme, see the Appendices to the Convention on the deepening of the navigation channel in the Western Scheldt.

Table 7.2 *Implementation of the 48'/43'/38' deepening programme (1/1995-6/1997)*

Round	Start situation	Major actors	End situation
1 (1/1995-11/1995)	Convention on the deepening of the navigation channel in the Western Scheldt	Flemish region Ministry of Transport, Public Works, and Water Management Environmental NGOs Lower level governments	Covenant between the Ministry and environmental NGOs
2 (11/1995-6/1996)	Decision of ZMF to continue its appeal to court	Ministry Lower level governments NGOs Dutch Council of State	Decision of Dutch Council of State to cancel the WVO Permit
3 (6/1996-6/1997)	Decision of Dutch Council of State	Dutch government Dutch Parliament	Dutch Parliament approves special legislation

The Flemish region needs the following permits to implement the deepening programme[201]:

- A permit based on the Dutch Surface Water Act (WVO Permit), which was issued by the Dutch Minister of Transport, Public Works, and Water Management (See Round XI).
- A dredging permit based on the Dutch River Act, the Landfills Act, and the dredging regulations, to be issued by the Dutch Minister of Transport, Public Works, and Water Management.[202]
- A permit based on the Dutch Act on Nature Conservation, to be issued by the Dutch Minister of Agriculture, Nature Management and Fisheries.
- Permits based on the Dutch Act on Land-Use Planning, to be issued by several Dutch municipalities along the Western Scheldt.[203]

The WVO Permit the Dutch government has granted to the Flemish region for the period between 1995 and 2000 is controversial for various reasons, which were discussed before (See Round XI). The main objection, however, concerns the absence of an EIA of the 48'/43'/38' deepening programme. In spite of the decision of the Dutch ministers that an EIA would neither be obligatory nor desirable for the deepening programme (See Section 6.2,

---

201. Files TSC-meeting 39, 28 March 1995, Schuddebeurs.

202. Rivierenwet, Ontgrondingenwet en Baggerreglement.

203. The Flemish region needs permits for dredging works and/or the storage of dredged material from the municipalities of Flushing, Reimerswaal, and Hontenisse. The Dutch government needs permits from the municipalities of Terneuzen and Hulst for the construction of works to protect the channel ('geulrandverdediging').

Round IX), environmental NGOs insist on an EIA. After the WVO Permit is issued (See Round XI), seven Belgian and Dutch environmental NGOs appeal to the Dutch Council of State (*Raad van State*). In addition, they demand the European Parliament to take action against the Dutch government. Because of this initiative the Dutch government has to answer questions put by the European Commission (EC). The Flemish and Dutch governments do not expect that the Dutch Council of State will decide that an EIA is obligatory. Therefore, they decide not to start an EIA procedure voluntarily, the more because such a procedure would take several years.[204] Furthermore, it is expected that the EC will not demand an EIA, since ample research on the impacts of the deepening programme has been carried out, and that the EC will confine itself to the control of the implementation of the EC Habitat-Directive.

To reduce the procedural complexity, the Dutch Ministry of Transport, Public Works, and Water Management aims to issue one combined permit for the implementation of the dredging works, which is based on the Dutch River Act, the Landfills Act, and the dredging regulations for the Western Scheldt. In the past, the Dutch government always granted the dredging permit to the Belgian and later the Flemish governments unconditionally. Because of the increasing knowledge on the impacts of the dredging works on the morphology (and ecology) of the Western Scheldt, however, the Dutch government decides to formulate conditions to the dredging permit. Although the Flemish government opposes the formulation of conditions in the dredging permit, the Dutch draft a permit in which the licensee will be obliged to carry out research on the impacts of the dredging works on the ecology of the Western Scheldt, and possibilities to reduce infrastructure and maintenance dredging works.[205] According to a Dutch respondent, in future dredging permits the Dutch government might formulate conditions concerning the amount of material to be dredged as well.[206]

Whereas the permits discussed above have to be granted by the Dutch national government, the permits based on the Dutch Act on Land-Use Planning have to be issued by the municipalities along the Western Scheldt. In January 1995, just before the conclusion of the Convention on the deepening of the Western Scheldt, the Dutch Ministry of Transport, Public Works and Water Management has started deliberations with the five municipalities along the Western Scheldt that have to issue a permit based on the Dutch Act on Land-Use Planning. As mentioned before, some municipalities do not intend to issue the necessary permits unconditionally, and formulate several conditions. The Ministry faces most problems with the municipalities of Flushing, Hontenisse, and Reimerswaal. First, these municipalities want to have more information on the compensation of nature losses in the Western Scheldt. The Ministry promises to prepare a discussion report with projects for nature compensation that will be discussed with the municipalities (See next section).[207] A second main issue introduced by the municipalities is the safety of navigation on the Scheldt.[208] For the

---

204. Report of a meeting between representatives of Flemish and Dutch national government agencies to discuss the implementation of the 48'/43'/38' deepening programme, Middelburg, 18 May 1995.

205. Files TSC-meeting 40, 27 September 1995, Cleydael.

206. Interview anonymous.

207. Anonymous (1995a, p. 5).

208. Research has shown that the implementation of the deepening programme may shift the so-called risk contours, and therefore has an impact on the safety at some places along the Western Scheldt, among which some areas in the municipalities of Flushing and Hontenisse. There is, however, uncertainty about the actual impact of the implementation of the deepening programme on the safety along the Western Scheldt, since the

municipality of Reimerswaal the uncertainty concerning the safety along the Western Scheldt is a good reason to demand an EIA for the 48'/43'/38' deepening programme.[209] The municipalities of Flushing and Reimerswaal do explicitly link the issuing of the permit based on the Dutch Act on Land-Use Planning to the safety of navigation on the Western Scheldt.[210] A third issue introduced by the municipality of Flushing is the erosion of some beaches along the Western Scheldt. The municipality wants the Ministry to pay for the sand suppletions that are necessary to protect the shores.[211]

The preceding clearly shows that the implementation of the 48'/43'/38' deepening programme is a highly complicated administrative process. The Dutch Ministry of Transport, Public Works, and Water Management tries to reach an agreement with the parties opposing the issuing of the necessary permits. As regards the opposition of environmental NGOs, the Ministry is rather successful. On 15 November 1995 it is able to conclude a covenant with several environmental NGOs, which, among other things, states that the NGOs will no longer appeal to court as regards the issuing of permits needed for the implementation of the deepening programme.[212]

Round 2: More obstacles by-passed, and the unexpected decision of the Dutch Council of State (11/1995-6/1996)

The executive committee of the ZMF, which has concluded the covenant with representatives of the Dutch Ministry of Transport, Public Works, and Water Management, is unable to defend its policy, when the members of the environmental association oppose its decision not to continue the appeal to court concerning the issuing of the WVO Permit. Therefore, the

---

implementation of the deepening programme would reduce the chance on shipping accidents, and consequently have a positive impact on the safety of navigation as well (Ibid).

209. Letter from the Municipality of Reimerswaal to the Ministry of Transport, Public Works, and Water Management of 19 January 1995.

210. Letter from Daamen and van Heteren, regional Directorate Zeeland of Rijkswaterstaat, to the Director-General of Rijkswaterstaat, of 22 March 1995. Report of a meeting between representatives of Flemish and Dutch national governments to discuss the implementation of the 48'/43'/38' deepening programme, Middelburg, 18 May 1995.

211. Anonymous (1995a, p. 5).

212. In this covenant the environmental NGOs promise that they will (List of joint decisions approved by representatives of the Main Directorate of the Ministry of Transport, Public Works, and Water Management, the regional Directorate Zeeland of the same Ministry, the 'Zeeuwse Milieufederatie' (ZMF), the 'Stichting het Zeeuwse Landschap', and the 'Vereniging Natuurmonumenten' on 15 November 1995, in Goes):
- accept a decision of the Dutch parliament and of the EC concerning the obligation to make an EIA-study on the 48'/43'/38' deepening programme;
- not appeal to court as regards the issuing of permits needed for the implementation of the deepening programme;
- convince their members and other environmental NGOs that they should act in the same way.
In turn, the Dutch Ministry of Transport, Public Works, and Water Management promises that (Ibid.):
- the ZMF will be invited to participate in preparatory meetings of the Dutch delegation in relevant working groups of the ICPS.
- Representatives of the Ministry and environmental NGOs will meet every three months to discuss the (impacts of the) implementation of the deepening programme.
- Representatives of the Ministry and environmental NGOs will cooperate on the development of alternative projects for nature restoration in the Western Scheldt, among which projects to compensate the nature losses caused by the implementation of the deepening programme, and nature restoration projects for a medium long and long period.

ZMF decides to continue its protest to the issuing of the WVO Permit.

In spite of the good intentions of the Dutch national government, some Flemish actors are worried about the progress of the implementation of the deepening programme. Because of the appeal to court of environmental NGOs, and the statements made by representatives of Dutch municipalities that they will not issue the necessary permits unconditionally, the Flemish Prime Minister, Van den Brande, and Minister of Public Works, Transport, and Physical Planning, Baldewijns, send a letter to the Dutch Prime Minister, Kok, in which they demand an extra effort to issue the necessary permits as soon as possible.[213] The city of Antwerp that knows about its dependence on the Dutch municipalities, invites representatives of the municipalities to visit Antwerp, and to look at the Antwerp efforts to protect the environment.[214]

On 19 January 1996, when the EC answers the questions put by members of the European Parliament concerning the EIA-obligations for the deepening programme, another potential obstacle is by-passed. As the Dutch government did expect, the EC states that according to EC-Law an EIA is not obligatory for the implementation of the deepening programme.[215] At about the same time, however, a new potential problem arises. Flemish environmental NGOs appeal to the Dutch Council of State against the issuing of the dredging permit.[216] The NGOs state that they will repeal their request if the Flemish government is prepared to make a similar agreement with the NGOs as the Dutch government has made with Dutch environmental NGOs. As a consequence, like the Dutch government, also the Flemish government concludes a covenant with Flemish environmental NGOs.[217]

On 14 June 1996, however, there is a hitch. The Dutch Council of State takes an unexpected decision, because it cancels the WVO Permit for the Flemish region, and decides that an EIA has to be made for the deepening programme.[218] The decision to cancel the WVO-Permit does not only make it impossible to implement the deepening programme, but to continue the maintenance dredging works in the Western Scheldt as well.

Round 3: The preparation of special legislation (6/1996-6/1997)

For the Dutch and Flemish governments the decision of the Dutch Council of State comes

---

213. Letter from the Flemish Prime Minister, Van den Brande, and the Flemish Minister of Public Works, Transport, and Physical Planning, Baldewijns, to the Dutch Prime Minister, Kok, of 13 December 1995. The Dutch Prime Minister, Kok, replies that the Dutch government makes every effort to facilitate the implementation process (Letter from the Dutch Prime Minister, Kok, to his Flemish colleague, Van den Brande, of 22 January 1996).

214. Report of a meeting between representatives of Flemish and Dutch national governments to discuss the implementation of the 48'/43'/38' deepening programme, Middelburg, 18 May 1995.

215. TK 56-4096.

216. Letter from the 'Bond Beter Leefmilieu' (BBL) to the Dutch Council of State of 22 January 1996.

217. Letter from Strubbe to Zijlmans of 24 January 1996.

218. Files TSC-meeting 42, 1 October 1996, Ghent.
The Dutch Council of State does not address the objections formulated by the ZMF and the other environmental NGOs, but decides on formal grounds that for the arrangement of storage locations in the Western Scheldt, the Minister of Transport, Public Works, and Water Management has to apply for a permit based on the Dutch Act on Environmental Management (Wet Milieubeheer). Consequently, an EIA becomes obligatory all the same.

like a bolt from the blue.[219] According to the Dutch Ministry of Transport, Public Works, and Water Management it is unacceptable to stop the maintenance dredging works, as that would be at the expense of the safety of navigation[220] Furthermore, a delay of the implementation of the deepening programme would certainly worsen the Flemish-Dutch relationship. Therefore, on 12 July 1996 it is decided that[221]:

- the maintenance dredging works in the Western Scheldt will be tolerated[222];
- special legislation will be drafted for the implementation of the deepening programme and the related maintenance dredging works;
- all regular procedures will be continued, for the case that the special legislation would not be accepted by the Dutch Parliament.

With that the Dutch government follows two tracks.[223] First, it starts to prepare special legislation, which sets aside all necessary licensing procedures for the implementation of the deepening programme, and tries to get this special legislation accepted by the Dutch parliament. Secondly, it starts to prepare EIAs for all storage locations in the Western Scheldt, which are needed in case the Dutch government or parliament would not approve the special legislation.[224] On 23 August 1996, the Dutch government approves the special legislation with a term of validity until 2001. With that the Dutch government sets aside, beside the WVO and WM-procedures, also the necessary procedures to obtain the permits based on the Dutch Act on Land-Use Planning and the Act on Nature Conservation. In the meanwhile, the Flemish government has reached an agreement with the Flemish environmental NGOs, and the dredging permit is granted.[225] Although several Dutch advisory bodies formulate fundamental criticism on the legislation[226], the Dutch Parliament is of the opinion that this very specific case justifies the issuing of special legislation. After the special

---

219. According to the Minister of Transport, Public Works, and Water Management, Jorritsma-Lebbink, the Ministry has not been able to anticipate this decision, because the created situation would not be caused by procedural mistakes, but by a new interpretation of existing law (EK 1995-1996, Report of the 36th meeting, 25-6-1996). Others, however, are surprised by the reaction of the Minister, and argue that the Ministry could have known that the Dutch Council of State would not approve the WVO-Permit, because the Ministry has not taken into account the Dutch Act on Environmental Management (Van Buuren, 1997, p. 122). For a discussion on the Permit Act for the Western Scheldt, see also Stroink (1998, pp. 56-60).

220. EK 1995-1996, 36th meeting, 25 June 1996, Minister Jorritsma-Lebbink.

221. Files TSC-meeting 42, 1 October 1996, Ghent.

222. Maintenance dredging works are continued on the basis of a WVO- and WM- 'gedoogbeschikking'.

223. Van Buuren (1997, p. 122).

224. After the Dutch parliament has approved the special legislation, these EIA-studies are not obligatory anymore. The studies, however, provide useful information for the preparation of the permits and for EIA-studies that are necessary after the term of validity of the special legislation is expired (Anonymous, 1997b, p. 7).

225. On 17 July 1996, the Flemish Administration Water Infrastructure and Marine Affairs and Flemish environmental NGOs conclude a covenant ("overeenkomst over de integrale benadering van de werken in het Scheldebekken") in which the NGOs state that they will no longer oppose the issuing of permits needed for the implementation of the deepening programme. In turn, the Flemish Ministry invites the NGOs to participate in decision making on nature restoration along the Sea Scheldt, and to participate in the meetings of the Flemish delegation in the TSC.

226. For more information on the legal details of the (criticism on the) special legislation, see van Buuren (1997).

legislation is carried on 17 June 1997[227], and the Permit Act for the Western Scheldt enters into force on 19 June 1997[228], all legal procedures are set aside, and nothing prevents the Flemish government any longer to deepen the navigation channel in the Western Scheldt.

A distortion of competition?
Beside the granting of the necessary permits, also a possible distortion of competition caused by the implementation of the 48'/43'/38' deepening programme is a topic of discussion in the Netherlands. Some Dutch MPs argue that, unlike in the port of Rotterdam, in the port of Antwerp the costs of infrastructure works are not passed on the users of the port. Therefore, they expect that the 48'/43'/38' deepening programme will cause a distortion of competition between Antwerp and Rotterdam.[229] Furthermore, the contracting out of the infrastructure dredging works is a topic of discussion. The Flemish authorities are of the opinion that the Flemish dredging companies that carry out the regular maintenance dredging works in the Western Scheldt do also have the right to carry out the non-recurring infrastructure dredging works. The Dutch dredging companies, however, are of the opinion that this is not in accordance with EC regulations.[230]

Potential future conflicts
Although Flanders and the Netherlands have reached an agreement on the implementation of the 48'/43'/38' deepening programme, it is most likely that the conclusion and implementation of the Convention on the deepening of the navigation channel in the Western Scheldt only marks a temporary end to the Flemish-Dutch disagreement on the management of the navigation channel in the Western Scheldt. A main reason for this is that the two neighbours still have not reached an agreement on a joint interpretation of the Scheldt Statute. Most likely, this issue will play a role in decision making on the implementation of the second stage of the deepening programme (the 50' deepening programme) or a further deepening of the navigation channel of the Western Scheldt, either in combination with the construction of a second maritime access to the Waasland harbour or not.

Decision making on nature restoration
Simultaneous to the implementation of the 48'/43'/38' deepening programme, the Dutch government has to develop a plan for nature restoration. Table 7.3 produces information on the rounds distinguished in decision making on nature restoration, which will be described below.

---

227. At this date the First Chamber of the Dutch Parliament approves the special legislation (Files TSC-meeting 43, 19 June 1997, Flushing).

228. Stb. 1997 No. 258.

229. The Municipal Executive of Rotterdam sends letters to several Dutch MPs, and argues that the implementation of the deepening programme contributes to the competitive position of the port of Antwerp, and may cause a distortion of competition between Antwerp and Rotterdam (TK 1995-1996, Report of the 56th meeting, 15-2-1996, pp. 4074, 4079).

230. TK 1995-1996, Report of the 56th meeting, 15-2-1996, pp. 4074.

Table 7.3 *Decision making on nature restoration (1/1995-8/1997)*

Round	Start situation	Major actors	End situation
A (1/1995- 1/1996)	Convention on the deepening of the navigation channel in the Western Scheldt	BOW Working group nature restoration Western Scheldt	Working group nature restoration Western Scheldt issues a study report with alternative projects for nature compensation
B (1/1996- 5/1996)	BOW starts participation rounds	BOW Members BOW NGOs Public	The BOW is not able to present a plan for nature restoration that is supported by the region, and therefore advises the Dutch Minister of Transport, Public Works, and Water Management to carry out additional research
C (6/1996- 8/1997)	The Minister installs a Commission of wise persons	Commission Western Scheldt Members of the BOW NGOs	The Commission Western Scheldt issues its advice to the Dutch Minister

Round A: Research on alternatives of nature restoration (1/1995-1/1996)

Beside the implementation of the 48'/43'/38' deepening programme, implementation of the Convention on the deepening of the navigation channel in the Western Scheldt entails the planning and implementation of works to restore nature losses.[231] The Dutch Minister of Transport, Public Works, and Water Management, who is responsible for the development and implementation of these works, demands the Core working group for the Western Scheldt (BOW) to prepare a plan for nature restoration to be approved by the Minister. On 27 March 1995 the BOW decides to organize a so-called 'open planning process'[232], and installs a working group with representatives of national, regional, and local government agencies, and representatives of the Flemish region, which has to study alternatives of naturerestoration.[233] Under the authority of this working group, which starts on 8 May 1995, the RIKZ and the Dutch consultancy Heidemij carry out research on the impacts of the implementation of the deepening programme on the ecology of the Western Scheldt, alternative nature restoration projects, and possible criteria for the evaluation of these projects. During the preparation of this report several parties are contacted. In the mean

---

231. Verdrag inzake de verruiming van de vaarweg in de Westerschelde, Artikel 2.2 sub c.

232. Although one may discuss the exact definition of an 'open planning process', central elements of this approach are that all relevant actors are given the opportunity to participate in the decision making process, and that decision making aims at the creation of consensus among these actors.

233. Letter from the chairman of the project group Nature Restoration Western Scheldt, Kop, to the members of the BOW of 26 June 1995.

while, the chairman of the BOW and representative of the Province of Zeeland, van Zwieten, has stated that he is prepared to coordinate the planning process in the region.[234]

By using the research results of the project OostWest (See Round Xa) the RIKZ and Heidemij are able to distinguish three categories of nature restoration projects:

- 'Ontpolderen';
- Nature restoration projects situated on the landside of the dikes along the estuary;
- Nature restoration projects situated outside the dikes along the estuary.

During the discussions in the BOW the experts often emphasize that 'ontpolderen' undoubtedly contributes most to nature restoration along the Western Scheldt, and that alternatives to 'ontpolderen' will not be sufficient to restore nature losses entirely.[235] Representatives of the municipalities along the Western Scheldt, however, emphasize the lack of regional support for the policy alternative 'ontpolderen'.[236] In addition, a Flemish representative in the BOW, Belmans, states that 'ontpolderen' will be most effective in the upstream parts of the estuary.[237] For this reason other experts criticize the articles dealing with nature restoration in the Convention on the deepening of the navigation channel in the Western Scheldt. According to them, the 44 million Dutch guilders could be spent more effectively on Flemish territory, because 'ontpolderen' on Flemish territory would yield more environmental benefits than the implementation of similar projects on Dutch territory.[238]

In January 1996 the project group Nature restoration Western Scheldt issues a report with projects for nature restoration.[239] It is estimated that the implementation of the deepening programme causes a loss of 500 hectares of nature areas, i.e. saltings, mud flats, and shallow waters.[240] In addition, the report contains policy alternatives of the three categories discussed above. A main conclusion of the report is that: "[...] *the policy measure 'ontpolderen' is indispensable in the development of the estuary, because this measure contributes on all fronts (morphodynamics, habitats and species, and shallow waters)."*[241] Although the BOW has the task to develop a plan for nature restoration in the Western Scheldt, the report does contain projects on Flemish territory as well. According to the project team, these projects could be taken into account in a long term plan for nature restoration in the entire Scheldt estuary.[242] In the BOW representatives of the municipalities demand whether the plan for nature restoration concerns the restoration of nature losses

---

234. Report Meeting of the executive committee of the BOW, 6 July 1995, Middelburg.

235. Reports Meetings BOW of 10 October and 18 December 1995.

236. Ibid.

237. Ibid.

238. See, for example, the presentation of C. Heip on the third Scheldt symposium (Heip, 1995, pp. 11-14). Whilst decision making on nature restoration is a Dutch concern, Flemish and Dutch government agencies have started cooperation in the LIFE-MARS-project (Marsh Amelioration along the River Schelde) (Niesing *et al.*, 1996). In this project some pilot projects on nature restoration are carried out. In addition, Flemish and Dutch government agencies and NGOs publish a joint report describing the ecological values of the Scheldt (Meire *et al.*, 1995).

239. Ministerie van Verkeer en Waterstaat, Directoraat-Generaal Rijkswaterstaat, Directie Zeeland *et al.* (1996a and b).

240. Ministerie van Verkeer en Waterstaat, Directoraat-Generaal Rijkswaterstaat, Directie Zeeland *et al.* (1996a, p. 7).

241. Ibid., p. 9.

242. Ibid., Preface.

caused by the implementation of the deepening programme only, or the restoration of nature losses caused by activities in the past as well. The BOW[243], just like the Dutch parliament[244], does not give a definite answer, mainly as it does not want to anticipate the participation rounds that will be organized concerning nature restoration along the Western Scheldt.

Round B: The participation meetings (1/1996- 5/1996)

The second stage of the open planning process is a discussion with the municipalities, water boards, NGOs, and the public in the province of Zeeland on the policy alternatives presented in the study report. In February and March 1996 the BOW organizes several participation meetings.[245] At these meetings it turns out that some parties, such as the waterboards and farmers associations are strong opposers to the policy alternative 'ontpolderen', but are prepared to discuss the other alternatives presented in the study report. The main reason for this is that they do not want to give up land that they have gained from the sea in the past. This standpoint should also be seen in its context. A few years back farmer associations have agreed with the implementation of the Nature policy plan, which implies that the farmers have to give up about 7,600 hectares of agricultural area.[246] Any additional loss of land is unacceptable for the farmer associations. The opposers also state that the costly dikes, which have been built after the flood disaster of 1953, should be maintained. Finally, they challenge the assumption of the research on the possibilities to compensate nature losses, and think that 'ontpolderen' will increase flood risks, and therefore will have a negative impact on the safety along the Western Scheldt. "*In a sometimes very probing way it has been made clear, how strongly many in the region are feeling about the water management, and related safety aspects. The latent fear for the sea, the hard-won Delta-safety, and the marks of 1953 have created so many emotions, that for many even thinking about 'ontpolderen' was a taboo*".[247] Some experts of the Dutch Ministry of Transport, Public Works, and Water Management, however, continue to argue that 'ontpolderen' will increase the safety along the Western Scheldt, because it increases the tidal volume in the Western Scheldt. At the participation meetings these experts strongly plead for the policy alternative 'ontpolderen', which would be the only serious option for nature restoration. Beside the HID of the regional Directorate Zeeland of Rijkswaterstaat, Saeijs, the Royal Commissioner, van Gelder, reveals himself as an important advocate of 'ontpolderen'.[248]

The participation meetings, however, clearly show that public and administrative support for the policy alternative 'ontpolderen' in the province of Zeeland is lacking. In March and April 1996 the Provincial Council of Zeeland and the various Local Councils involved formulate their standpoints. Subsequently, the BOW formulates an advice to the Minister of Transport, Public Works, and Water Management. The members of the BOW take the opinion of their inhabitants seriously, and all do conclude that there is no public support for

---

243. Report meeting BOW, 31 January 1996, p. 3.

244. TK 1995-1996, Report of the 56th meeting, 15-2-1996, pp. 4074-4114.

245. For more information on the planning of the participation meetings, see the *Schelde nieuwsbrief*, Vol. 2, January 1996.

246. Report of the meeting of the BOW of 8 May 1996, Middelburg.

247. Letter from the executive committee of the BOW to the Minister of Transport, Public Works and Water management of 8 May 1996.

248. PZC, 2-3-1996; PZC, 5-3-1996.

the policy alternative 'ontpolderen'.[249] Because the project group Nature restoration Western Scheldt in fact has stated that a full compensation of the nature losses caused by the deepening programme cannot be realized without 'ontpolderen', the BOW concludes that it has not been able to develop a plan for nature restoration in the Western Scheldt that is supported by the parties in the Province of Zeeland. Therefore, it advises the Dutch Minister of Transport, Public Works, and Water Management to carry out additional research on possibilities of nature restoration, in which the results of the existing study report and the participation meetings should be taken into account.[250]

Round C: The Commission of wise persons (6/1996-8/1997)

On 13 September 1996 the Minister of Transport, Public Works, and Water Management installs an advisory commission, the 'Commission Western Scheldt', which has three members, and has to advise on the Dutch compliance with the provisions concerning nature compensation in the Convention on the deepening of the navigation channel in the Western Scheldt.[251] After an orientation on the situation that has developed in the province of Zeeland, the Commission plans meetings with representatives of national, regional, and local governments, and several NGOs. In these meetings, each party is asked to propose projects for nature restoration. On the basis of these proposals the Commission drafts a plan for nature restoration. In a second round of deliberations the parties are given the opportunity to comment on the draft plan for nature restoration. Taking into account these comments, the Commission prepares a final draft, which is discussed once more with the Zeeland farmers associations and Nature protection organizations.[252] In addition to these deliberations, a professor in Public Management is asked to analyze decision making on the compensation of nature losses in the Western Scheldt.[253]

On 29 August 1997 the Commission Western Scheldt issues its advice to the Minister, which is based on the results of several meetings, and the scientific analysis of the decision making process.[254] The Commission proposes to distinguish between the short and the long term as regards decision making on the Western Scheldt. In the short term (5 years) nature restoration projects that are needed to comply with the Convention on the deepening of the navigation channel in the Western Scheldt should be implemented.[255] A possible compensation

---

249. Stenographic report meeting BOW, 8 May 1996, Middelburg.

250. Letter from the executive committee of the BOW to the Minister of Transport, Public Works and Water management of 8 May 1996.

251. Decision of the Dutch Minister of Transport, Public Works, and Water Management of 13 September 1993 on the installation of a Commission Western Scheldt, Letter from the Minister to the chairman of the BOW of 13 September 1996. The members of the Commission Western Scheldt are the Royal Commissioner in the Province Overijssel, J.A.M. Hendrikx, the chairman of the product boards for agricultural products ('Produktschappen voor Akkerbouwprodukten'), J.H.M.Kienhuis, and the Royal Commissioner in the Province Zuid-Holland, J.M. Leemhuis-Stout.

252. Commissie Westerschelde (1997, pp. 9-10).

253. This analysis is made by Prof.dr. A.B. Ringeling, professor in Public Management at the Erasmus University, Rotterdam; Ringeling (1997).

254. Commissie Westerschelde (1997), Letter from the chairman of the Commission Western Scheldt, Hendrikx, to the Minister of Transport, Public Works, and Water Management, Jorritsma-Lebbink, of 29 August 1997.

255. Commissie Westerschelde (1997, p. 3).

of the nature losses caused by activities in the past may be taken into account in the development of policies for the long term. According to the Commission, for the development of both the short term and the long term policies, cooperation between the national government and lower level governments is essential.[256] The point of departure of all plans should be that the protection against floodings should be guaranteed. According to the Commission, a 'partial landward movement of the sea walls' can only be taken into account (in the long term planning) if there are ponderous arguments for that, which are based on a solid technical-scientific foundation.[257]

For the nature restoration plan for the short term the Commission recommends several projects[258], and advises the Minister to discuss these plans with the governments involved and the land-owners, and to conclude a covenant.[259]

For the long term the Commission Western Scheldt advises the Minister to develop a long term vision, which preferably should relate to the Flemish and Dutch parts of the Scheldt estuary. This vision could be developed in the ICPS, the TSC, or by an *ad hoc* steering group with Flemish participants. It should, among other things, take into account recent developments in water management. The Commission explicitly relates the issue of the safety along the estuary to the policy alternative 'ontpolderen', and states: "*With that respect, the partial movement of the sea walls, should not be precluded in the long term*"[260] For the development of a long term vision extra research would be needed. The final reports of the project OostWest could be a first input for the discussions on the development of a long term vision.[261] An interesting challenge concerns the selection of the right platform(s) for preparing this long term vision. Although the ICPS and TSC certainly are relevant, a transboundary basin committee for the Scheldt estuary might be a good alternative.[262]

---

256. Ibid.

257. Ibid., p.3. The Commission Western Scheldt introduces a new term for 'ontpolderen', namely the 'partial landward movement of sea walls' ('partieel-landinwaards verplaatsen van de zeewering'). A main reason for this is that the term 'ontpolderen' is associated with drastic policy measures, whilst experts are of the opinion that in many cases a partial landward movement of seawalls would have limited consequences only.

258. For a short description of these projects, see Commissie Westerschelde (1997, Appendices I and II).

259. Ibid., p.4.

260. Ibid., p.5.

261. Ministerie van Verkeer en Waterstaat, Directoraat-Generaal Rijkswaterstaat, RIKZ (1996, 1997), Ministerie van Verkeer en Waterstaat, Directoraat-Generaal Rijkswaterstaat, Directie Zeeland (1997).

262. In recent years both Flemish and Dutch water managers plead for the installation of a basin committee for the Scheldt estuary, and for the development of a nature restoration plan or integrated water management plan for the Scheldt estuary. See for example Saeijs *et al.* (1995), Meire (1996), and Adriaanse (1996). Such a transboundary basin committee may be created by extending the BOW with Flemish participants. After the conclusion of the water conventions, the Flemish participation in the BOW has become a topic of discussion again (See Section 6.2, Round VII). The Flemish region expresses its willingness to intensify its participation in the BOW (Report of a meeting between representatives of Flemish and Dutch national governments to discuss the implementation of the 48'/43'/38' deepening programme, Middelburg, 18 May 1995). It is, however, reserved as regards the signing of the covenant ('bestuursovereenkomst') concerning the management of the Western Scheldt (See Section 6.2, Round IX) (Files TSC-meeting, 41, 15 Februari 1996, Slot Moermont). The main reason for this is that the Flemish administrators are not familiar with the phenomenon 'bestuursovereenkomst'. The BOW and Flemish government agencies do reach an agreement on the joint publication of a newsletter concerning the developments and activities in the Scheldt estuary (Covenant signed by the chairman of the BOW, and the Directors-General of the Flemish Administration Water Infrastructure and Marine Affairs, and the Administration Environment, Nature- Land- and Water management of 15 May 1997, Temse).

7.3   Analysis and explanation of the decision making rounds Xa to XIIb

In this section, the strategic interactions described in the preceding section will be analyzed and explained. Section 7.3.1 focuses on the relationship between the perceptions and strategies of national and lower level governments and NGOs, and with that gives an answer to research question 3 for the period between 1993 and 1997. Subsequently, Section 7.3.2 discusses the strategic interactions and institutionalization processes, and with that presents the answer to research question 4 for the period indicated. Section 7.3.3 analyzes the strategic and cognitive learning processes that developed in decision making on international Scheldt issues, and gives an answer to research question 5 for the period studied in this chapter. Section 7.3.4 discusses the influence of contextual variables on the perceptions and strategies of the actors involved in the decision making process, and gives an answer to research question 6 for the last case study period. To conclude, in Section 7.3.5 decision making on nature restoration is addressed shortly.

7.3.1 Influence of perceptions of actors involved in the decision making rounds Xa to XIIb on their strategies

Perceptions and strategies of the Scheldt basin states and regions
Table 7.4 produces information on the main strategies of the Scheldt basin states and regions in each decision making round. In the following, the strategies of these actors are related to their perceptions of (inter)dependence, which provides insight in the action rationality of the individual basin states and regions.

The Flemish region: desperate needs lead to desperate deeds
The Flemish government, which had been waiting for a Dutch permission to deepen the navigation channel in the Western Scheldt for more than ten years, started to get nervous, and cut capers. First, it proposed a parallel discussion of the 48'/43'/38' deepening programme and the alignment of the HSL, and with that tried to make an international agreement on the implementation of the deepening programme more attractive for the Dutch government (See Table 7.5). Shortly after the Dutch and Flemish governments had reached an agreement to deal with these issues simultaneously, however, decision making on the alignment of the HSL turned out to be laborious, whilst the Dutch and Flemish delegations could reach an agreement on the texts of the Conventions on the deepening of the navigation channel in the Western Scheldt and the distribution of the Meuse water relatively easily. Because of the Dutch interest in decision making on the alignment of the HSL, and the Flemish-Dutch agreement to deal with the issues simultaneously, the Dutch government did not want to sign a convention that would enable the Flemish region to implement the deepening programme until decision making on the alignment of the HSL would have been concluded.[263] At that time the Flemish region knew that it had made a tactical mistake, and decided to re-establish the   linkage between bilateral and multilateral Scheldt and Meuse issues it had opposed for so many years. The Flemish region did not want to sign a convention on any of the Scheldt or Meuse issues at stake, unless the Dutch government would sign the Convention on the deepening of the navigation channel in the Western

---

263. One of the problems the Flemish region faced was that for decision making on the alignment of the HSL it was dependent on the Federal State of Belgium.

Table 7.4 *Main strategies of Scheldt basin states and regions in decision making rounds Xa to XIIb*

Round	Actor	Type of strategy	Strategic actions
Xa (1993-1994)	Flemish government	Issue-linkage	Proposes parallel discussion of deepening programme and alignment of the HSL
		Compensation, Mitigation	Is willing to contribute financially to the compensation of negative side-effects of the implementation of the 48'/43'/38' deepening programme
	Dutch government	Issue-linkage	Holds on to Flemish-Dutch agreement on a parallel discussion of the 48'/43'/38' deepening programme and the alignment of the HSL, and does not want to sign the convention on the deepening programme until a Flemish-Dutch agreement on the alignment of the HSL is reached
Xb (1993-1994)	Dutch government	Offensive interactive	Takes the initiative for informal deliberations Proposes conventions with a broad scope
	French government	Offensive interactive	Takes the initiative for formal negotiations
	Brussels/Walloon governments	Reactive interactive	Do want to cooperate on the water quality of Scheldt and Meuse
		Reactive aimed at maintaining autonomy	Do want cooperation to be restricted to the water quality of the main course of the rivers
	Flemish government	Reactive interactive	Does want to cooperate on conventions with a broad scope
		Issue-linkage	Links all Scheldt and Meuse issues to the deepening of the navigation channel in the Western Scheldt

Round	Actor	Type of strategy	Strategic actions
XI (1994-1995)	Flemish government	Compromising	Is prepared to reach a procedural agreement concerning decision making on the alignment of the HSL
	Dutch government	Reactive interactive	Is prepared to unlink decision making on the alignment of the HSL and the deepening programme
		Reactive interactive	Formulates additional conditions in a new WVO Permit
XIIa (1995-1997)	Flemish and Dutch governments	Trust building	Do accept that the first SAP will not contain policies that go beyond existing regional and national policies in the short term
	Walloon, Brussels, and French governments	Reactive interactive Reactive aimed at maintaining autonomy	Do aim at a modest first SAP that does not go beyond existing regional and national policies.
XIIb (1995-1997)	Flemish government	Offensive interactive	Tries to get necessary permits for implementation of the deepening programme
	Dutch government	Trust building Cooperative	Helps Flemish government with application of necessary permits
		Reactive interactive	Formulates conditions in dredging permit

Scheldt. Therefore, unlike all other basin states and regions, the Flemish government did not sign the multilateral conventions on the Scheldt and the Meuse in 1994, and did even delay the conclusion of a bilateral convention that had never been part of the negotiations on the water conventions: the Convention on the Revision of the Scheldt regulations. This unexpected Flemish decision is a real grand finale of the negotiations on the water conventions. Later, the Flemish and Dutch governments were able to reach an agreement on the procedures to be followed in decision making on the alignment of the HSL, and the Dutch government declared that it was willing to sign a Convention on the deepening of the navigation channel of the Western Scheldt, if the Flemish government would sign the other water conventions. At that moment, the Flemish region decided to sign the water conventions.

In the negotiations on the deepening programme the Flemish region used the strategy of compensating losses, and was willing to contribute financially to the restoration of nature losses.

In the multilateral negotiations and the first meetings of the provisional ICPS the Flemish delegation had about the same points of view as the Dutch delegation, and aimed at the implementation of integrated river basin management. The region did not want to disturb the process, and therefore did not demand too much in the multilateral negotiations. The Dutch and Flemish delegations formed a rather stable coalition in the negotiations on the development of international institutional arrangements and policies to combat Scheldt pollution. Also in the informal study group preceding the installation of the provisional ICPS, the ISG, the Flemish and Dutch delegations had already about the same points of view. The Flemish water managers used the ISG and the LIFE-DSS project for the water quality of the Scheldt to intensify their contacts with the upstream parties, in particular the Walloon region.

After the conclusion of the Convention on the deepening of the navigation channel in the Western Scheldt, the Flemish region on the one hand controlled precisely the Dutch efforts to implement the deepening programme, and on the other hand expressed its appreciation for the efforts made by the Dutch national government to facilitate the implementation of the deepening programme.

Table 7.5 *Issue-linkages in decision making on international Scheldt issues (1993-1997)*

Year	Actor proposing issue for negotiation	Proposed issues	Linking actor	Linked issues
1993	Flemish government Dutch government	Deepening programme Water quantity Meuse	Dutch government Flemish government	Water quantity Meuse Deepening programme
1993	Flemish government Dutch government	Deepening programme HSL	Dutch government Flemish government	HSL Deepening programme
1993	Flemish government	WVO permit	Dutch government	Extraction of polluted sediments from the Lower Sea Scheldt
1994	Dutch government	Water quality of Scheldt and Meuse Water quantity of Meuse Revision of Scheldt regulations	Flemish government	Deepening programme

The Dutch government: increased attention for the process

Bilateral negotiations. The Dutch government agreed with the Flemish proposal to deal with the deepening of the navigation channel in the Western Scheldt and the alignment of the HSL simultaneously. After the Dutch and Flemish delegations had reached an agreement on a Convention on the deepening of the navigation channel in the Western Scheldt, the Dutch government refused to sign this convention unless an agreement would have been reached on the alignment of the HSL. In spite of the severe criticism, both in Flanders and in the Netherlands, the Dutch government continued the linkage, because it perceived a big interest in decision making on the alignment of the HSL, and, after all, the Flemish region had proposed to deal with the two issues simultaneously.

When the Flemish government refused to sign the multilateral Scheldt and Meuse conventions, and decided to re-establish the linkage across bilateral and multilateral Scheldt and Meuse issues, the Dutch were astonished, but not really impressed. After all, the Dutch had never made a secret of their interest in the water quality of the Meuse rather than the water quality of the Scheldt. Therefore, for the Dutch it was most important that Wallonia would sign the Meuse convention. Furthermore, the Dutch knew that the delay had a purely tactical reason, and expected the Flemish Region to sign the Scheldt and Meuse Conventions in the near future.

Only after a procedural agreement concerning decision making on the alignment of the HSL had been made, the Dutch were prepared to sign the Convention on the deepening of the navigation channel in the Western Scheldt, and the water conventions could be concluded. One may discuss whether it was a wise decision of the Dutch government to continue the linkage between decision making on the alignment of the HSL and the deepening of the navigation channel in the Western Scheldt, and with that to punish the Flemish region for its tactical blunder. On the one hand the Dutch indeed had a main interest in decision making on the alignment of the HSL, and the linkage possibly has contributed to a Flemish-Dutch agreement on that issue. On the other hand, the Dutch delay of the approval of the Convention on the deepening of the navigation channel in the Western Scheldt seems to have increased the Flemish distrust of the Dutch, which had grown during history, even more, and caused a delay of joint decision making on several other international issues. After the conclusion of the water conventions, however, the Dutch government made every effort to implement the 48'/43'/38' deepening programme, and did even issue special legislation to enable the Flemish region implementing this programme.

In 1994 the Dutch government continued the linkage between the WVO permit for the Flemish region and the extraction of polluted sediments from the Lower Sea Scheldt. Compared with the preceding permits, the licensee was obliged to provide some additional information. Until the provisional installation of the ICPS, the WVO permit certainly was the most important resource the Dutch government could use to exert influence on the Flemish region to reduce the transboundary load of pollutants in the Scheldt basin. Partly because of the good experiences with the formulation of conditions in the WVO permit, the Dutch started to use a similar strategy in decision making on the dredging permit for the Flemish region. Whereas in the past the dredging permit had been granted unconditionally, in the last case study period the Dutch started to formulate conditions in this permit as well. This copying of a successful strategy may be interpreted as a strategic learning process.

Multilateral negotiations. In the multilateral negotiations and the first meetings of the provisional ICPS the Dutch favoured integrated river basin management, and with the Flemish region formed a rather stable coalition. Nevertheless, the Dutch (and Flemish) delegations did accept multilateral conventions, which provide for a joint management of the water quality and ecology of the surface waters only, and which do not contain concrete water quality objectives. Whilst at the end of 1991 the chairman of the Dutch delegation had stated that the Dutch would not accept an 'empty' convention, since by that the problems of the negotiation commission would be shifted entirely to the river commissions (See Section 6.2, Round VIII), in 1993 the Dutch were prepared to sign multilateral conventions, even though negotiations on concrete policies still would have to take place in the international river commissions to be installed. There are several possible alternative or supplementary explanations for the Dutch strategic change.

The first possible explanation is as simple as it is clear, namely that there was no chance to formulate a more ambitious convention, because the other basin states and regions did not support the proposals that had been made by the Dutch delegation. Because, with the exception of the Flemish Region, the Dutch did not compensate the other basin states or regions financially or otherwise, the contents of the multilateral conventions reflect the greatest common denominator. External pressure on the Dutch government to conclude the water conventions, however, had increased tremendously, and the Dutch could hardly afford a further delay of the conclusion of these conventions.

A second explanation may be that gradually international emission-based policies had been developed in various international platforms, such as in the EC, and at the North Sea Ministerial Conferences. Therefore, the emphasis of decision making on river basin policies had gradually shifted from the development of emission-based policies to the coordination and joint implementation of existing international and national policies, and the joint formulation of water quality objectives. The so-called combined approach addresses emission and immission-based policies simultaneously.

A third possible explanation are the developments in international law, which tend to the conclusion of framework conventions enabling the basin states to deal with the issues at stake in a flexible manner, rather than conventions with detailed policy objectives and/or measures. The Rhine basin states, for example, had gained positive experiences with the development of Action Programmes.

Finally, the Dutch may have learned that careful process management, trust building, and a step by step approach probably contribute more to an improvement of the water quality in the Scheldt and the Meuse than lengthy political negotiations on the development of ambitious policies. After the conclusion of the multilateral conventions and the installation of the provisional ICPS, the Dutch tried to stimulate trust building between the parties, and did primarily aim at good relations with the Belgian regions and the upstream basin state France.

Research and information exchange. The Dutch also continued to stimulate the exchange of information and joint research. They put a lot of effort in research on the impacts of dredging and dumping in the Scheldt estuary (OostWest). Research results indicating the negative impacts of dredging and dumping on the ecology of the Western Scheldt strengthened the Dutch position in the negotiations with Flanders on infrastructure and maintenance dredging works in the Western Scheldt, and, among other things, enabled the Dutch to demand a (financial) compensation for nature losses.

Together with the VMM, Rijkswaterstaat was the main initiator of the ISG, where the first regular contacts between experts of the Scheldt basin states and regions were established. Rijkswaterstaat was a main initiator of the LIFE-DSS project as well, because it wanted to extend a Dutch water quality model. Although Rijkswaterstaat seems to have had good substantial reasons for its refusal to cooperate on the extension of a Walloon model, one may discuss whether it was a wise decision to pass an opportunity to intensify the contacts with the upstream parties.[264] After all, all parties knew well what should be done to clean-up the

---

264. This seems to be true, all the more, since time after time the establishment of contacts between on the one hand the downstream parties Flanders and the Netherlands and on the other hand the upstream parties France, Wallonia, and Brussels, appeared to be difficult. This was the case in the ISG, in the workshop ISG-DES, and in the case of the organization of the third Scheldt symposium.

river Scheldt, and did not need a new model for that.[265] Because of the distrust that had grown between the basin states and regions, however, intensified interactions between the water managers of the basin states and regions, which could foster mutual understanding and learning processes, probably would have contributed more to the achievement of the Dutch objective, i.e. an improvement of the water quality of the Scheldt, than a perfect model.

The Walloon and Brussels Capital Regions: toward more interactive strategies
Whilst on the one hand the Flemish and Dutch governments and delegations formed a rather stable coalition in the negotiations on the multilateral Scheldt and Meuse conventions and the first SAP, on the other hand the Walloon and Brussels Capital regions formed a rather stable coalition. Although these two regions started to use more interactive strategies in the period between 1993 and 1997, they had more reservations as regards the development of international institutional arrangements and policies than the Flemish region and the Netherlands.

Evidence for their use of more interactive strategies are the participation in the ISG and LIFE-projects, their willingness to conclude multilateral conventions on the Scheldt and the Meuse, and their active participation in the ICPS and the ICPM. There are several possible (supplementary) explanations of the strategic change of these regions. First, the Dutch (and Flemish) governments had met two important wishes of these regions: the Dutch had decided to unlink the bilateral and multilateral Scheldt and Meuse issues, and were prepared to involve France in the negotiations on the Scheldt and the Meuse. Secondly, because Flanders and the Netherlands had started bilateral negotiations on the water quantity of the Meuse, the Walloon region would not have to discuss water quantity issues with the Dutch and Flemish governments any longer. Since the discussions on the construction of storage reservoirs in the Walloon Ardennes, water quantity management had been a very sensitive issue. Thirdly, because of the attribution of treaty-making competencies to the Belgian regions, the regions were able to negotiate directly with the basin states France and the Netherlands, and therefore could formulate their own standpoints clearly. Fourth, the strategic change may be interpreted as a strategic learning process as well. The regions may have learned that after the rapid development of international water management policies and legislation in the eighties and the nineties, they were hardly able to continue their strategies aimed at maintaining autonomy. Finally, after the Flemish region had raised investments in water quality policies, also in the Walloon and Brussels Capital regions investments were raised. Consequently, international water quality policies to be developed by the basin states and regions would have less extra financial consequences for these regions than before.

Despite the use of more interactive strategies, the Walloon and Brussels Capital regions were more reserved than the Dutch and Flemish governments as regards the development of international institutional arrangements and policies. Unlike the Flemish and Dutch delegations, they preferred the conclusion of a convention dealing with the water quality of the surface water of the main course of the river only, and opposed the conclusion of an integrated convention addressing the management of the water quality and quantity of both surface water and ground water basins in the Scheldt river basin. Furthermore, these parties opposed the formulation of strict water quality policies in the short term. This may be

---

265. All parties knew that waste water treatment plants would have to be constructed, just like new sewage systems. In addition, they were convinced of the necessity to develop institutional arrangements, such as systems for environmental permitting and taxation.

explained partly by the bad economic situation, in particular in the Walloon part of the basin. According to the representatives of the Walloon and Brussels Capital regions they needed time to develop regional water quality policies. Furthermore, the Walloon and Brussels Capital regions had to cope with a lack of personnel capacity. Finally, as regards the water pollution in the river Meuse, the Walloon Minister was of the opinion that the Dutch complaints were unjust (See Round Xb).

As regards the management of the water quantity in the Scheldt and the Meuse, the Walloon and Brussels Capital regions continued to use reactive strategies aimed at maintaining autonomy. The Walloon objections to the extension of the scope of the ICPS and the ICPM with water quantity issues may be explained partly by the distribution of water management competencies among two ministers.

The Belgian Federal state: not involved directly, but still influential
In Section 6.3 it was already concluded that in the period between 1985 and 1993 the power of the Federal state of Belgium was gradually declining. The Belgian Federal state is not a contracting party to most of the water conventions concluded, and therefore was no longer directly involved in decision making on international Scheldt issues. The Belgian Federal state, however, was involved in decision making on the alignment of the HSL, and for that reason influenced decision making on international Scheldt issues indirectly. Furthermore, the Federal State of Belgium was a contracting party to the Convention on the revision of the Scheldt regulations. Finally, because the Federal state is no contracting party to the Scheldt and Meuse conventions, but still has the competencies concerning nuclear policies, the Belgian regions were unable to discuss these policies in the ICPM and the ICPS.

The French government, the mediator
France was the new party involved in decision making on international Scheldt and Meuse issues. According to some respondents, France more or less played a mediating role between the Netherlands and the Belgian regions, which had so many conflicts in the past. With respect to some issues the French points of view were very similar to the Dutch and Flemish ones. Like the Dutch and Flemish, the French aimed at an integrated convention dealing with water quality and quantity issues simultaneously. As regards the development of water quality policies, however, France was more reserved, and preferred a gradual, step by step, approach. The French seem to perceive the pollution issues in the Scheldt as typical commons problems (See Section 2.3.2). The French statements made in the ISG and the ICPS indicate that the French parties were willing to clean-up the river Scheldt, but were prepared to raise investments in the development of water quality policies only if the basin states and regions situated downstream, among which the Brussels Capital region, would do the same. This French perception and strategy may be explained partly by the bad economic situation in the French part of the Scheldt basin, which makes it more difficult to impose charges on polluting industries and households. A second reason may be the organization of French water management. In the French basin committees, the water users, among which industries and households, do have a big say in the development of water quality policies, and decision making on charges.

Perceptions and strategies of lower level governments and NGOs
   Provinces. The three southern Dutch provinces, Zeeland, Northern-Brabant, and Limburg, formed a coalition, and jointly opposed the linkage between decision making on the deepening programme and the alignment of the HSL. According to them, the linkage

would mainly benefit the Dutch 'Randstad'. For the province of Zeeland, which was mainly interested in the water quality of the river Scheldt, this was the second linkage, after the linkage across Scheldt and Meuse issues-areas, that would serve national Dutch interests rather than the interests of the province of Zeeland. The provinces criticized the Dutch refusal to sign the Convention on the deepening of the navigation channel in the Western Scheldt, because a further delay of the implementation of the deepening programme would frustrate the cooperation between the three southern provinces and the Flemish region, and the cooperation in the RSD. The provinces also pleaded for the establishment of a sanitation fund and for a direct involvement of the provinces in the international negotiations. The Dutch minister, however, did not want to establish a sanitation fund for the reasons discussed in Section 6.3.1, and was of the opinion that a direct involvement of the provinces in the negotiations would not be useful. After the conclusion of the water conventions a representative of the province of Zeeland became member of the Dutch delegation in the ICPS. From that moment the province of Zeeland was directly involved in decision making on the Scheldt water quality issues.

Municipalities. The Dutch municipalities along the Western Scheldt opposed the linkage between decision making on the deepening of the navigation channel and the alignment of the HSL for the same reason as the three southern provinces. To increase their influence on decision making on the various Scheldt issues, several municipalities along the Western Scheldt used their competence to issue a permit based on the Dutch Act on Land-Use Planning, and made a linkage between the issuing of this permit and the safety of navigation on the Western Scheldt. In addition, the municipalities demanded an EIA for the deepening programme. Because of its interest in the water quality of the Western Scheldt, the Task group for the Western Scheldt pleaded for the formulation of extra water quality conditions in the WVO-Permit for the Flemish region. Furthermore, the municipalities established cross-border contacts with Flemish and Walloon municipalities to transfer their knowledge about environmental management. After the conclusion of the water conventions a representative of the Task group for the Western Scheldt became member of the Dutch delegation in the ICPS. From that moment, like the provinces, also the municipalities were directly involved in decision making on the Scheldt water quality issues.

Seventeen municipalities along the river Meuse, among which the city of Rotterdam, and drinking water production firms, which suffered from water and sediment pollution in the river Meuse, did not want to wait any longer on the results of the international negotiations on the water quality of the Meuse. They passed over the Dutch national government and ministries, and took action themselves. In the project POM these parties threatened with litigation against polluting industries in the Meuse basin, and with this big stick tried to conclude covenants on pollution reduction with these industries.

NGOs. Like the provinces and the Task group of the Western Scheldt, also several NGOs opposed the linkage between decision making on the deepening programme and the alignment of the HSL, and the Dutch refusal to approve the Convention on the deepening of the navigation channel in the Western Scheldt. Representatives of the Rhine Scheldt Delta (RSD), the chairmen of the Chambers of Commerce of Antwerp and Rotterdam, and the Dutch national Port Council all did argue that a further delay of the implementation of the deepening programme would obstruct cooperation between the ports of Antwerp and Rotterdam. The AGHA got enough of the continuous Dutch refusal to sign the deepening programme, and threatened with litigation several times, as the Dutch policy would not be in accordance with the Scheldt Statute.

On the other hand, Flemish and Dutch environmental NGOs strongly opposed the implementation of the deepening programme, demanded an EIA, and eventually appealed to court. In addition, they contacted members of the European Parliament, who were asked to control the Dutch and Flemish implementation of EC-policies. Flemish environmental NGOs tried to influence decision making by publishing a CBA-study, which stated that implementation of the deepening programme would not be cost-effective. Furthermore, they started litigation against the dredging permit issued by the Dutch government.

Both Dutch and Flemish environmental NGOs concluded a covenant with their respective Ministries concerning the implementation of the deepening programme, and the compensation of nature losses caused by the implementation of that programme. The Dutch environmental NGO ZMF, however, was unable to live up to the agreements made in the covenant, because its members wanted to continue litigation against the WVO Permit granted by the Dutch government.

Like the Dutch lower level governments, also Dutch environmental NGOs demanded participation in the ICPS and the ICPM. Unlike the government agencies, however, the NGOs were not invited to become a member of the Dutch delegation in these commissions. The covenant mentioned above, however, included the right to participate in the preparatory meetings of the Dutch delegations of the working groups.

In the project 'Scheldt without frontiers', environmental NGOs tried to establish transboundary contacts, and with that to build transboundary coalitions.

7.3.2 Strategic interactions and institutionalization processes

In the period between 1993 and 1997 the relationship between the Scheldt basin states and regions did get more elements of cooperation, but continued to have elements of conflict as well. On an abstract level the basin states and regions were able to reach an agreement on the principles of cooperation in the Scheldt and Meuse basins. The joint decision making on concrete policy objectives and measures, however, clearly had conflicting elements.

Strategic interactions and the role of issue-linkage

An interesting conclusion of the case study is that the conclusion of all water conventions did involve issue-linkage. The Flemish and Dutch governments reached an agreement on an extensive package deal including all international Scheldt and Meuse issues. Decision making on the content of the WVO-permits for the Flemish region did involve issue-linkage as well. These results are in accordance with the theory that in situations where the decision making rule is unanimity, and there are heterogenous preference intensities across issue-areas, the probability that movement away from the status quo will involve tactical issue-linkage is high. The statement made by the Dutch Minister after the conclusion of the water conventions, that she got enough of the linkages and that she, just like her Flemish colleague, does not want to hear the word linkage anymore, is comprehensible but nevertheless unrealistic (See Section 7.2, round XI). The conclusion of the lengthy and laborious negotiations on the water conventions should not be that the strategy of issue-linkage as such is wrong, but that a wrong application may be disastrous for the course of a decision making process. Diverse statements have been made concerning the linkages that were made in the negotiations. Some argued that the linkage between issues in two basins was the main mistake of the Dutch negotiators, whilst others were of the opinion that a linkage between infrastructure issues and water management issues caused the main problem. The case analysis, however, has shown that not a specific type of linkage should be excluded as a strategy promoting international cooperation, but all linkages that do not make all parties better off. The linkage between the deepening of the navigation channel in the Western

Scheldt and the water quality of the Meuse did not benefit all parties, because the Walloon region did not perceive an interest in any of the issues at stake, and the linkage therefore turned out to be largely unsuccessful. The linkage between the deepening of the navigation channel in the Western Scheldt and the water quality of the Scheldt, however, for example in the WVO Permit, did create a win-win situation, and therefore appeared to be successful. Also the linkage between Flemish-Dutch decision making on the water quantity of the Meuse and the deepening of the navigation channel in the Western Scheldt was a wise one, because the same parties were involved in decision making on both issues.

The development of international institutional arrangements and policies
The multilateral agreements between France, the Walloon and Brussels Capital regions, and the Netherlands, did neither involve issue-linkage nor any financial compensation. As argued before, these conventions reflect the greatest common denominator, and certainly do not go beyond existing national policies of the individual basin states and regions. Nevertheless, the conventions provide for important institutional arrangements, such as agreements on the installation of international river commissions, the exchange of information, joint monitoring, the development of an alarm system for cases of accidental pollution, and the preparation of Action Programmes with joint water quality policies.

Like in the period studied in the previous chapter, also in the period between 1993 and 1997 the conditions formulated in the WVO Permit for the Flemish region are the most important international water quality policies in the Scheldt basin.

The Convention on the deepening of the navigation channel in the Western Scheldt entails the implementation of the 48'/43'/38' deepening programme and the development of a plan for nature restoration, and therefore, unlike the two water quality conventions, does contain concrete policies.

Beside the conclusion of important international conventions in 1994 and 1995, which entailed the development of international institutional arrangements and policies, also transboundary research networks further developed in the period between 1993 and 1997. Water management experts of the basin states and regions exchanged information and carried out joint research in the international projects OostWest, ISG, ISG-DES, and the two LIFE-projects. Until the installation of the provisional ICPS, however, the establishment of contacts between the upstream and downstream basin states and regions went rather laboriously.

Limits to the development of international institutional arrangements
In spite of the increasing intensity of interactions between the Scheldt basin states and regions in the period between 1993 and 1997, national boundaries remained important obstacles for the development of institutional arrangements or policies on a basin scale. First, the basin states did not establish a sanitation fund, whilst some investments that are made in the downstream parts of the rivers Scheldt and Meuse would certainly yield more benefits if they would be made in the upstream parts of these rivers. Secondly, partly because of the Convention on the deepening of the navigation channel in the Western Scheldt, the Dutch had to spend the 44 million Dutch guilders on Dutch territory, whilst experts preferred to develop one nature restoration plan for the Scheldt estuary.

Gradually improving relationships
After the Dutch decision to unlink the bilateral and multilateral Scheldt and Meuse issues, and the involvement of France in the negotiations on the water quality of the Scheldt and the Meuse, the deep distrust that had grown between the parties declined gradually. The careful

start of the multilateral negotiations with two informal meetings, and the strategies used by the parties contributed to a gradually improving negotiation climate. Nevertheless, some events caused a disturbance of the relations. The Dutch POM-project, some cases of accidental pollution in the Meuse, and the statements made by the Dutch Minister of Transport, Public Works, and Water Management, caused Walloon-Dutch conflicts at the beginning of the multilateral negotiations. In the bilateral Flemish-Dutch negotiations, the Dutch delay of the approval of the Convention on the deepening of the navigation channel in the Western Scheldt seems to have reinforced the Flemish distrust of the Dutch. The Flemish decision to re-establish the linkage between bilateral and multilateral issues illustrates the high degree of distrust that had grown between the parties. Only after the conclusion of the water conventions, international relations in the Scheldt (and the Meuse) basin could really improve. Whilst the expectations concerning the content of the first SAP did not run high, in 1996 most respondents judged positively on the relations between the parties, and the negotiation climate.

7.3.3 Learning processes
Like in the other case study periods, also in the period between 1993 and 1997 some strategic changes of the Scheldt basin states and regions could be interpreted as the result of a strategic learning process. These examples were discussed in Section 7.3.1, where the dynamics of the perceptions and strategies of the individual basin states and regions were discussed.

Beside examples of strategic learning, some interesting examples of cognitive learning processes could be found. As was argued in Section 6.3.3, the research project OostWest had provided valuable information on the functioning of the Scheldt estuary, among which information on the impacts of dredging works, and the dumping of sediments on the morphology and ecology of the Scheldt estuary. Furthermore, in this project the policy alternative 'ontpolderen' was introduced. In the period between 1993 and 1997 these research results were going to play an important role in the decision making process. First, the information on the consequences of dredging and dumping in the Scheldt estuary provided more insight in the nature losses that would be caused by the implementation of the deepening programme, which could be used in the discussions on the compensation of these nature losses. Secondly, because of the increased knowledge on the negative impacts of dredging and dumping in the Scheldt estuary, the Dutch started to formulate conditions in the dredging permit for the Flemish region, which aimed at a reduction of the dredging intensity in the Western Scheldt. Thirdly, the policy alternative 'ontpolderen' played an important role in decision making on a plan for nature restoration. In spite of the objections in the Dutch Province of Zeeland to 'ontpolderen', and the advice of the commission of wise persons not to use this policy alternative in the short term, research on this policy alternative will continue. Fourthly, the review team, which studied the research results of OostWest, concluded that the research results had shed a new light on the construction of the Baalhoek Canal (See Round Xa). Possibly, these considerations will be taken into account in future decision making on the possible construction of a second maritime access to the Waasland harbour, and the development of a joint Flemish-Dutch long term policy plan for the Scheldt estuary.

Whilst in the issue-area of the maritime access to the port of Antwerp research clarified the intellectual relations between issues, such as dredging, dumping, and nature restoration, and seems to have had a considerable influence on decision making, in the issue-area of water and sediment pollution, research seems to have been less influential. The ISG and LIFE-DSS projects, however, contributed to the development of personnel networks between the parties. Since multilateral Scheldt water quality policies have not yet been developed, the influence of these research projects on the content of such policies could not be assessed.

7.3.4 Influence of the context on the perceptions and strategies of the Scheldt basin states and regions

Underlying hydrological structure of the issues at stake
There is quite some evidence that the underlying hydrological structure of the international Scheldt and Meuse issues continued to influence decision making in the period between 1993 and 1997. The Dutch government used its relative power advantage in decision making on the deepening programme to exert influence on the Flemish region in decision making on the water quantity of the Meuse and the alignment of the HSL. In turn, the Flemish region used its relative power advantage in decision making on the water quantity of the Meuse and in decision making on the alignment of the HSL to exert influence on the Dutch government in decision making on the deepening programme.[266]

Beside the influence of the power asymmetries caused by the underlying hydrological structure of the issues at stake on the strategies that were used in the international negotiations on these issues, there is another indication of the influence of the underlying hydrological structure of the issues at stake. Most initiatives for projects related to the water quality or ecology of the river Scheldt, such as the ISG, ISG-DES, the Benelux and Scheldt symposia, and the LIFE-project, were taken by the downstream parties Flanders and the Netherlands. This, however, may be explained by the limited personnel capacity in the Walloon and Brussels Capital regions, or the relative limited mandates of civil servants in these regions as well (See the second next section).

Although there still are indications that the underlying hydrological structure of the issues at stake has influenced decision making, there also is some evidence that this contextual factor did play a less dominant role in decision making on the pollution issues in the rivers Scheldt and Meuse than in the periods studied in the preceding chapters. As mentioned before, the new party France perceived the pollution issues more as commons problems than as real upstream-downstream problems. In spite of its upstream position in the basin, and its relative power advantage, the French favoured joint action to clean-up the rivers Scheldt and Meuse. Furthermore, also the Walloon and Brussels Capital regions, in spite of their upstream positions, started to use more interactive strategies, and to cooperate on a clean-up of these rivers.

International context
Decision making on international Meuse issues formed a main international context of decision making on the international Scheldt issues. In the period between 1993 and 1995, however, because of the Flemish-Dutch linkage of the deepening programme and the alignment of the HSL, decision making on the alignment of the HSL was an even more influential contextual development. Only after procedural agreements on this issue had been reached, the road was clear for the signing of the water conventions.

---

266. In principle, as regards decision making on the alignment of the HSL the parties are mutually dependent. Nevertheless, because the Dutch government perceived a bigger interest in rapid decision making on this issue than the Flemish government, the Dutch government was more dependent on the Flemish government than vice versa.

The development of international water quality policies in the EC, i.e. the development of emission and water quality directives, the North Sea Ministerial Conferences, and the UN-ECE convention clearly influenced the multilateral negotiations on the Scheldt and Meuse conventions. Partly because of the development of emission-based and water quality policies in these international organizations, along with the experience gained with the Rhine Action Programme, the Dutch did no longer aim at detailed policy agreements, but favoured the conclusion of so-called framework conventions.

Apart from the influence of the EC-water policies, the EC influenced decision making on international Scheldt issues in some other respects. First, the EC took important decisions concerning the obligation of an EIA on the 48'/43'/38' deepening programme, and the compensation of nature losses that would be caused by the implementation of this programme. Secondly, beside its intervention position, the EC had an incentive position in decision making on the international Scheldt issues, and stimulated transboundary interactions with the LIFE-programme (two LIFE-DSS projects and the LIFE-MARS-project). Some respondents did expect the EC to play an even more important role in future decision making on international Scheldt issues. According to them, negotiations on the Scheldt and the Meuse will have to be reopened after the adoption of the EC Framework Directive for Water, because the present Scheldt and Meuse conventions are no integrated conventions, but deal with water quality and ecological issues only.

The Benelux Parliament continued to exert pressure on the Scheldt (and Meuse) basin states and regions, and, among other things, adopted a resolution demanding a rapid implementation of the deepening programme, and the installation of river commissions.

Cultural context
Cultural background variables may provide part of the explanation of the perceptions and strategies of the actors involved in the decision making process.

First, some examples seem to illustrate the relatively large influence of politics on policy making in Belgium, and the relatively limited mandates of civil servants. Dutch civil servants, for example, faced difficulties with the establishment of contacts with Walloon water management experts, and for a long time experienced that only contacts on the (sub)political level were possible.

Secondly, the basin states and regions clearly have different traditions concerning the openness of decision making processes. The Dutch and Flemish delegations in the multilateral negotiations preferred to invite NGOs to become a member of the river commissions to be installed. In particular the Walloon region opposed the involvement of NGOs in the discussions of the commissions. Whilst the Dutch are used to relatively open decision making processes and the participation of interest groups, Belgian decision making tends to be more hierarchical. Interesting, however, is that in spite of its high score on the cultural dimension 'Power Distance', the Flemish region was in favour of the involvement of NGOs. Furthermore, it should be noted that at a more local level, the Walloon region does involve NGOs in decision making on the so-called 'contrats de rivière'.[267]

Thirdly, the basin states and regions seem to have a different perception of the (hierarchical) relationship between the national government, lower level governments, and NGOs. This may explain why the Walloon region reacted furiously on the POM-project, whilst the Dutch national government had to accept this initiative, because of the autonomy

---

267. See Appendix 6.

of Dutch local government agencies. Possibly, the Dutch national government was more used to initiatives of the relatively autonomous lower level governments than the Walloon government. Another example concerns the implementation of the deepening programme. In first instance, the Flemish government did only want to discuss the implementation of the deepening programme with the Dutch Ministry, and refused to speak with lower level governments. Later, however, the Flemish learned about the Dutch decision making procedures, and contacted the municipalities to apply for the necessary permits.

Intranational context
After the conclusion of the Sint Michiels agreements and the attribution of treaty-making competencies to the regions, intranational Belgian relations became more transparent, and negotiations on the water conventions went smoother. Nevertheless, the relation between the Belgian federal state and the regions still had an impact on international decision making. First, as regards decision making on the alignment of the HSL, beside the Flemish region also the federal state of Belgium had important competencies. Secondly, in the ICPS and the ICPM it turned out that the Belgian regions are not able to discuss radioactivity policies, because the Belgian federal state, which is no contracting party to the Scheldt and Meuse conventions, has the competencies in this field. Also the intranational organization of water management in the Walloon region seems to have influenced the development of international institutional arrangements and policies. One of the reasons why the Walloon region was not inclined to discuss water quantity issues in the international Meuse commission, was that two ministers are responsible for water management, one for the water quality and one for the water quantity management, and that only one of them was involved in the negotiations on the Scheldt and Meuse water management issues.

Apart from the internal Belgian state organization, also the different stages of development of water quality policies influenced the decision making process. These differences may explain why the delegations of the Walloon and Brussels Capital regions in the multilateral negotiation commission and the ICPS preferred a more gradual development of international policies than the Dutch and Flemish delegations. The Belgian regions did also have less personnel capacity than the Dutch government agencies involved in decision making on international Scheldt issues. On the other hand, the gradual development of water quality policies in the Belgian regions may explain why they were prepared to sign the multilateral water quality conventions, and to start cooperation on the clean-up of the rivers Meuse and Scheldt at all.

The last part of the case study also provided quite some evidence for the influence of economic background variables. The economic problems and high unemployment rates in the upstream parts of the Scheldt basin, i.e. France and the Walloon region, make it rather difficult to impose costly water quality policies on industries and households in these parts of the Scheldt basin. This may also explain why the French emphasized the need for comparable efforts in all parts of the basin.

7.3.5 Decision making on nature restoration
Although in Section 7.2 decision making on nature restoration was described shortly, a detailed analysis of this decision making would go beyond the scope of this thesis, which after all focuses on the *international* interactions between the Scheldt basin states and regions. Nevertheless, like the implementation of the deepening programme, the case of nature restoration along the Western Scheldt is an interesting illustration of the multi level character of decision making on international river issues. On the international level the Netherlands

and the Flemish region had reached an agreement on the compensation of nature losses caused by the implementation of the deepening programme. The implementation of this provision in the Convention on the deepening of the navigation channel in the Western Scheldt, however, was an *intra*national Dutch concern. The Dutch national government, which had concluded the convention, faced the problem of a lack of regional support for the proposed policy measure to compensate nature losses, namely 'ontpolderen'.

Some main conclusions of the analysis of decision making on nature restoration made by Ringeling are that[268]:

- the decision making process did not have a clear organization, mainly because two problem definitions were added to the original one. The original problem discussed was the compensation of nature losses caused by the implementation of the deepening programme. A first problem definition added by Rijkswaterstaat was the restoration of nature losses caused by activities in the past. A second problem definition added was the development of a long term vision for the Scheldt estuary;
- there was a lack of communication between the parties;
- a good process manager that would have to organize and facilitate this communication was lacking;
- Rijkswaterstaat sometimes played two roles in the decision making process simultaneously. It tried to manage the decision making process as a more or less independent party, whilst some representatives of Rijkswaterstaat were main protagonist of the policy alternative 'ontpolderen';
- the psychological aspects related to the policy alternative 'ontpolderen' were underestimated.

In the theoretical chapter, it was stated that it generally takes a period of a decade or more until research results do have an impact on decision making, and that it is the cumulative effect of findings from different studies and from ordinary knowledge that has the greatest influence on policy (See Section 2.2.7). Therefore, continued research on the dynamics of the Scheldt estuary, and possibilities of a 'partial landward movement of the sea walls', may influence decision making in the long run. This on the one hand may be a comfort for the advocates of 'ontpolderen', but on the other hand be a threat to the opposers of this policy alternative.

7.4 Conclusions

This chapter addressed the third case study period, which extends from 1993 to 1997. In the first part of the chapter a detailed reconstruction of the decision making process was presented. This reconstruction is based on the analysis of written documents and reports of interviews with participants in the decision making process. The results of this analysis are a long list of decisions taken by the various actors involved in the decision making process, and a coherent description of the interactions between them. Like in the periods studied in the preceding chapters, also in the period between 1993 and 1997 several crucial decisions, and decision making rounds could be distinguished. The first decision making round described in this chapter (Round Xa) concerned the Flemish-Dutch negotiations on the deepening of the navigation channel in the Western Scheldt, the distribution of the Meuse

---

268. Ringeling (1997).

water, and the alignment of the HSL. Simultaneous to these bilateral negotiations, the Scheldt and Meuse basin states and regions negotiated multilateral conventions for the water quality of the Scheldt and the Meuse (Round Xb). After the Flemish decision to re-establish the linkage between the bilateral and multilateral Scheldt and Meuse issues, it took another round of Flemish-Dutch deliberations until the Flemish and Dutch governments were able to reach an agreement on the conclusion of the water conventions (Round XI). Negotiations between the Scheldt basin states and regions on the international pollution issues were continued in the provisionally installed ICPS (Round XIIa). Simultaneous to these multilateral negotiations, Flanders and the Netherlands made every effort to implement the convention on the deepening of the navigation channel in the Western Scheldt (Round XIIb). The Flemish and Dutch governments tried to implement the 48'/43'/38' deepening programme, and Dutch government agencies tried to develop a plan for nature restoration.

The case analysis focused on the action rationality of the individual basin states and regions, and addressed the dynamics of their perceptions and strategies. In the period between 1993 and 1997 the Dutch government continued to use the issue of the deepening of the Western Scheldt as exchange in the negotiations on issues the Dutch were interested in, such as the distribution of Meuse water, and the alignment of the HSL. In the negotiations on the water quality issues, however, the Dutch government gradually changed its strategy, and aimed at the conclusion of framework conventions and the installation of basin commissions, rather than at detailed agreements on water quality objectives. In the provisionally installed ICPS the Dutch did primarily aim at trust building between the parties. The Flemish strategy was characterized by the expression 'desperate needs lead to desperate deeds'. Because of the continuous Dutch refusal to cooperate on the implementation of the deepening programme unconditionally, the Flemish region cut capers. First, it made a tactical mistake by linking the issue of the alignment of the HSL to the deepening programme. Later, the region re-established the linkage across bilateral and multilateral issues it had always opposed. In the negotiations on the water quality of Scheldt and Meuse the Flemish region formed a rather stable coalition with the Netherlands, and aimed at integrated conventions dealing with water quality and quantity management of groundwaters and surface waters. The Walloon and Brussels regions had about the same points of view, and with that formed a rather stable coalition as well. On the one hand, these regions began to use more interactive strategies, and were prepared to discuss the international water quality issues in the Scheldt and the Meuse. On the other hand, these regions preferred a relatively slow and gradual development of international institutional arrangements and policies, and did not want to formulate policies that would go beyond existing national and regional policies. Finally, the new party France perceived the international pollution issues in the Scheldt basin as commons problems. The French wanted to cooperate on a clean-up of the river Scheldt, but emphasized more often the need for comparable efforts in all parts of the basin. After the long history of conflicts between the Netherlands and the Belgian regions, the French parties did play a mediating role in the multilateral discussions on the water quality of the Scheldt and the Meuse.

The analysis of the perceptions and strategies of lower level governments and NGOs clearly illustrated the two-level character of decision making on international river issues, and the danger of the occurrence of an implementation gap. Lower level governments and NGOs used their resources, such as the competence to grant permits, or the right to appeal to court, to introduce their problem perceptions, and to demand concessions from national government agencies. The issuing of special legislation, however, also showed that, unlike in most international decision making processes, formal hierarchy may be an important instrument in *intra*national decision making.

As regards the linkages that were made in the period between 1993 and 1997, an interesting conclusion is that almost any agreement that has been reached on any Scheldt or Meuse issue did involve issue-linkage. Consequently, issue-linkage seems to be a useful means for the development of international institutional arrangements and policies in cases where decision making is based on unanimity, and the parties indeed do have different preference intensities in different issue-areas.

The conclusion of the water conventions, which was a significant contribution to the development of international institutional arrangements and policies, indicates that the international relations did get important cooperative elements. On an abstract level the basin states and regions were able to reach an agreement on the need for international joint action to clean-up the rivers Scheldt and Meuse. Furthermore, transboundary professional networks were extending gradually. Equally important is that after the Dutch decision to unlink bilateral and multilateral Scheldt and Meuse issues, a culture of trust had been growing gradually. In spite of these cooperative elements, however, after the conclusion of the water conventions the interactions between the basin states and regions continued to have conflicting elements as well. The Scheldt basin states and regions still had different and conflicting preferences concerning the water quality policies to be developed, and in spite of the good relations between Flanders and the Netherlands, these parties still have not reached an agreement on the interpretation of the Scheldt Statute.

Some strategic changes could be interpreted as the result of a strategic learning process. In particular in the issue-area of the maritime access to the port of Antwerp, beside strategic learning processes, also cognitive learning processes have developed. The diverse linkages that were made in the negotiations make it very plausible that the distribution of interests and resources caused by the underlying hydrological structure of the issues at stake did largely influence the strategies of the basin states and regions. The issuing of EC and North Sea policies, and the experiences gained with the Rhine Action Programme may partly explain the preference of the Scheldt basin states and regions to conclude framework conventions for the water quality of the Scheldt and the Meuse, and to prepare Action Programmes for the clean-up of these rivers in the international river commissions to be installed. In some cases perceptions and strategies could be related to cultural characteristics of the basin states and regions, and the influence of cultural background variables on the decision making process could be made plausible. Finally, like in the periods studies in the previous chapters, also in the period between 1993 and 1997 the intranational context of the actors seems to have strongly influenced their perceptions and strategies. Several examples illustrated that the intranational organization of water management shaped relations of (inter) dependence that influenced the strategies used in international decision making. In addition, the bad economic situation in the French and Walloon parts of the Scheldt most likely were important explanatory variables for the French and Walloon perceptions of the international pollution issues, and the strategies these parties used in decision making on these issues.

Part III:
Conclusions,
Recommendations, and
Final Discussion

CHAPTER 8

CONCLUSIONS

8.1 Introduction

Part II of the thesis contains the description and analysis of decision making on international Scheldt issues between 1967 and 1997. In this chapter the main conclusions from this case study are summarized. In Section 8.2 the answers to the descriptive research questions are presented, whereas in Section 8.3 the answers to the four explanatory research questions are discussed. Finally, in Section 8.4 some additional conclusions, which are not related directly to any of the research questions formulated beforehand, are drawn.

8.2 Answers to descriptive research questions

The two descriptive research questions formulated in Chapter 3 were:

1. *How did decision making on international Scheldt issues develop?*

2. *What were the characteristics of the context of decision making on international Scheldt issues?*

The first research question was answered in the Sections 5.2, 6.2, and 7.2. The analysis of written documents, and the information collected by forty interviews enabled the researcher to make a chronological list of decisions made by the actors involved in the decision making process (See Appendix 12). Some of these decisions appeared to have had more influence on the decision making process than others, and to have changed the composition of the international arena or the issues on the international agenda. Therefore, in the period of 30 years studied in this thesis, fourteen decision making rounds could be distinguished. Two times decision making rounds took place simultaneously. Table 8.1 gives on overview of the decision making rounds distinguished in the case study of decision making on international Scheldt issues.

The second research question was answered in Chapter 4. The main conclusions of the analysis of the context of decision making on international Scheldt issues were summarized in Section 4.9.

Table 8.1 *Decision making rounds* The Table lists the decision making rounds distinguished in the case study of decision making on international Scheldt issues between 1967 and 1997. In the periods from 1993 to 1994, and from 1995 to 1997 two decision making rounds took place simultaneously. These rounds are indicated a and b.

Round		Title round
I	(1967-1968)	Dutch linkage of Scheldt and Meuse issues
II	(1969-1975)	Preparation of three draft conventions
III	(1975-1983)	Internal Belgian disagreement
IV	(1983-1985)	Dutch linkage of the 48'/43'/38' deepening programme to Meuse water management issues
V	(1985-1987)	Modification of the declaration of intent of 1985
VI	(1987-1989)	Negotiation commission Biesheuvel-Davignon
VII	(1989-1991)	Negotiation commission Biesheuvel-Poppe
VIII	(1991-1992)	Dutch and Belgian draft conventions
IX	(1992-1993)	The break-through
Xa	(1993-1994)	Bilateral negotiations
Xb	(1993-1994)	Multilateral negotiations and Flemish decision to re-establish the linkage between bilateral and multilateral issues
XI	(1994-1995)	A Flemish-Dutch agreement and the conclusion of the water conventions
XIIa	(1995-1997)	Preparation of the first Scheldt Action Programme
XIIb	(1995-1997)	Implementation of the Convention on the deepening of the navigation channel in the Western Scheldt

8.3   Answers to explanatory research questions

The four explanatory research questions formulated in Chapter 3 were:

3. *How did the perceptions of the actors involved in decision making on international Scheldt issues influence their strategies?*

4. *How did the strategic interactions between the actors involved in decision making on international Scheldt issues influence structural and cultural network characteristics?*

5. *Did learning processes develop?*

6. *How did the context of decision making on international Scheldt issues influence the perceptions and strategies of the actors involved in decision making on these issues?*

In the Sections 8.3.1 to 8.3.4 the answers to these questions are summarized.

8.3.1 Influence of perceptions of the actors involved in decision making on international Scheldt issues on their strategies

Representatives of national governments and their bureaucracies are the main players in the international arena. Nevertheless, in some cases lower level governments or NGOs have had a considerable impact on the decision making process as well. This section presents the main conclusions concerning the relation between the perceptions of the actors involved in the decision making process and their strategies.

The Belgian state
The Belgian national government perceived a dependence on the Netherlands as regards the realization of several infrastructure projects to improve the maritime access to the port of Antwerp: the construction of the Baalhoek Canal, the straightening of the bend near Bath, and the 48'/43'/38' deepening programme. Therefore, the Belgian state used offensive interactive strategies, and tried to place these issues on the international agenda. Although the Belgian state opposed formally the Dutch linkages between the deepening programme and Meuse water management issues, because this linkage would not be in accordance with the Scheldt Statute, the Belgian national government aimed at a package deal with the Dutch including several Scheldt and Meuse issues. The Belgian national government, however, continuously faced the problem of internal Belgian disagreement. Whilst the Scheldt infrastructure issues were of the utmost importance for the economic development of the Flemish region, the Walloon region did not perceive an interest in any of the Scheldt or Meuse issues at stake. On the national level several attempts were made to reach a consensus between the regions, but all failed. This may be explained partly by the ongoing federalization process, which entailed the attribution of most water management competencies to the Belgian regions. After the attribution of treaty-making competencies to the Belgian regions, the Belgian federal state would only play a minor role in decision making on the international Scheldt (and Meuse) issues. Because of its involvement in decision making on the alignment of the HSL, the approval of some international conventions, and its competencies as regards the development of nuclear policies, however, the Belgian federal state continued to influence decision making on international Scheldt issues indirectly.

The Flemish Region
As mentioned above, in particular the Flemish Region perceived an interest in the infrastructure projects aiming at an improvement of the maritime access to the port of Antwerp. For the realization of these projects the region did not only perceive a dependence on the Dutch, but, after the Dutch had made linkages between the Scheldt infrastructure and Meuse water management issues, on the Walloon Region as well. Although the Flemish Region opposed the linkages that were made to the deepening programme, and the conditions formulated to the WVO permit needed for the dumping of dredged material in the Western Scheldt, because these Dutch policies would not be in accordance with the Scheldt Statute, the region had to accept the Dutch linkages, and aimed at a package deal with the Dutch. Because of its dependence on the Dutch authorities, the Flemish region tried to build up a reservoir of goodwill. Anticipating the attribution of water management competencies to the Belgian regions, the Flemish Region started with the development of water quality policies. Furthermore, the region was prepared to conclude bilateral Flemish-Dutch agreements on the water quantity of the Meuse, and the water quality of the Scheldt and the Meuse, in case the

Walloon region would continue its refusal to cooperate on these issues internationally. The Dutch decision to unlink the bilateral and multilateral Scheldt and Meuse issues was a major relief for the Flemish Region, because it would no longer depend on the Walloon Region as regards the realization of its infrastructure projects. Shortly after this decision, however, the Flemish region made a tactical mistake by linking the deepening issue to the discussions on the alignment of the HSL. This linkage would cause another delay of the Dutch approval of the deepening programme, and caused the Flemish region to re-establish the linkage between bilateral and multilateral issues the region had always opposed. In the multilateral negotiations on the water quality of the Scheldt, and in the provisional ICPS the Flemish Region had about the same points of view as the Dutch, and therefore formed a rather stable coalition with them.

The Walloon Region

Unlike the Flemish Region and the Netherlands, the Walloon Region did not perceive an interest in any of the Scheldt or Meuse issues at stake. On the contrary, the Walloon politicians were of the opinion that they would have to pay for the benefit of the Flemish Region and the Netherlands. The region did not want to impose costly water quality policies, neither was it willing to construct storage reservoirs on its territory to regulate the flow of the river Meuse. Consequently, the region used reactive strategies aimed at maintaining autonomy. The Walloon objections to the water quality conditions formulated in the Meuse and Bath conventions of 1975 may be explained partly by the relatively bad economic situation in the region. Other reasons for the Walloon objections, however, were that the draft conventions did only contain obligations for Belgium, that the upstream basin state France was no contracting party, the linkage across Scheldt and Meuse issues-areas, and last but not least that the region had been insufficiently involved in the negotiations on the water conventions. Because neither the Dutch nor the Flemish parties changed their strategy fundamentally, continued to aim at a package deal, and did not involve France in the negotiations, the Walloon region continued its reactive strategies aimed at maintaining autonomy until 1993. Among other things, the region discontinued the discussions in the SWC and the TMC. Only after the Dutch decision to unlink the bilateral and multilateral Scheldt and Meuse issues, and the involvement of France in the negotiations on the water quality of the Scheldt and the Meuse, the region started to use more interactive strategies. Nevertheless, the region wanted to restrict international cooperation to the water quality of the main course of the river only, and did prefer a gradual development of water quality policies. The region was not inclined to discuss water quantity policies in the ICPS, which may be explained partly by the fragmented organization of water management in the Walloon Region.

The Brussels Capital Region

Unlike the Flemish and Walloon Regions, the Brussels Capital Region only received water management competencies after the constitutional amendment of 1988. Even more than the Flemish and Walloon Regions, the Brussels Capital Region had to eliminate a backlog as regards the development of water quality policies. When the region became involved in the negotiations on the water conventions in 1989, it had about the same points of views as the Walloon region, and mainly used reactive strategies aimed at maintaining autonomy. Although since 1993 the Brussels Capital Region started to use more interactive strategies, and was willing to conclude multilateral agreements on the management of the Scheldt and the Meuse, it preferred the international policies to be restricted to the water quality of the main course of the river only. It should be noted that the Brussels Capital Region is situated in the basin of the river Senne, which is a sub-tributary of the river Scheldt.

The Netherlands
The Dutch national government perceived most interest in the water quality and quantity of the river Meuse. Some main reasons for this are the use of Meuse water for drinking water production in the Dutch 'Randstad', and the high costs of the clean-up of contaminated water beds in the port of Rotterdam. Therefore, the Dutch used the opportunity to link the Scheldt infrastructure issues to these Meuse water management issues. This strategy is in accordance with the theory that in cases where decision making is based on unanimity, and the actors have different preference intensities in different issue-areas, the probability that movement away from the status quo will involve tactical issue-linkage is high.[1] In spite of the Belgian objections, the Dutch did not want to cooperate unconditionally on the implementation of the 48'/43'/38' deepening programme as well, and linked the approval of this programme to the Meuse water management issues. The Dutch faced a negotiation partner that was unable to reach an internal consensus on the strategies to be used in the negotiations with the Netherlands, and speculated on a Flemish compensation of the Walloon 'losses'. Because the negotiations reached an impasse, the Dutch gradually changed their strategy. First, they did no longer aim at the construction of storage reservoirs in the Walloon Ardennes. Secondly, the Dutch conventions drafted in the early nineties did no longer contain obligations for Belgium only, but for the Netherlands as well. Thirdly, the Dutch started bilateral discussions on the water quality of the Scheldt in the TSC, where it used the Flemish obligation to apply for a WVO-permit to exert influence on Flemish water quality policies. Recently, the Dutch began to formulate conditions to the dredging permit for the Flemish region as well. The Dutch did stimulate international research continuously, and put a lot of effort in international research projects, such as the ISG and the OostWest-project. Only in the beginning of the nineties, however, the Dutch changed their strategy fundamentally. First, they contacted the French parties, and aimed at the conclusion of multilateral conventions for the Scheldt and the Meuse. This strategic change may, among other things, be explained by the conclusion of the UN-ECE Convention. Secondly, the Dutch decided to unlink the bilateral and multilateral Scheldt and Meuse issues. This, among other things, may be explained by the involvement of France in the negotiations, and the attribution of treaty-making competencies to the three Belgian regions, which enabled the Dutch to negotiate directly with these regions. In the bilateral negotiations, the Dutch accepted the Flemish proposal to deal with the deepening programme and the alignment of the HSL simultaneously, and took the opportunity to use this linkage to exert influence on the Flemish government in decision making on the alignment of the HSL. In the multilateral negotiations and the first negotiations in the provisional ICPS the Dutch mainly aimed at trust building between the parties, and formed a rather stable coalition with the Flemish region.

France
The French parties reacted positively on the invitation to participate in the multilateral negotiations on the water quality of the Scheldt and the Meuse. Because of the conflicts between in particular the Netherlands and the Walloon region, the French parties more or less played the role of mediator. In the Scheldt basin, the French seem to perceive the

---

1. See Section 2.3.2.

international pollution issues as typical commons problems.[2] France is willing to clean-up the river Scheldt on its territory, only if the other basin states and regions, among which the Brussels Capital Region, will do the same. If not, the French do not want to impose costly water quality policies on its industries and households. This perception and strategy may be explained partly by the relatively bad economic situation in the French part of the basin.

Perceptions and strategies of lower level governments and NGOs

Lower level governments. Interests and resources were not only distributed among the basin states, but among national and lower level governments, and NGOs within the individual basin states as well. Because within individual basin states different actors had different problem perceptions, they also had different preferences concerning the strategies to be used in the international decision making process. The different strategic preferences of the Belgian regions probably are most striking. The three regions clearly had diverging problem perceptions, and therefore preferred different strategies to be used in the negotiations with the Netherlands. Within the Netherlands, however, there were interesting differences between national, regional, and local government agencies as well. National government agencies, serving the national Dutch interest, for a long time were in favour of a linkage between Scheldt infrastructure and Meuse issues, and later favoured a linkage between the deepening of the navigation channel in the Western Scheldt and the negotiations on the alignment for a new high speed train between Antwerp and Rotterdam. Regional and local actors in the Province of Zeeland, however, did not perceive a direct interest in the Meuse issues and the alignment of the high speed train. These parties were in favour of a linkage between Scheldt infrastructure and Scheldt water quality issues, because such a linkage would be beneficial for the Province of Zeeland. Because regional cross-border cooperation in the RSD, between the three southern provinces and the Flemish region, and in the BOW was hindered by the continuous disagreement on the political level, lower level governments in the region of Zeeland tried to influence *intra*national decision making in the Netherlands on the strategies to be used in the international negotiation, and pleaded for intervention by the EC in these negotiations. Dutch municipalities along the Western Scheldt founded the Task group for the Western Scheldt, and with that formed a coalition. They tried to establish cross-border contacts with municipalities in the other basin states and regions as well. Because the Dutch national government apparently was unable to solve the international water quality problems in the Meuse basin, Dutch municipalities along the river Meuse, and drinking water production firms using Meuse water for drinking water production passed the Dutch government, and contacted polluting industries in the upstream parts of the Meuse basin. Threatening with litigation, they tried to conclude covenants on a reduction of waste discharges with these industries. Provincial and local governments demanded direct participation in the negotiations on the Scheldt and Meuse issues as well. The Dutch national government did not want to involve representatives of lower level governments in the negotiations on the water conventions, but in 1995 representatives of these governments were invited to participate in the provisional river commissions for the Scheldt and Meuse.

The distribution of interests and resources among national, regional, and local government agencies caused interdependencies between them, which enabled the regional and local actors to exert influence on the national actors. Municipalities along the Western

---

2. See Section 2.3.2.

Scheldt possessed indispensable resources for the implementation of the 48'43'/38' deepening programme, and therefore were able to make a linkage between the deepening of the navigation channel in the Western Scheldt and the safety of navigation on the Western Scheldt. Furthermore, the national government needed regional support for nature restoration projects along the Western Scheldt. The implementation of the deepening of the navigation channel in the Western Scheldt, and nature restoration along the Western Scheldt clearly illustrate the importance of regional and local support for international policies, and the risk of an implementation gap.

NGOs. Whilst the perceptions and strategies of lower level governments may be explained by their (limited) geographical scope, those of NGOs involved in the decision making process may be explained by their specific interests. Obviously, environmental NGOs tried continuously to get environmental issues on the international agenda, and to keep these issues there. They started litigation against Belgian polluting industries, organized protest marches, and published research reports indicating the bad water quality of the rivers Scheldt and Meuse. Environmental NGOs established cross-border contacts, and by that formed transboundary coalitions as well. Furthermore, they monitored the progress made with the implementation of international policies. They, for example opposed the second and third WVO-Permits the Dutch government granted to the Flemish Region, because the Flemish Region would not have complied with the conditions formulated in the preceding permits, and an EIA would be obligatory for the 48'/43'/38' deepening programme. Furthermore, they stated that the conclusion of the multilateral conventions on the water quality of the Scheldt and the Meuse only is a small first step toward the clean-up of these two rivers, because the conventions do not contain concrete water quality objectives and policies. Like the Provinces and municipalities, Dutch environmental NGOs would like to be involved in the discussions in the international river commissions for the Scheldt and the Meuse. The Dutch national government actually invited these NGOs to participate in the preparatory meetings of the Dutch delegations to the working groups of the ICPS.

Flemish industrial and port associations, among which the AGHA, threatened with litigation against the Dutch state, because the Dutch would not live up to the Scheldt Statute. Representatives of the port of Rotterdam tried to link the discussions on the improvement of the maritime access to the port of Antwerp to a discussion on the supposed distortion of competition between the ports of Antwerp and Rotterdam.

8.3.2 Strategic interaction and institutionalization processes
During the strategic interactions between the Scheldt basin states and regions so-called institutionalization process developed, which influenced structural and cultural characteristics of the network shaped around international Scheldt issues. The strategic interactions did influence the:
- international institutional arrangements;
- international policies;
- trust or distrust between the parties.

Development or erosion of international institutional arrangements and policies
The three draft water conventions of 1975 contained many concrete policies, and provided for elaborate international institutional arrangements to regulate the interactions between Belgium and the Netherlands. The proposed package deal, however, failed because it did not create a situation in which all relevant parties perceived that they would be better off with the agreement proposed than without this agreement. The Walloon region did not perceive

any of the proposed agreements as beneficial. Because this Walloon perception did not change in the period between 1975 and 1985, and neither of the other parties involved did change its strategy, there was a lengthy deadlock between 1975 and 1985. The linkages made in the declaration of intent signed by the Belgian and Dutch Ministers in 1985, failed for exactly the same reason as the linkages that had been made in the negotiations between 1969 and 1975: they did not create a win-win situation. Because neither of the parties fundamentally changed its strategy, but all wanted or expected the others to change their strategies, the negotiations reached an impasse. Consequently, no institutional arrangements or policies were developed. On the contrary, existing international platforms, such as the SWC and the TMC did erode.

In the beginning of the nineties, with the projects ISG, OostWest and several international symposia, transboundary professional network extended gradually. The establishment of contacts between on the one hand the downstream parties the Netherlands and the Flemish region and on the other hand the upstream parties France, and the Walloon and Brussels Capital Regions, appeared to be rather difficult. The scope of the bilateral Flemish-Dutch platform TSC was extended gradually. Flemish and Dutch water managers started to discuss the morphological and ecological aspects related to the dredging works and the dumping of dredged material in the Scheldt estuary, and water quality policies as well. In the TSC Flanders and the Netherlands were able to reach an agreement on concrete policies, such as the extraction of large amounts of polluted sediments from the Lower Sea Scheldt.

The Scheldt water quality convention concluded in 1994 was an important step in the development of institutional arrangements to combat Scheldt river pollution. The Convention, among other things, provided for the installation of an international river commission that has to prepare a Scheldt Action Programme. The convention, however, did not contain concrete water quality policies. One of the conclusions of the case study is that negotiations on international policies were delayed until the installation of the international river commission for the Scheldt, and delegated to this commission. In the commission the parties have to reach an agreement on the first Scheldt Action Programme with the first basin-wide water quality policies. Unlike the multilateral water quality convention, the bilateral Flemish-Dutch Convention on the deepening of the navigation channel in the Western Scheldt does contain concrete policies. Appendix 13 contains an overview of the main international commissions and working groups where international Scheldt issues were discussed in the case study period.

(Dis)trust
The long Belgian-Dutch history of conflicts on the management of the navigation channel in the Western Scheldt, and the continuous Dutch refusal to cooperate on the improvement of the maritime access to the port of Antwerp unconditionally, have given the Dutch a reputation of being unwilling to cooperate on the further development of the Flemish part of Belgium. Because of the linkages made across Scheldt and Meuse issues, the Dutch got a bad reputation in the eyes of Walloon politicians as well. The Belgian government, and the governments of the Belgian regions, however, did not want to cooperate on the Scheldt and Meuse water management issues unconditionally. As regards environmental management the Belgian parties had a bad reputation in the eyes of the Dutch. This in particular goes for the Brussels Capital Region, because until 1997 the capital of Europe had not yet started treating its waste water.

Rigid mutual strategies causing lengthy deadlocks in the decision making process, reinforced mutual bad reputations, and caused a profound distrust between the parties. This distrust may explain why the parties did hardly use the strategy of building up a reservoir of goodwill. Furthermore, mutual distrust seems to have hindered international information exchange and research, and transboundary cooperation on regional or local scales. Although since the Dutch decision to unlink the bilateral and multilateral Scheldt and Meuse issues, and the start of the multilateral negotiations on the water quality of the Scheldt and the Meuse, relations between the Scheldt basin states and regions were improving gradually, the Dutch POM-initiative, and the reaction of the Dutch minister on some accidental discharges in the river Meuse caused another Dutch-Walloon conflict. Only since the conclusion of the water conventions, and the provisional installation of the ICPS, relations are really improving.

Issue-linkage and the development of international institutional arrangements and policies
The probably most important conclusions of the case study of decision making on international Scheldt issues relate to the many linkages that have been made in the decision making process.

First, the case study results are in accordance with the theory that in cases where the decision rule is unanimity, and the parties do have different preference intensities in different issue-areas, the probability that movement away from the status quo will involve tactical issue-linkage is high. Almost any agreement that has been reached on any Scheldt or Meuse issue did involve issue-linkage. Examples are the Dutch approval of the Convention on the deepening of the navigation channel in the Western Scheldt, the Flemish approval of the Convention on the water quantity of the Meuse, the Flemish approval of the Conventions on the protection of the Scheldt and the Meuse, and the Revision of the Scheldt regulations, and the conditions formulated to the WVO-permits for the Flemish Region. Therefore, the case study seems to confirm Martin's statement that in cases where the decision rule is unanimity and there are different preference intensities in different issue-areas, tactical issue-linkage often is the key to understanding international cooperation.[3]

A second conclusion of the case study is that tactical issue-linkages did cause lengthy deadlocks in the international decision making process, and consequently hindered the development of institutional arrangements and policies, in case these linkages did not create a situation in which all parties perceived that they would be better off with the proposed package deal than without one (a win-win situation). The linkage between Scheldt infrastructure and Meuse water quality and quantity issues, which lasted for more than 25 years, does not seem to have contributed to the development of international institutional arrangements and policies. On the contrary, this linkage was accompanied with lengthy deadlocks in the decision making process, which reinforced bad reputations of the parties involved, contributed to a bad negotiation climate, and frustrated cross-border interactions in other (regional) platforms.

Apart from the many issue-linkages, another possibility of compensating potential losers from an international agreement was discussed: the establishment of a sanitation fund. Although such a fund might be attractive from an economic point of view, it has never been established. The main reason for this is that the establishment of a sanitation fund is not in accordance with an important principle of international environmental law, namely the polluter pays principle. In addition, one may argue that such a fund does not stimulate economic efficiency, because polluters would not bear the full (environmental) costs of their

---

3. See Section 2.3.2.

activities, and therefore be unable to make efficient decisions.

8.3.3 Learning processes
During the games that are played the actors may gain new knowledge on the perceptions of the other parties, and the (inter)dependencies in the policy arena, and consequently change their objectives and/or strategies. In addition, the actors may gain new knowledge on the intellectual relationships in an issue-area.[4] In this section the main conclusions concerning these strategic and cognitive learning processes in decision making on international Scheldt issues are summarized. A main problem of drawing these conclusions is that there mostly are multiple rival or supplementary explanations for strategic change. Therefore, in general it is difficult to attribute strategic change to a specific cause, such as strategic or cognitive learning. Nevertheless, the case study contained several examples of strategic change, where the occurrence of learning processes could be made plausible. In the following sections, these examples are discussed.

Strategic learning
A first example of strategic change, which may be interpreted as strategic learning, is the Dutch decision not to insist on the construction of storage reservoirs in the Walloon Ardennes. During the negotiations on the water conventions, the Walloon region had opposed continuously the construction of these reservoirs, and this issue turned out to be a major bottleneck in the international negotiations. Partly for that reason the Dutch decided to skip the issue from the international negotiation agenda. The Dutch decision may be interpreted as the result of a cognitive learning process as well, because research had indicated the negative impacts of the construction of storage reservoirs on the environment. Because of the difficulties involved in the negotiations with the Walloon Region on the distribution of the water of the river Meuse, the Dutch later aimed at the conclusion of a bilateral Flemish-Dutch agreement on a saving scheme for cases of a low river flow.

Secondly, during the negotiations the Dutch learned about the objections to the draft conventions formulated by the other parties, and partly for that reason changed the contents of their draft conventions. For example, the conventions drafted in the early nineties did no longer contain obligations for Belgium only, but contained obligations for the Netherlands as well.

A third example of strategic change, which may be interpreted as strategic learning, is the Dutch decision to unlink the bilateral and multilateral Scheldt and Meuse issues. Although there are several possible explanations for this strategic change (See Section 6.3.1), strategic learning seems to be part of the explanation. At the end of the eighties and the beginning of the nineties more and more Dutch experts started to question the effectiveness of the Dutch linkage, because the linkage the Dutch had made for more than 20 years did not seem to have been very successful. After intensive discussions between those who were in favour of a strategic change and the 'hardliners', who wanted to continue the linkage, the former group won in 1993. Because the Dutch decision to unlink bilateral and multilateral Scheldt and Meuse issues came rather late, it can be concluded that the Dutch government and delegation lacked learning capacity.

A fourth example of strategic learning by the Dutch is the decision to formulate conditions to the WVO Permit granted to the Flemish region for dumping dredged material

---

4. See Section 2.2.7.

in the Western Scheldt. The Dutch had experienced continuously that the upstream basin states and regions did not succeed in a clean-up of the river Scheldt, and that the strategies the Dutch had used so far apparently were not effective. Therefore, they decided to make a linkage between their permission to dump dredged material in the Western Scheldt, and the improvement of the water quality of this river, for example by the extraction of polluted sediments from the Lower Sea Scheldt. Because the Dutch had gained positive experiences with the conditions formulated to the WVO Permit, they started to use the same strategy in the negotiations on the dredging permit. The conditions formulated to this permit, however, show a clear intellectual coherence with the content of the permit itself, and therefore should be interpreted as a substantive rather than as a tactical linkage.

Some Flemish strategic changes may be interpreted as strategic learning processes as well. The Flemish parties learned that somehow they had to come up to the demands of the Netherlands, since otherwise the Dutch would not be willing to cooperate on the implementation of infrastructure projects aiming at an improvement of the maritime access to the port of Antwerp, such as the deepening of the navigation channel in the Western Scheldt. This may explain partly the development of water quality policies in the Flemish region, and why some Flemish parties opposed the bringing into use of the Tessenderlo pipeline.

Furthermore, until 1993 the Flemish experienced a continuous Walloon unwillingness to cooperate on the international Scheldt and Meuse water management issues. This may explain partly why the Flemish government gradually changed its strategy aiming at an internal Belgian agreement, and started bilateral discussions with the Netherlands on both the water quantity of the Meuse, and the water quality of the Scheldt.

Finally, the recent strategic changes of the Walloon and Brussels Capital Regions may be interpreted as strategic learning processes. Apart from the attribution of treaty-making competencies to the Belgian regions, the continuous demand for international action to clean-up the rivers Scheldt and Meuse, which was made by the other basin states and regions, international organizations and several environmental NGOs may have taught these regions that they have hardly any choice but to use more interactive strategies, and to cooperate internationally.

Strategic learning is a continuous (and ongoing) process, and numerous small strategic changes of the Scheldt basin states and regions are the result of strategic learning as well. The main examples discussed above showed that strategic learning indeed played an important role in decision making on international Scheldt issues, and mostly contributed to international agreement.

Cognitive learning

Strategic change may not only be explained by strategic learning, but by learning about the cognitive or 'factual' aspects of the issues at stake as well. The case study contained various interesting examples of cognitive learning.

Since the end of the seventies Flemish and Dutch researchers have carried out research on the impacts of the 48'/43'/38' deepening programme of the Western Scheldt in the Technical Scheldt Commission (TSC). The results of a first extensive study, which was finished in 1984, played an important role in the discussions on the content of the Convention on the deepening of the navigation channel in the Western Scheldt. Because the main result of this research was that the implementation of the deepening programme would hardly affect the safety along or the environment of the Western Scheldt, the Dutch did not insist on the preparation of an EIA for this project. However, as the research had indicated that implementation of the deepening programme would cause shore erosion, the Dutch did ask for measures to prevent or restore this shore erosion.

In the research project OostWest, which was carried out under the authority of the TSC as well, research on the morphology and ecology of the Scheldt estuary, and the impacts implementation of the deepening programme would have on the morphology and ecology of the Scheldt estuary, was continued. Because of increased knowledge on the dynamics of the estuary, this research indicated that the dredging activities and the dumping of dredged material in the Scheldt estuary did have a negative impact on the ecology of this estuary. Unlike the TSC-report issued in 1984, the Pilot Study OostWest issued in 1991, and several succeeding research reports indicated that implementation of the 48'/43'/38' deepening programme would cause a considerable loss of tidal areas and marshlands along the Western Scheldt. The research results of the long term research programme OostWest seem to have had a considerable influence on the perceptions and strategies of both the Flemish and Dutch parties involved in decision making on international Scheldt issues. First, the new knowledge on the impacts of the deepening programme played a role in the negotiations on a provision for nature restoration in the Convention on the deepening of the navigation channel in the Western Scheldt. Secondly, the policy alternative 'ontpolderen', which was introduced in the international project OostWest, did have a considerable influence on the discussions concerning nature restoration and flood protection in both Flanders and the Netherlands. Thirdly, the research results induced a Flemish-Dutch discussion on the dredging strategy in the Western Scheldt, and explain why the Dutch launched the idea to formulate conditions to the Flemish dredging permit. Finally, the researchers participating in the project OostWest argued that, in addition to the negative consequences, the construction of the Baalhoek Canal could have positive consequences for the ecology of the Western Scheldt as well. The construction of this canal would contribute to an increase of the tidal volume, and make it possible to decrease the dredging intensity in the eastern part of the Western Scheldt. Possibly, this new knowledge on cause-effect relations and the consequences of policy alternatives will play a role in future decision making in the issue-area of the maritime access to the port of Antwerp.

Cognitive learning processes seem to have played a less prominent role in the issue-area of water and sediment pollution. This may be explained by the fact that the main cause-effect relations in this issue-area are relatively simple compared to the complexity of the morphodynamics and ecology of the Scheldt estuary. Policy measures are rather obvious. To solve the pollution problems the basin states and regions have to reduce their discharges of pollutants. They, for example, should construct waste water treatments plants, and develop and implement systems for environmental permitting and taxes. International research carried out in this issue-area, however, played an important role in the decision making process. During the deadlocks in the negotiation process, the ISG-project, the LIFE-project, and various international symposia where Scheldt water quality, navigation and ecological issues were discussed, seem to have contributed to trust building and mutual understanding between the parties.

8.3.4 Influence of the context on the perceptions and strategies of the Scheldt basin states and regions

The in depth study of perceptions and strategic behaviour made it possible to formulate alternative or supplementary explanations for most strategic changes. Therefore, it turned out to be difficult to draw 'hard' conclusions concerning the explanatory variables for the course and outcome of the decision making process. Nevertheless, the detailed reconstruction of decision making, and the 'thick descriptions' enabled the researcher to make certain influence relations more plausible than others, and the case study showed ample evidence for the influence of specific contextual variables on the decision making process. In this section the main conclusions concerning the influence of the contextual dimensions discussed in Chapter 4 are summarized.

Underlying hydrological structure of the issues at stake
The underlying hydrological structure of the issues at stake was a constant variable, which seems to have been a very important, if not the most important contextual variable that has influenced decision making on international Scheldt issues. The underlying hydrological structure of the various issues at stake influenced the distribution of interests and resources among the actors involved in decision making on these issues, and consequently their perceptions of (inter)dependence. Because of its downstream position in the basins of the Scheldt and the Meuse, the Dutch perceived a dependence on all upstream basin states and regions as regards the supply of a reasonable amount of water of a reasonable quality. On the other hand, in the issue-area of the maritime access to the port of Antwerp, the parties having a relative upstream position, i.e. the Belgian state and Flemish Region, perceived a dependence on the Dutch as regards the maintenance or improvement of the navigation channel in the Western Scheldt. Section 8.3.1 discussed the influence of these perceptions on the strategic behaviour of these parties, and consequently on the course and outcome of the decision making process. Only in recent years, the influence of the underlying hydrological structure seems to have become less influential. First, the new party France, in spite of its relative upstream position in the basins of the Scheldt and the Meuse, under certain conditions, was willing to cooperate on a clean-up of these basins. Secondly, the Walloon and Brussels Capital regions, which mainly have a relative upstream position in the Scheldt and Meuse basins, started to use more interactive strategies.

International context
In the case of decision making on international Scheldt issues, the international context seems to have been almost as influential as the underlying hydrological structure of the international Scheldt issues. The first and by far most important reason for this are the diverse linkages made between international Scheldt issues, Meuse issues and even a non-river issue, namely the alignment of the High Speed Train. Because of the many linkages made between Scheldt and Meuse issues, one may even discuss whether decision making on the international Meuse issues is not an integral part of the phenomenon studied rather than an important element of the context.

Other influential contextual developments were the negotiations on and the conclusion of the UN-ECE Convention for the Protection and Use of Transboundary Watercourses and International Lakes. Some persons involved in the Scheldt and Meuse negotiations were involved in the negotiations on the UN-ECE Convention as well. For these persons the deadlocks in the negotiations on the Scheldt and Meuse formed an important incentive to stimulate the conclusion of the UN-ECE Convention. Therefore, also here the distinction between the context and the decision making process studied is rather vague. Since the conclusion of the UN-ECE convention basin-wide cooperation between the Scheldt basin states and regions was a necessity, and the upstream basin state France had to be involved in the negotiations on the water quality of the Scheldt and the Meuse.

Other contextual developments that have influenced decision making on international Scheldt issues were the issuing of several EC Directives, the North Sea Ministerial

Conferences, and decision making in the Rhine basin. Because of the absence of multilateral and bilateral agreements on the water quality of the Scheldt, EC and North Sea policies were the main international water policies applying to this river. The case study research did not address the influence of these policies on the development of water policies in the individual basin states, but addressed the influence of these policies on the decision making on international Scheldt issues. As regards this influence two conclusions can be drawn. First, these policies enabled the parties suffering from transboundary water and sediment pollution, such as the state of the Netherlands and several environmental NGOs, to substantiate their demand for international joint action to clean-up the rivers Scheldt and Meuse. These parties referred frequently to the success of the clean-up of the river Rhine as well. The case study, however, has shown that not only the downstream basin state the Netherlands, but also the Walloon region tried to legitimate its proposals by referring to existing international policies. During the international negotiations the Walloon Region expressed more often its willingness to comply with the EC Directive for cyprinids, but did not want to formulate more ambitious objectives in bilateral or multilateral conventions between the Scheldt and Meuse basin states.

Beside the formulation of emission- and immission-based water policies, the EC stimulated the development of cross-border networks by granting subsidies to several transboundary research projects in the Scheldt basin, among which the LIFE-DSS and the LIFE-MARS-project.

If the Framework Directive for water that presently is being developed by the EC, will be issued, negotiations on the Scheldt and the Meuse most likely will have to be reopened, because the Scheldt and Meuse conventions do not address the water quantity management in these basins.

Cultural context
The in depth study of strategic behaviour, and the 'thick descriptions' presented in Part II of the thesis made it possible to address the influence of the 'soft' factor culture as well. Because of the considerable difference between on the one hand the Latin cultures of France and Belgium, and on the other hand the Nordic culture of the Netherlands (See Sections 4.6 and 4.9), it has been worthwhile to include this factor in the case analysis.

In several cases the diverging perceptions and preferences of the actors may be explained by the difference between the Dutch consensus culture and the more hierarchical French and Belgian decision making cultures characterized by a relatively large influence of politics on decision making. Whilst the Dutch Minister of Transport, Public Works, and Water Management preferred to consult an advisory council, and other interested parties as regards the implementation of the 48'/43'/38' deepening programme, the Belgian government preferred rapid decision making on that issue without complex procedures. In addition, for a long time the Belgian government did want to negotiate the deepening programme with representatives of the Dutch national government only, and did not accept the relative autonomy of the Dutch municipalities on that issue. This may be explained partly by a different perception of the relationship between national and lower level governments, and of international relations between sovereign states. Later, however, the Flemish region would contact these municipalities to apply for the necessary permits. Another example of the different perceptions of the relation between national and lower level governments concerns the Dutch POM-initiative. Whilst Walloon politicians blamed the Dutch national government for this initiative, the Dutch national government was of the opinion that it was unable to prevent this initiative, because of the relative autonomy of local government agencies.

The case study provided quite some evidence for the relatively large influence of politics on decision making in Belgium compared with the Netherlands. Dutch experts faced difficulties with the establishment of cross-border contacts on the administrative level, and complained that for a long time they had been able to establish contacts on the subpolitical level of the ministerial cabinets only. Furthermore, in the ISG the Walloon delegation stressed that the activities of the ISG should not go beyond the exchange of information, and that in the ISG no Scheldt policies could be prepared. The case study seems to confirm that Dutch civil servants have more mandate than their Belgian colleagues

The basin states and regions clearly had different traditions concerning the openness of decision making processes as well. For a long time the Walloon region did not want to provide technical information concerning the permits it had issued for waste water discharges by industries. Furthermore, the Scheldt basin states and regions had different preferences concerning the accessibility of the paper work produced by the river commissions for the Scheldt and the Meuse. In addition, the parties had different preferences concerning the participation of NGOs in the river commissions to be installed. Unlike the Flemish and Dutch delegations, the Walloon delegation opposed an observer status for NGOs in the International Commissions for the Protection of the Scheldt and the Meuse against pollution.

The relatively low Dutch score on the dimension Uncertainty Avoidance may explain why Dutch representatives generally had less problems with informal international initiatives than Belgian and French representatives, who preferred formalized cross-border cooperation. Some examples concern the organization of the last meetings of the ISG, when the political negotiations were restarted, and the difficulties related to the organization of the third Scheldt symposium.

The Belgian clientilism seems to have influenced the decision making process as well. According to one respondent the Dutch have tried to prepare a package deal with the Walloon region. In such a package deal, Walloon efforts to clean-up the river Meuse or to guarantee a minimum flow of the river Meuse would be compensated by Dutch cooperation on infrastructure works improving navigation possibilities to the port of Liège. After the appointment of another Walloon minister, however, who did not live in the region of Liège, discussions on this package deal were discontinued.

The case study provides some evidence for the Dutch characteristic that they sometimes tend to overestimate the value of their ideas. Although they possibly had good reasons, in the ISG the Dutch initially had a strong preference to work with a Dutch GIS-system, and in the LIFE-DSS project the Dutch experts (like the Walloon experts) strongly favoured the model they had developed. Furthermore, some statements made by the Dutch Minister of Transport, Public Works, and Water Management concerning the Meuse water pollution, and the reactions of the Walloon Minister on these statements, seem to illustrate the Dutch overestimation of their ideas, their directness, and the sometimes rather negative perception of this behaviour by representatives of other cultures.

Beside the decision making cultures, it is most likely that language differences between the Scheldt basin states and regions have influenced the development of cross-border networks as well. The use of the same language may explain partly why contacts between Flemish and Dutch water management agencies developed earlier and presently are more intensive than contacts between Dutch and Walloon, or Flemish and Walloon water management agencies.

Finally, in the Sections 4.7.4 and 4.9 it was argued that the different scores of the Scheldt basin states and regions on the cultural dimensions Femininity and Uncertainty Avoidance may explain partly the different stages of development of environmental policies.

In spite of the explanatory power of the scores on the four cultural dimensions discussed in Section 4.6, some case study results seem to contradict these scores. First, French decision making on water policies is less hierarchical than one may expect on the basis of the French cultural scores, and does even look rather similar to the Dutch consensus decision making. Already since 1964 the subnational basin committees and agencies are at the centre of French water management, and in the basin committees user groups are involved in the development of basin policies (See Appendix 6). Secondly, although the Walloon region opposed the participation of NGOs in the international river commission, on a regional scale the Walloon region organizes open decision making processes on the conclusion of so-called river contracts. Thirdly, although the Flemish region has about the same scores on the cultural dimensions as the Walloon region, the Flemish delegation, unlike the Walloon delegation, was in favour of the participation of NGOs in the international river commissions, which indicates a relatively open decision making culture. Furthermore, the conclusion of a covenant between the Flemish Ministry and environmental NGOs on nature restoration along the Lower Sea Scheldt is an interesting example of consensus decision making.

Although probably less influential than the factors discussed in the two preceding sections, there is ample evidence for the influence of cultural factors on the perceptions and strategies of the actors involved in decision making on international Scheldt issues. The conclusions presented above seem to confirm Soeters' statement that it is the 'soft' factor of the national culture, that can be very 'hard' in its consequences.[5]

Intranational context
The fourth contextual dimension distinguished concerns the intranational developments in the basin states and regions. Undoubtedly, the most influential intranational development was the Belgian federalization process. This process, which consisted of three stages, entailed a continuous redistribution of competencies among the federal state of Belgium and the Walloon, Flemish, and Brussels Capital Regions, and affected the distribution of competencies in the policy fields of water and environmental management, and infrastructure planning as well. The general pattern was one of a gradually declining competence of the federal state of Belgium, and gradually increasing competencies of the three regions. The main problem caused by this federalization process was that the neighbouring states France and the Netherlands formally had to negotiate with the state of Belgium, whilst there actually were four Belgian parties involved in decision making on international Scheldt (and Meuse) issues. Internal Belgian disagreement between these parties complicated decision making. Only since 1993, when the three Belgian regions received treaty-making competencies, France and the Netherlands were able to start direct negotiations with the three Belgian regions (and vice versa). Among other things, this intranational Belgian development caused the Dutch to unlink the bilateral Flemish-Dutch and multilateral Scheldt and Meuse issues. Although the completion of the third stage of the Belgian state reforms in 1993 has made the internal Belgian organization more transparent, the distribution of competencies between the federal state and the regions continued to influence the international decision making process. Some examples of this influence are the Belgian approval of the Convention on the revision of the Scheldt regulations, the Belgian ratification of the UN-ECE convention, and some discussions concerning the scope of activities of the international commissions for the Scheldt and the Meuse.

Apart from the Belgian federalization process, the different stages of development of

5. Soeters (1993, p. 651).

water management policies in the Scheldt basin states and regions were an important explanatory variable for the course of the decision making process. These differences may explain the different levels of ambition of the basin states and regions. During the last years, however, the Belgian regions are eliminating the backlog they have compared with the downstream basin state the Netherlands rapidly, and some do even argue that the dialectics of progress apply to the case. As argued in the Section 4.7.4 these different stages of development in turn may be explained partly by the different socio-economic circumstances in the different parts of the basin. Although the relationship can hardly be proven, the relative low political priority of environmental policies, and the lack of financial resources of the parties in the southern parts of the Scheldt basin may be explained partly by their problematic economic situation.

Historical context
The long history of conflict on the main Scheldt and Meuse issues caused a profound distrust between the parties, and bad mutual reputations. For the conclusions concerning the influence of this history on the decision making process, see Section 8.3.2.

8.4 Additional conclusions

Several research findings could not be related directly to the research questions formulated beforehand, and therefore are presented separately. Four additional conclusions, which are presented in the Sections 8.4.1 to 8.4.4, concern the effectiveness of international river commissions, the different geographical scales for dealing with international river issues, the role of individual persons in decision making, and the relation between international and *intra*national decision making processes.

8.4.1 The effectiveness of international river commissions
Although the ICPS is relatively new, and this commission was not even installed formally when this chapter was written, it is possible to draw some preliminary conclusions concerning its effectiveness.

A first observation is that in the last years before the conclusion of the Convention on the protection of the Scheldt against pollution important developments took place in the Scheldt basin states and regions as regards water quality management. The Flemish government developed ambitious plans to clean-up the river Scheldt, and assigned large budgets for the construction of waste water treatment plants. In the Walloon region investments in waste water treatment infrastructure increased rapidly as well. Finally, in the Brussels Capital region the construction of waste water treatment plants was started eventually. Because of the development of water quality policies, in the mid-nineties the water quality of the river Scheldt started to improve gradually. These facts should be taken into account in a first evaluation of the effectiveness of the ICPS.

A second observation is that in 1996 neither of the respondents did expect the first Scheldt Action Programme (SAP) to go beyond existing national policies. Most respondents expected the first SAP to be a summary of existing national policies.

These two observations may lead to the rather negative conclusion that the ICPS does hardly influence the development of water quality policies, and therefore is hardly effective. This hypothesis, however, may be challenged for several reasons.

First, the Scheldt basin states and regions may have anticipated the conclusion of the

Convention on the protection of the Scheldt, and the (provisional) installation of the river commission. Possibly, the prospect of the development of international policies in an international river commission stimulated the development of water quality policies in the Belgian regions. This anticipatory effect, however, should not be overestimated. The Dutch linkage between infrastructure and water quality policies, and the development of international water quality policies on the EC-level, probably were more important incentives for the Flemish region to clean-up its watercourses than the prospect of the installation of an international river commission.

Secondly, the interactions in the commission may induce cognitive learning processes. The continuous exchange of information enables the parties to reach an agreement on the factual aspects of the international issues at stake. Furthermore, the parties may inform each other on their experiences with policies and policy alternatives.

Thirdly, although the commission possibly does not have a large impact on the development of national water quality policies, it does enable the parties to better coordinate their policies. The parties may coordinate the main river functions or immission standards on each side of a border. Furthermore, they may develop a joint alarm system for cases of accidental pollution.

Fourthly, the basin states may generate economies of scale. As an example, it may be cheaper to construct one joint monitoring station at the border, than constructing two stations on each side of a national border. Furthermore, in transboundary (sub)basins the joint construction and exploitation of waste water treatments plants may generate economies of scale.

Fifthly, the establishment of an international river commission may support national environmental ministries in the national competition on the division of government budgets. Because of the (provisional) installation of the Scheldt commission, water management departments in the Belgian region, which face the problem of a lack of financial resources and personnel capacity, may receive extra money and personnel capacity.

For the five reasons discussed above, some empirical evidence could be found in the case study. There are, however, two more reasons why an international river commission could be effective. For these reasons, however, no empirical evidence could be found yet.

First, the interactions in the commission may induce strategic learning processes, which enable the basin states to recognize win-win situations, and to reach an agreement on mutually beneficial policies.

Secondly, international joint action theoretically may solve the classical commons problem discussed in Section 2.3.2. As argued in Section 8.3.1, the French problem perception typically requires international joint action on a basin scale.

For the reasons discussed above, it is most likely that the benefits of the ICPS do exceed its costs, such as the costs of the secretariat, and the costs made by the individual basin states and regions.[6]

8.4.2 Different geographical scales for dealing with international river issues

The case study has shown that different international river issues emerge on different scales and that the solution of these issues were found on different scales.

The causes and effects in an issue-area sometimes are restricted to part of the basin,

---

6. Similar conclusions about the effectiveness of international river commissions were drawn by Bernauer (1996), and Durth (1996).

whilst the geographical scope of another issue-area may easily exceed the boundaries of a river basin. The issue of sand-mining in the Scheldt estuary is a river issue, but nevertheless does not emerge on a basin-scale. The actors in the upstream riparian state France probably are not interested in a solution of the problems some actors in the issue-area perceive, neither do they have problem solving capacity. Consequently, from a network perspective it is rather illogical to involve France in the discussions on this issue. Obviously, water and sediment pollution is an issue that emerges on a basin scale, and therefore should be discussed with the upstream basin states and regions. A second example is the issue-area of the maritime access to the port of Antwerp. Causes and effects in this issue-area are increasingly conceived of as being spread over the whole Rhine-Scheldt Delta, including the port of Rotterdam. The increasing interdependence between the ports in the Rhine-Scheldt Delta demands for an approach of this issue on a scale that is different from the basin scale. In this issue-area the relation between Antwerp and Zeebrugge is an interesting one as well, because more intense cooperation between these ports would make a further deepening of the Western Scheldt less urgent. In the latter case, the interdependencies in a major international issue-area go beyond the basin scale. Whilst for the international pollution and related ecological issues one may advocate a river basin approach, for the infrastructure and related ecological issues one may equally plead for a 'delta approach'.

Apart from the fact that different issues emerge on different geographical scales, the case study has shown that most agreements on international Scheldt issues did involve issue-linkage (See Section 8.3.2). Consequently, issues were addressed on scales exceeding the scales of the single issues at stake.

8.4.3 The role of individual persons in decision making

Another additional conclusion concerns the role of individual persons in decision making. Although the importance of this factor should not be overestimated, the case study showed ample evidence for the important role individual persons may play in a decision making process. Some persons did influence decision making in a positive way, whilst others had a rather negative influence on the course of the international decision making process. A role is defined as positive when the person was able to initiate or stimulate international interactions and agreement, whereas a negative role is characterized by the use of reactive strategies aimed at maintaining autonomy, or rigid strategic behaviour.

The case study contains several examples of a positive role of individuals. An interesting observation is that some of these persons worked for some time in one basin state, and subsequently moved to another basin state to start working there. These persons stimulated the development of transboundary networks, and were aware of cultural and structural differences between the basin states and regions. The same goes for those who worked for an international organization, such as the EU. The researcher has the impression that individuals playing a rather negative role in the international decision making process were present in all basin states and regions. Characteristic to these persons was their inflexibility, incapability of (strategic) learning, or lack of empathy.

8.4.4 The relation between international and *intra*national decision making

In Section 2.3.1 it was argued that decision making on international river issues should be conceived of as a multi-level game. On the international level the international decision making game is linked to various *intra*national games. The case study of decision making on international Scheldt issues contained several examples of such linkages, and made it possible to distinguish four ways in which decision making on international policies was linked to *intra*national decision making processes.

First, the national parliaments of the basin states may disapprove conventions and/or policies drafted in an international negotiation commission. The history of international Scheldt negotiations contained an interesting example, namely the decision of the First Chamber of the Dutch parliament not to approve an international Treaty in 1927 (See Section 4.8).

Secondly, whilst the national government may agree with developed international policies, in other governmental layers objections may be formulated. The Walloon opposition to the proposed Belgian-Dutch package deal is a perfect illustration of this (See Sections 5.3.1, 6.3.1, and 8.3.1).

Thirdly, the judiciary may oppose the implementation of international policies. This can be illustrated with the decision of the Dutch Council of State to cancel the WVO Permit granted to the Flemish region (Section 7.2, Round XIIb).

Fourthly, NGOs and the public may oppose the implementation of international policies. The opposition to the plans for nature restoration in the Dutch province of Zeeland is a clear example of this (Section 7.2, Round XIIb).

CHAPTER 9

SOLVING UPSTREAM-DOWNSTREAM PROBLEMS

9.1   Introduction

The aim of this chapter is to answer research question seven, and to indicate strategies that can be used to contribute to the solution of upstream-downstream problems in international river basins.

In Section 9.2 three ways in which this study may make a contribution to the practice of international river basin management are distinguished. Subsequently, in Section 9.3, entitled 'lessons learned', several recommendations are formulated on the basis of the conclusions of the case study presented in Chapter 8. Thereafter, in Section 9.4 a typology of strategies is presented that may be used by actors involved in decision making on upstream-downstream issues in international river basins. This section successively addresses strategies of problem solving that may be used by individual basin states, the European Union, and intermediaries.

9.2   Possibilities of prescription

This thesis research may contribute to the practice of international river basin management in three different ways.

First, it may stimulate those involved in decision making on international river issues to look at decision making from a pluricentric perspective, and to develop strategies of network management themselves.[1] The theory of decision making presented in Chapter 2, the 'thick descriptions' of decision making, and the case analysis may contribute to a better awareness and understanding of the complexity and dynamics of decision making processes on international river issues. The study emphasized that different actors do have different perceptions and employ strategic behaviour, and illustrated that actors should take into account the action rationality of others, if they want to achieve their objectives.

Secondly, on the basis of the conclusions of the case study presented in Chapter 8 some lessons can be learned that may be relevant for cases of decision making on upstream-downstream issues in other international river basins as well.

Thirdly, the theory presented in Chapter 2 and the lessons learned from the case study enabled the author to develop a typology of strategies that may be used to contribute to

---

1. Termeer (1993b, p. 299).

the solution of upstream-downstream problems in international river basins. This typology, which will be presented in Section 9.4 of this chapter, may serve the purpose of a checklist by those involved in decision making on these issues.

9.3   Lessons learned

This doctorate thesis research is a so-called single case study. Consequently, possibilities to generalize the case study results to other cases by definition are limited. The case study has clearly illustrated the large impact contextual variables may have on the course and outcomes of a decision making process. Because in different international river basins the characteristics of the context may be very different, one should be careful with generalizing the case study results to other international river basins. In spite of these considerations, on the basis of the conclusions of the case study presented in Chapter 8 some lessons can be learned that may be relevant for cases of decision making in other international river basins as well.

1. Strategic learning formed an important explanation for international agreement on the development of institutional arrangements and policies (See Section 8.3.3). The parties involved in decision making on international river issues should be able and prepared to adapt a strategy in case this strategy has proven to be ineffective. In addition, they should monitor carefully developments in their interorganizational environment to recognize opportunities for reaching mutually beneficial agreements.

2. Issue-linkage was the key to understanding the development of international institutional arrangements and policies in the Scheldt basin (See Section 8.3.2). Therefore, in cases where the decision rule is unanimity, and the parties do have different preference intensities in an issue-area, the parties should investigate possibilities of issue-linkage.

3. Linkages that were not accepted by the actors involved because they did not create a win-win situation caused the growth of a culture of distrust, and influenced negatively decision making on other international issues (See Section 8.3.2). Therefore, actors should be careful with continuing linkages that are not accepted by all relevant parties, i.e. the parties possessing indispensable resources.

4. The presence of different platforms where representatives of the basin states and regions discussed the same or similar issues appeared to be very useful. When negotiations in the international negotiation commissions went laboriously, the TSC could be used to start discussions on the water quality of the river Scheldt, the ISG could be used to start the exchange of information, and the UN-ECE meetings could be used to reach an agreement on the need for basin-wide cooperation (See Section 8.3.2). Therefore, despite the often expressed wish for coordination and integration of existing platforms, it seems to be recommendable to maintain a certain redundancy of international contacts in different international platforms.

5. International Scheldt issues emerged on different scales, and the solutions of these issues were found on different scales (Section 8.4.2). Therefore, one should not put a

one-sided emphasis on the basin scale, and neglect the many interdependencies that go beyond the boundaries of a basin. By not taking into account these interdependencies one may easily overlook possibilities of cooperation and problem solving. Effective problem-solving requires flexibility of the actors to be involved in the decision making process and the issues to be addressed, and consequently flexibility of scale. Each issue demands for another network to be activated that will rarely exactly coincide with the basin scale. Nevertheless, some kind of structuring of international relations in river basins is very useful, and may stimulate the development of international policies.[2] Furthermore, the basin scale is an appropriate scale as indeed many international hydrology related river issues do emerge on a basin scale.

6. The results of the Flemish-Dutch research project OostWest contributed to a convergence of the perceptions of the issues of dredging and dumping in the Scheldt estuary, and stimulated agreement on these issues (See Section 8.3.3). Actors involved in decision making on upstream-downstream issues in international river basins, which do involve uncertainties, should start joint research, because this may contribute to an agreement on the cognitive aspects of the issues at stake (See also Section 9.4.1).

7. Even in a relatively small international river basin, such as the river Scheldt basin, there were considerable differences between the national and decision making cultures of the basin states, which influenced decision making in various ways (See Section 8.3.4). Therefore, those involved in decision making on international river issues should be aware of cultural differences, and pay attention to the problem of intercultural communication (See also Section 9.4.1).

9.4   A typology of strategies that may contribute to the solution of upstream-downstream problems in international river basins

The theory of decision making on international river issues treated in Chapter 2, and the case study of decision making on international Scheldt issues presented in part II of the thesis made it possible to develop a typology of strategies that may contribute to the solution of upstream-downstream problems in international river basins.

In the theoretical framework, three different positions shaped by the distribution of resources among the actors were distinguished: the interaction position, intervention position, and the incentive position.[3] The possibilities of an actor to exert influence on decision making not only depend on his position, but on the possibility to influence variables at all as well.[4] For example, the underlying hydrological structure of the issues at stake, or the (decision making) cultures of the basin states cannot be influenced by any of the actors involved in a decision making process, no matter which position they have. The next sections successively discuss strategies that may be used by individual basin states, the European Union, and intermediaries. Figure 9.1 contains a schematic overview

---

2. See also the conclusions concerning the effectiveness of international river commissions presented in Section 8.4.1.

3. See Section 2.2.3, Table 2.1.

4. Ellemers (1987, pp. 229-230).

of the types of strategies discussed below.

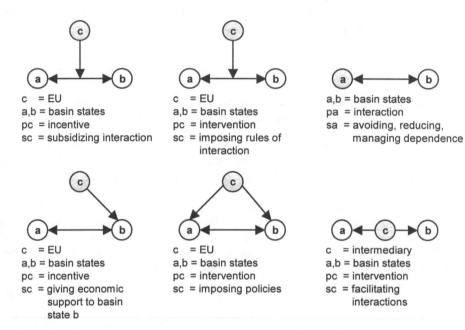

c   = EU
a,b = basin states
pc = incentive
sc = subsidizing interaction

c   = EU
a,b = basin states
pc = intervention
sc = imposing rules of
      interaction

a,b = basin states
pa = interaction
sa = avoiding, reducing,
      managing dependence

c   = EU
a,b = basin states
pc = incentive
sc = giving economic
      support to basin
      state b

c   = EU
a,b = basin states
pc = intervention
sc = imposing policies

c   = intermediary
a,b = basin states
pc = intervention
sc = facilitating
      interactions

Figure 9.1 *Strategies for solving upstream-downstream problems* In this figure the strategies of problem solving that may be used by individual basin states, the EU, and intermediaries are depicted. a,b, and c are actors, p is the position of the actor, and s is the strategy that is used by this actor. The upper left corner of the figure shows the European Union (c) which grants a subsidy to stimulate interactions between the basin states a and b. The upper middle part of the figure shows the European Union (c) which imposes rules of interaction on the basin states a and b. The upper right corner of the figure shows an individual basin state a that tries to avoid, reduce, or manage its dependence on another basin state b. The lower left corner of the figure shows the European Union (c) which gives economic support to a basin state b. The lower middle part of the figure shows the European Union (c) which imposes policies on the basin states a and b. Finally, the lower right corner of the figure shows an intermediary (c) facilitating interactions between the basin states a and b.

9.4.1 Individual basin states
Representatives of individual basin states involved in decision making on upstream-downstream issues in an international river basin do have an interaction position. They only are able to exert influence on other basin states in international interaction processes. A basin state that perceives an international upstream-downstream problem may use two

broad strategies of problem solving.[5] First, it may try to avoid or reduce its dependence on the other basin states. Secondly, it may try to manage its dependence on them. The following sections successively discuss these types of strategies. In addition, some general recommendations concerning (cross-cultural) communication and negotiation are made. Table 9.1 produces information on the types of strategies that can be used by representatives of individual basin states involved in decision making on upstream-downstream issues in international river basins.

Table 9.1 *Strategies that can be used by individual basin states involved in decision making on upstream-downstream issues in international river basins* Individual basin states do have an interaction position. They can choose alternative strategies of problem solving from two broad categories of strategies: the strategies aiming at the avoidance or reduction of dependence, and the strategies aiming at the management of dependence

AVOID OR REDUCE DEPENDENCE	Lower level of ambition	
	Increase of (control of) resources	*Stimulate development of supranational policies*
		Build coalitions
		Carry out (advocative) research
		Develop legal resources
	Develop substitutes	
MANAGE DEPENDENCE	Arrange interactions	*Set agenda*
		Create platform for interaction
	Influence perceptions	*Stimulate consensus on cognitive aspects of issues at stake*
		Persuade
		Influence background variables
	Compensate potential losers	*Issue-linkage*
		Compensate financially
		Enlarge the shadow of the future

Strategies aiming at the avoidance or reduction of dependence
Generally, representatives of individual basin states do have three possibilities to reduce their dependence on other actors. They can:
1. lower their ambitions;
2. increase their (control of) resources;

---

5. See Section 2.2.3.

3. develop substitutes.

Lower ambitions
Because a problem perception is defined as a discrepancy between a perceived actual situation and a desired situation, individual basin states are able to solve their problems by formulating less ambitious objectives. Although this strategy of problem solving is an academic one, it undeniable is one of the strategies that may be used by actors having an interaction position, and therefore should be included in the typology presented in this section. Although this strategy may be useful in situations where actors have formulated unrealistic objectives, generally actors should not be recommended to lower their ambitions, but to be flexible as regards the strategies they use to achieve ambitious objectives.

Increase of (control of) resources
Interdependencies are caused by the distribution of interests and resources among the network of actors in an issue-area. Actors may reduce their dependence on others by increasing their control of resources.

A specific type of resource, which enables actors to use authoritative or coercive strategies, is authority. Generally, actors having an interaction position are unable to acquire this type of resource. They may, however, stimulate that international organizations receive such resources, and consequently are able to impose rules of interaction or policies. Downstream basin states in the European Union, for example, which suffer from upstream waste discharges, may stimulate the development of supranational European water policies, because these policies may contribute to the solution of the problems they perceive.

Another possibility to increase the control of resources is to build coalitions with actors perceiving (about) the same problems. The pooling of resources may contribute to the solution of their problems.

Furthermore, knowledge on the cognitive aspects of the issues at stake may be an important resource for problem solving. Research may contribute to more knowledge on causes and effects in an issue-area, policy alternatives, and impacts of these policy alternatives, and therefore enables the basin states to develop more effective strategies. In addition, advocative research may be used to found preferences concerning policy objectives and alternatives.

The creation of legal resources may reduce dependencies as well. The introduction of an obligatory WVO-Permit in the case study of decision making on international Scheldt issues is an interesting example of this strategy.

Develop substitutes
A third possibility to reduce dependence on the other basin states is to develop substitutes for resources that are controlled by other parties. For example, an upstream basin state that depends on a downstream basin state as regards the maintenance or improvement of the maritime access to its main ports, may decide to dig a canal to the sea on its territory, and a downstream basin state that extracts drinking water from a polluted river may decide to extract drinking water from a less polluted river or from a ground water basin.

Because actors naturally strive towards maintenance of autonomy, they tend to use strategies aiming at the avoidance or reduction of dependence first. In many cases, however, these strategies are insufficient for problem solving, and basin states have to manage their dependence on other basin states.

Strategies aiming at the management of dependence
Once actors have recognized and accepted their dependence on other actors, they can use a combination of three types of strategies to manage dependencies. They can:
1. arrange interactions;
2. influence perceptions;
3. compensate potential losers.

Arrange interactions
Basin states that want to manage dependencies have to create an opportunity of interaction first, because this enables them to communicate and negotiate with the basin states having indispensable resources. Two different ways to create an opportunity of interaction are agenda setting, and the creation of a new platform for interaction. In the first situation an actor tries to place his problem on the agenda of an existing organization, such as an international river commission. Opportunities to place new issues on the international agenda, for example, may be a disaster (See Section 2.3.2), or a policy proposal made by other parties, as this makes it possible to apply a strategy of issue-linkage.

If no suitable platform for interaction exists, the basin state may try to create a new one. In the latter case the basin state has to establish contacts with the other basin states, and to start negotiations on the development of institutional arrangements to regulate their interactions.

Influence perceptions
If actors depend on others as regards the solution of their problems, they may try to influence the perceptions of these actors. A convergence of the problem perceptions of the actors involved in decision making on an international river issue increases the chance that an agreement will be reached on the development of institutional arrangements and policies to deal with that issue. Strategies that can be used to influence problem perceptions are strategies that aim at reaching consensus on the cognitive aspects of the issue at stake, persuasive strategies, and strategies that aim at influencing background variables.

Differences between problem perceptions may be caused partly by a lack of consensus on the cognitive or 'factual' aspects of an upstream-downstream issue. First, the actors may disagree on the cause-effect relations in a river. As an example, two basin states may disagree on the impact waste discharges in the upstream part of the basin have on the population of a certain species in the downstream part of the basin. Secondly, the actors may disagree on the consequences of policy alternatives. They may, for example, disagree whether the construction of a specific type of waste water treatment plant is sufficient for the achievement of a certain water quality objective. Because research on cause-effect relations, and the consequences of policy alternatives inevitably comprises many subjective elements, such as the assumptions made, and the selection of research and measurement methods, joint research generally contributes more to a consensus on the cognitive aspects of an issue than research that is carried out by a single participant in the decision making process. Two main strategies that may be used to further consensus on the cognitive aspects of international river issues are the exchange of data and expertise, and joint research. These strategies contribute to a consensus on the cognitive aspects of the issue at stake, and therefore mostly to a *convergence* of problem perceptions. They do, however, not lead automatically to a *shared* problem perception, because preferences concerning the desired situation may continue to diverge.

If the actors involved in decision making on an international river issue do have different values and interests, the actor perceiving a problem may try to influence the perceptions of the other actors by using persuasive strategies. For example, an upstream polluter may know about his waste discharges, and know the figures about the impacts these discharges have on the ecosystems along the river. Nevertheless, his problem perception may change when he is confronted directly with the negative impacts of his activities. Therefore, actors responsible for the management of wetlands in the downstream part of a river basin may invite upstream polluters to visit the sites along the river, which may lead to an increased understanding of the problems caused by the upstream waste discharges. Possibly, the problem perceptions and behaviour of the upstream polluters will change.

Whilst the strategies discussed above may change problem perceptions in the short run, influencing background variables may influence problem perceptions in the long run. For environmental issues, such as transboundary water and sediment pollution, environmental awareness and economic prosperity are important background variables. Environmental awareness increases the willingness to invest in environmental policies, whereas economic prosperity enables the parties to allocate financial resources to make such investments.[6] Basin states perceiving transboundary environmental problems may try to raise environmental awareness in the upstream basin states. They may, for example, subsidize the activities of environmental NGOs. Secondly, they may support basin states having economic problems. Such support may be given either directly or indirectly.[7] In the latter case the basin states may plead for economic support by an international organization, such as the European Union.[8]

Compensate potential losers
If actors are unable to reduce their dependence fully, and strategies that aim at influencing problem perceptions are insufficient or only effective in the long run, compensation of potential losers may be an effective strategy of problem solving. This strategy aims at the creation of a situation in which all actors are better off with an agreement on joint problem solving than without one, a so-called win-win situation. Three possibilities to compensate potential losers are issue-linkage, to compensate financially, or to enlarge the shadow of the future.

Issue-linkage may change a situation of dependence into a situation of interdependence. As an example, an agreement with a neighbouring country on an international river scheme this country prefers most may be used to gain concessions on other issues, either related to river basin management or not (See Section 2.3.2). The

---

6. This recommendation relates to the fundamental dilemma of environmental policy, which was also mentioned shortly in Chapter 1 of this thesis. The dilemma is that on the one hand economic growth enables actors to invest in environmental policies, but on the other hand economic growth often causes additional environmental pollution or degradation.

7. This support, however, may be perceived as contradicting with the internationally accepted polluter pays principle.

8. Below, the strategies of problem solving that may be used by the European Union are discussed.

case study presented in this book has lead to ambiguous conclusions concerning the use of issue-linkage in the case study (See Sections 8.3.2.). On the one hand, issue-linkage appeared to be a useful strategy for the development of international institutional arrangements and policies if it created a situation in which all parties involved perceived that they would be better off with an agreement than without one. On the other hand, issue-linkages that did not create such a 'win-win situation' caused lengthy deadlocks in the decision making process, and distrust between the parties.

Sometimes actors do only oppose a proposed policy programme because this programme entails negative side-effects. In these cases mitigation measures may compensate potential losers. As an example, the negative impacts the construction of an infrastructure project may have on the ecology of a water system may be compensated by the creation of new nature areas.

A second possibility to compensate potential losers from an international agreement is to compensate these parties financially. An example of financial compensation in river basin management concerns the establishment of a sanitation fund. The idea of a sanitation fund is that all basin states contribute to the fund, and that the money available is invested in the projects yielding most benefits. In the Scheldt case the establishment of a sanitation fund would imply a financial transfer from the downstream basin state, the Netherlands, to the upstream basin states and regions. The problems related to the establishment of a sanitation fund were discussed in Section 8.3.2

A third possibility to compensate potential losers is to enlarge the 'shadow of the future' (See Section 2.3.2). If actors do expect that they will meet each other more often in the future, the potential loser may be willing to build up a reservoir of goodwill he can draw on when he needs support or a concession from the other actors as regards policies of greater interest to him later. Axelrod distinguishes two types of strategies to enlarge the 'shadow of the future' (See Section 2.3.3). First, interactions can be made more durable. Secondly, interactions can be made more frequent. As an example, interstate cooperation in the European Union contributes to durable and more frequent interactions between the member states, which in turn increases the chance that member states are willing to build up a reservoir of goodwill.

Each of the strategies discussed above may contribute to problem solving, but in most cases neither of them separately will be sufficient. Some strategies only are effective in the long run, such as strategies aiming at influencing background variables. Others can be effective in the short run, such as strategies aiming at a compensation of potential losers. Effective problem solving generally requires a combination of these strategies.

Improve communication and negotiation skills
Most strategies discussed above have to be used in an international communication and negotiation process. Therefore, communication and negotiation skills are essential for effective problem solving.

Cross-cultural communication skills
Knowing the other parties is a prerequisite for effective international communication. The actors involved in decision making on international river issues should know about the institutional differences between the basin states. These differences concern both the different organization of water management, and the different (decision making) cultures of the basin states. Knowledge about the administrative organization in the other basin

states enables the actors to select the right organizations and persons to be contacted. Knowledge about cultural characteristics of the other parties makes it possible to better interpret their behaviour, and therefore to prevent misunderstandings. Although the national culture is a factor that can hardly be influenced, intercultural relations can and should be managed consciously.

Three stages of intercultural communication are: the awareness of cultural differences, the acceptance of these differences, and dealing with the observed differences. Hopefully, the cultural analysis presented in Section 4.6 contributes to the awareness of cultural differences between the Scheldt and (most) Meuse basin states. As regards the acceptance the most important recommendation is that the parties would better accept cultural differences, because cultures can hardly be influenced, and do only change in the long run. If parties do interact with each other for a long time, however, for example in an international river commission, a shared culture may develop gradually. As regards the management of intercultural relations several recommendations can be made. First, the actors involved in international decision making processes could be taught general cross-cultural communication skills, such as[9]:

1. the capacity to communicate respect;
2. the capacity to be nonjudgmental;
3. the capacity to accept the relativity of one's own knowledge and perceptions;
4. the capacity to display empathy;
5. the capacity to be flexible;
6. the capacity for turn-taking (letting everyone take turns in discussions);
7. tolerance for ambiguity.

Secondly, a cultural-specific training may be given.[10] For example, culture-specific training, focused on intercultural interaction between representatives of the Latin culture of France and Belgium, and the Nordic culture of the Netherlands, could contribute much to the awareness and acceptance of cultural differences, and the capacity to deal with these differences of those who are involved in decision making on international Scheldt and Meuse issues.

The case study has shown that transboundary job migration had a positive impact on the development of cross-border networks, partly because the persons involved are familiar with the cultural characteristics of two basin states. Therefore, transboundary job-migration may be an important instrument to improve intercultural communication and negotiation.

Negotiation skills and learning capacity

In addition to cross-cultural communication skills, those involved in decision making on international river issues should have negotiation skills and learning capacity.

First, those involved in negotiations should be aware that the main problem in a negotiation process mostly is not the objective reality, but what is in the minds of the people involved in the process. Although factual knowledge can be useful, the perceptions of the parties involved are most important for the course of a negotiation process, and finding solutions.[11] Therefore, the capacity to display empathy, to look at the issue at

9. Ruben (1977), cited in Hofstede (1980, p. 398).

10. Hofstede (1980, p. 397).

11. Fisher *et al.* (1981, p. 43).

stake from different perspectives, and to understand the point of view of the other parties involved is one of the most important negotiation skills.[12] The capacity to understand other parties points of view, however, does not imply that one should accept them.[13]

Secondly, the negotiators should be flexible as regards the policy programmes proposed, and strategies used to solve their problems. This flexibility should enable the negotiators to search for policy programmes and strategies that are acceptable for other parties as well.

Thirdly, because in an international context the dominant decision rule is unanimity, the parties should be aware of and accept the necessity for searching mutually beneficial agreements or win-win situations.[14] According to Fisher, one of the main obstacles to the creation of win-win situations is that negotiators tend to perceive negotiation problems primarily as distributive problems. Therefore, they may easily engage in a fight about the share of a fixed pie, whereas, they overlook possibilities to increase the pie, and to create a win-win situation.

Fourthly, although the parties do have different interests, it may be possible to reach an agreement on principles.[15] Agreement on principles may largely facilitate the negotiation process. For example, if the basin states of an international river disagree on the contents of international river policies, they may try to reach an agreement on certain principles, such as the polluter pays principle, first.

9.4.2 The European Union

This section focuses on the contributions the EU could make to the solution of upstream-downstream problems in river basins shared by EU-member states. Unlike the individual basin states, which have an interaction position, the European Union may have an intervention and an incentive position in decision making processes on international river issues. The following sections successively discuss the strategies the EU may use from these two positions.

The intervention position
Strategies that can be used from an intervention position are to:
1. impose policies;
2. impose rules of interaction;
3. facilitate interactions.

Impose policies
First, the EU can impose emission-based or immission-based policies on its member states. An example of this type of strategy is the EC Directive on the treatment of urban waste water. Generally, the EU does impose policy objectives, and no policy measures. Because EU-policies are largely based on the principle of subsidiarity, the development of strategies to reach the EU-policy objectives is a matter of concern of the individual basin states.

---

12. Ibid., pp. 43-44.

13. Ibid., p. 45.

14. Ibid., p. 78.

15. Ibid., p. 105.

Impose rules of interaction
Secondly, the EU is entitled to impose rules of interaction. An example of this type of strategy is the development of the EU Framework Directive for Water.[16] Part of the proposal for this Directive contains procedures for the development of river basin plans for international river basins and their subbasins. Furthermore, the directive would make obligatory the installation of river basin commissions to develop these plans. By doing so, this directive establishes rules of interaction between the basin states.

A key question is whether these types of authoritative strategies contribute to the solution of upstream-downstream problems in international river basins. Apart from the potential of authoritative strategies to influence water policies in the EU member states, and decision making on international river issues, these strategies do have several limitations as well.

First, decisions on the use of authoritative strategies are the result of a joint decision making process. The dominant decision rule in these processes is unanimity or at least a large majority. Consequently, decisions on the use of authoritative strategies often are compromises, and will rarely be able to change the status quo significantly.

Secondly, imposed policies or rules of interaction that are not supported by the basin states will cause implementation gaps in case provisions for effective monitoring and control are absent or insufficient. In general, one may expect more serious implementation gaps when decision making is not based on unanimity. Especially if basin states do not perceive the main river issues as commons problems, but are not interested in cooperation at all, the effectiveness of authoritative strategies strongly depends on the organization of monitoring and control of the imposed regulations.

Finally, on the European level it is more difficult to take into account regional differences than on the national level. Effective institutional arrangements and policies, however, should fit into the specific regional and local institutional context.

Facilitate interactions
Beside imposing rules of interaction or joint policies, the EU may facilitate interstate interactions, and play the role of intermediary (See below). The draft Framework Directive for Water does provide for such a role.

The incentive position
In addition to an intervention position, the EU may have an incentive position in decision making on international river issues. Although giving financial incentives is a kind of top down steering, its character is fundamentally different from the kind of steering discussed in the previous section. The main difference is that actors are free to use subsidies or accept financial support, whilst they have to live up to imposed regulations.

The EU may:
1. stimulate international interactions by granting subsidies to transboundary projects;
2. give economic support to economically weak regions.

---

16. Commission proposal for a council directive establishing a framework for community action in the field of water policy, and several amendments made afterwards. 26.02.1997.

The first steering option is to grant subsidies for the organization of international projects. The case study has shown two main examples of subsidized projects, the LIFE-project for the development of a decision support system, and the LIFE-MARS-project directed at the implementation of pilot projects for nature restoration. Both projects stimulated international exchange of information and research, which may have contributed to an agreement on the cognitive aspects of the issues at stake, and a convergence of the problem perceptions of the parties involved in decision making on these issues.

Secondly, the EU may give economic support to areas with economic problems. Because one of the conclusions of the case study research was that most likely economic prosperity was an important explanatory variable for the willingness to develop environmental policies, economic support may be an effective strategy of problem solving. This recommendation, however, touches the fundamental dilemma in environmental management mentioned earlier in Section 9.4.1.

9.4.3 Intermediaries

A final perspective for the formulation of recommendations is the perspective of an intermediary. An intermediary is an independent party facilitating interactions, and stimulating cooperation.[17] The role of an intermediary resembles the role assigned to government agencies in most studies of network management. Unlike the other parties involved in a decision making process, the intermediary does not have a specific interest. In an international context it is difficult to find independent parties, because the actors involved in the decision making process generally serve the interests of the respective basin states.

Examples of actors that may play the role of an intermediary in decision making on international river issues are:

- an independent policy analyst;
- an independent chairman of an international river commission;
- a representative of the European Union or another international organization, such as the World bank;
- a new party entering the international arena.

As indicated before, these intermediaries are only able to facilitate interactions, and stimulate cooperation if they are accepted by the parties involved.[18] If one or more of the parties do not accept the intermediary, for example as they prefer to keep on fighting for internal political reasons, it is unlikely that an intermediary will be successful.

The main contribution of the policy analyst is to stimulate agreement on the cognitive aspects of the issues at stake, i.e. on the relevant 'facts'. He may do so by stimulating the exchange of information between the parties involved and/or by initiating joint research projects. In this way policy analysts may contribute to the prevention or solution of cognitive conflicts.

An independent chairman or representative of an international organization does have many more possibilities to play a mediating role. Apart from stimulating agreement on the 'facts', an important task of these intermediaries is to make clear to all actors involved in a decision making process that a mutually beneficial agreement is a hard precondition for

---

17. For information on types of intermediaries and the strategies they may use, see for example Wessel (1988), Dieperink (1997), and Wolf (1997).

18. See Section 6.3.4.

any international policy or institutional change. This should stimulate the parties involved to search actively for win-win situations. Opportunities to create mutually beneficial agreements may be the generation of economies of scale, the solution of commons problem, or possibilities to gain from trade (package deals). To recognize these opportunities, the intermediary should stimulate exchange of information concerning problem perceptions and preferences of the parties involved. To facilitate the communication and negotiation process, the intermediary should stimulate the development of a common language, which may prevent misunderstandings caused by the use of a different jargon.[19] In addition, the intermediary should prevent misunderstandings caused by different cultural backgrounds. Therefore, the intermediary should be aware of these differences and their potential implications.[20] Furthermore, the intermediary should contribute to a climate in which doubts and times for reflection are allowed.[21] Another important contribution of the intermediary is to stimulate agreement on the rules of interaction (or game rules). Participants should, for example, know when major decisions will be taken, and how long the decision making process will continue.

Finally, the case study of decision making on international Scheldt issues has shown that new parties entering the international arena may play a mediating role as well.[22] If two parties have a history of conflicts, a new party acceptable to both parties may play an important role in the joint decision making process.

---

19. Termeer (1993a, p. 113).

20. Avruch & Black (1993, p. 131-145).

21. Termeer (1993a, p. 114).

22. See Section 7.2, Round Xb, and Section 7.3.1.

CHAPTER 10

FINAL DISCUSSION

10.1 Introduction

This chapter contains a reflection on the research process and results, and recommendations for further research. In the Sections 10.2 and 10.3 successively the methodological and theoretical choices that were made in this research are evaluated. Subsequently, in Section 10.4 the experiences with doing international research are discussed. Section 10.5 contains an evaluation of the research results. Section 10.6 concludes this chapter with some recommendations for further research.

10.2 Evaluation of methodological choices

The study of decision making on international Scheldt issues presented in this book entailed several methodological choices, which inevitably had an impact on the research process and results. In this section, the main methodological choices are evaluated. These choices concerned the choice for:

1. an in depth single case study covering a long time span;
2. the use of the rounds model of decision making;
3. a rather inductive approach and *ex post* explanation of decision making.

Because of the interest in a wide range of variables and their interrelations, the interest in real-life decision making, and the rather vague distinction between the phenomenon decision making and its context, the choice for a *case study strategy* was quite obvious. The choice for a *longitudinal* case study covering a period of 30 years enabled the researcher:
- to distinguish between more or less constant explanatory variables and dynamic ones;
- to study the sequence of decision making rounds, sometimes characterized by conflict, and sometimes more by cooperation;
- to study learning processes.

The focus on a *single case study* enabled the researcher to carry out an in depth study of decision making. The author knows about several comparative case studies of decision making on international river issues.[1] Although these studies without exception are excellent, they inevitably are less detailed than the single case study presented in this book. For some time the researcher aimed at the selection and study of multiple cases of international issues in the Scheldt basin. In depth study, however, learned that most of these issues were strongly related, either because of their intellectual coherence, or because of their strategic coherence. On the basis of the intellectual coherence, two main international issue-areas could be distinguished in the Scheldt basin: the issue-area of the maritime access to the port of Antwerp, and the issue-area of water and sediment pollution. Because of the many tactical linkages made between these issue-areas, there was a strategic coherence between all international Scheldt issues. For that reason, it turned out to be impossible to describe and analyze decision making on the various issues separately.

A second important methodological choice was the choice for using the rounds model of decision making. Although the selection of crucial decisions and the distinction between decision making rounds may be disputed easily, the rounds model appeared to be a useful model to structure the lengthy and complicated decision making process. The coherent description of the strategic interactions between the Scheldt basin states and regions, and the analysis of the action rationality of the actors involved contributed to a better insight in the course of the decision making process.

A third methodological choice was the choice for a rather inductive approach, and *ex post* explanation of decision making. By using this approach strategic interactions around international Scheldt issues were reconstructed, and case specific hypotheses concerning the relation between perceptions and strategies, institutionalization processes, learning processes, and the influence of contextual variables on the decision making process were formulated. During the early stages of the research the author has struggled with the choice for a hypothetico-inductive or a hypothetico-deductive research approach.[2] By now he has learned that the real problem of research on social phenomena, such as complex decision making, is the so-called cases-variables problem, which refers to the fact that the number of relevant explanatory variables often exceeds the number of cases. Therefore, it is always difficult to draw 'hard' conclusions, no matter which methodological choices are made.

10.3 Evaluation of theoretical choices

The main theoretical choice was the choice for using the theory on policy networks. Although many scientists doubt whether the theory on policy networks is a real theory, because it would not have much explanatory power, the case study has shown that the theory has *ex post* explanatory power at least.[3] Indeed, the theory on policy networks is not a predictive theory about the development of decision making. One may, however, discuss whether this is a shortcoming of the theory, or the consequence of the complexity

---

1. LeMarquand (1977), Faure and Rubin (1993), Durth (1996).

2. An interesting example of a hypothetico-deductive study of decision making on international river issues is the research of Dieperink (1997).

3. For a discussion on the status of the theory on policy networks, see Klijn and Koppenjan (1997), Pröpper (1997), and Klijn and Koppenjan (1997a).

of real life decision making.[4]

There were two main reasons to choose the theory on policy networks for the description, analysis, and *ex post* explanation of decision making on international Scheldt issues. First, the author expected the theory to be especially suitable for the study of international decision making. Characteristic to the theory on policy networks is its focus on communication, negotiation, and consensus building, rather than on hierarchical steering. The absence of possibilities of hierarchical steering, and the need for consensus exactly are the main characteristics of decision making on most international river issues. Secondly, like in most research, the choice was partly based on personal preferences of the author.

Although most theoretical concepts presented in Chapter 2 draw on the theory of policy networks, the focus on the study of *international river* issues made it necessary to consult part of the International Relations literature and the literature on international river basin management as well.[5] In addition, special attention was paid to Hofstede's research on cultural differences among nation states.[6]

10.4 Experience with doing international research

The international character of the research had important consequences for both the contents of the research, and the research methods applied.

The international character of the issues studied made it necessary to study the difference between institutional characteristics of France, Belgium, and the Netherlands. These differences, for example, concerned the administrative-organization of water management, and the decision making cultures in these countries.

As regards the research methods, the international character of the research made it necessary to collect information in three different countries. The research comprised a series of interviews with representatives of government agencies in each of the basin states and regions. A major advantage of doing research in the Scheldt basin was that hardly social research on decision making on international Scheldt issues has been carried out yet. Consequently, most respondents were interviewed on this topic for the first time, showed much interest in the study, and were willing to spend their valuable time.

In addition to visiting government agencies in France, The Walloon, Flemish, and Brussels Capital regions, and the Netherlands, the author attended several international meetings where international Scheldt issues were discussed.[7] This enabled him to observe real-life international interactions, to establish contacts with representatives of each basin state, and to develop a small international network himself.

---

4. See Klijn and Koppenjan (1997).

5. For example Axelrod (1984), Caldwell (1988, 1990), Frey (1993), LeMarquand (1977), Martin (1995), and Marty (1997).

6. See Section 4.6.

7. The author attended several meetings of RIZA-AMINAL, the workshop ISG-DES, a meeting of the Working group of the Western Scheldt, and several international conferences where international Scheldt issues were discussed.

10.5 Evaluation of research results

In this section the practical and theoretical relevance of the research results are discussed. Chapter 9 should have indicated the practical relevance of this research. There, it was argued that the research could contribute to the practice of international river basin management in three different ways. First, it may stimulate those involved in decision making on international river issues to look at decision making from a pluricentric perspective. Secondly, on the basis of the conclusions of the case study of decision making on international Scheldt issues some recommendations could be formulated. Thirdly, in Chapter 9 a typology of strategies was presented that may serve the purpose of a checklist by those involved in decision making on international river issues. This typology contained strategies that may be used by:
- representatives of individual basin states perceiving upstream-downstream problems in international river basins;
- representatives of the European Union;
- independent parties, such as independent policy analysts, independent chairmen of river basin commissions, or EU representatives, which play the role of mediator in a decision making process.

In addition to the practical relevance, the study has contributed to the theory of decision making on international river issues. Whilst the theory of policy networks generally is applied to cases of national decision making, this study has demonstrated that the theory can be used to analyze decision making on international issues as well.[8] Because of the international character of the issues at stake, special attention was paid to the influence of institutional differences between the basin states and regions on the decision making process.

No extensive study of decision making on international Scheldt issues was available yet. Fortunately, several overviews of the international negotiation process could be used, among which articles of Suykens and Strubbe[9], and the research of Planchar.[10] Furthermore, Ovaa's extensive description of the Scheldt basin and the organization of water management in the Scheldt basin states provided interesting introductory information.[11] Neither of these studies, however presented a complete overview of decision making on international Scheldt issues between 1967 and 1997, nor did they contain a theoretical interpretation. The Scheldt case appeared to be a very interesting one. It clearly demonstrates the difficulties the basin states of an international river face if they want to reach an agreement on the solution of upstream-downstream issues. The case study especially contributed to the knowledge on the use of the strategy of issue-linkage in international negotiations, and the influence of institutional differences between the basin states on the decision making process.

---

8. As regards the use of network concepts in an international context, the author only knows about the work of Soeters (1993). In his research, however, the focus is on the (change of) network characteristics, rather than on the course of decision making on specific issues.

9. Suykens (1995), Strubbe (1988a and b).

10. Planchar (1993).

11. Ovaa (1991).

10.6 Recommendations for further research

Like most research also this research evokes more new questions than it produces answers. Therefore, recommendations for further research can be formulated rather easily.

First, similar analyses could be made for international issues in other river basins.

Secondly, it could be interesting to design comparative studies of decision making on international river issues. Because of the different characteristics of the context of these decision making processes, however, such comparative research designs are highly complicated.

Thirdly, because the theory on policy networks has proven to be useful for the analysis and *ex post* explanation of decision making on international river issues, it may be worthwhile to apply the same conceptual framework to cases of decision making on other international issues, such as the construction of railways or highways.

Fourthly, in the Scheldt case there appeared to be considerable cultural differences between the basin states that seem to have influenced decision making in various ways. Therefore, it would be interesting to carry out additional research on the influence of cultural factors on international decision making processes.

Fifthly, in the theoretical framework Kingdon's theory of policy windows was used to explain the impact of the Sandoz-fire in the river Rhine on the Rhine river negotiations, and the impacts of the floods in the rivers Rhine and Meuse on decision making on flood protection and alleviation. It might be interesting to use this theory for a more detailed analysis of the impact of these and other disasters on decision making on international river issues.

Finally, it would be interesting to carry out research on *intra*national decision making processes on strategies to be used in international negotiations. How do interactions between politicians, civil servants, and NGOs develop, how is (democratic control of) this decision making organized, and how could it be improved?

References

References

A. Literature

Adriaanse, L. (1996); *Case-study (Wester) Schelde-estuarium op weg naar duurzaamheid. Economie, ecologie en sociologie*. Lezing Zomeruniversiteit Zeeland, Studiedag Verbreding Waterbeheer, 27 augustus 1996.

Alen, A., J. Billiet, D. Heremans, K. Matthijs, P. Peeters, J. Velaers (1990); *Rapport van de Club van Leuven, Vlaanderen op een kruispunt, Sociologische, economische en staatsrechtelijke perspectieven*. Leuven.

Anema, K. (1997); Geen kwestie van slechte wil, maar van diplomatie en geduld. *Platform*, Vol. 13, Nr. 7, 6-7.

Anonymous (1990); De rol van de Vlaamse polders en wateringen in het kwantiteits- en kwaliteitsbeheer van de onbevaarbare waterlopen. *Water*, Nr. 50, januari, februari 1990, 39-46.

Anonymous (1991); NVA/WEL-symposium over ecologisch herstel Schelde. Voor oplossing Scheldeprobleem zijn Nederlands-Belgisch Samenwerkingsverdrag en Integraal Schelde-Actieprogramma nodig. H_2O, Vol. 24, No. 23, 660-662.

Anonymous (1995a); Overleg met gemeenten over verdieping Westerschelde, Ook Reimerswaal mogelijk dwarsligger bij verstrekken aanlegvergunning. *Binnenlands Bestuur*, Vol. 16, No. 2, 5.

Anonymous (1995b); Het gaat iets beter met de Schelde. *Schelde nieuwsbrief*, Vol. 1, Nr. 2, 3.

Anonymous (1997a); Eerste kwaliteitsrapportage van de Internationale Commissie voor de Bescherming van de Schelde (ICBS), Schelde Actie Programma in aantocht. *Schelde nieuwsbrief*, Vol. 4, Nr. 13, p. 2.

Anonymous (1997b); Milieuaspectenstudie baggerstort gereed, Hoe de MER een MAS werd. *Schelde nieuwsbrief*, Vol. 4, Nr. 13, 7.

Argent, P. d' (1997); Les accords de Charleville-Mézières du 26 avril 1994, L'évolution historique du statut juridique de la Meuse et de l'Escaut. In: Proceedings of *"The Charleville-Mézières agreements on the Meuse and the Scheldt: New trends in the law of international rivers*. Tuesday 6 May 1997, Association belge pour le droit de l'Environnement, T.M.C. Asser Instituut. Palais des Académies, Brussels, Belgium.

Ast, J.A. van, L. Korver-Alzerda (1994); *SCHAR, De Schelde en handhaving regelgeving in Vlaanderen en Nederland*. Erasmus Studiecentrum voor Milieukunde. Rotterdam.

Avruch, K., P.W. Black (1993); Conflict resolution in intercultural settings: Problems and prospects. In: Sandole, D.J.D., H. van der Merwe (ed.) (1993); *Conflict resolution theory and practice: integration and application*. New York.

Axelrod, R. (1984); *The Evolution of Cooperation*. New York.

Axelrod, R. (1990); *De evolutie van samenwerking* (The evolution of Co-operation). Amsterdam.

Baakman, N.A.A., R. Maes, G. Bouckaert (1994); Besturen in Vlaanderen en Nederland. *Bestuurskunde*, Vol. 3, No. 6, 235-245.

Babbie, E. (1989); *The practice of social research*. Belmont.

Barraqué, B., J.-M. Berland, S. Cambon (1997); Frankreich. In: Correia, F.N., R.A. Kraemer (ed.); *Eurowater 1, Institutionen der Wasserwirtschaft in Europa, Länderberichte*. Berlin, 189-328.

Barraqué, B. (ed.) (1995); *Les politiques de l'eau en Europe*. Paris.

Bedet, M. (1993); Kan de Schelde net zo schoon worden als de Rijn? *ROM*, No. 1-2, Jan.-Febr. 1993, 26-29.

Belmans, H. (1995); De verdieping van de Westerschelde. *Water*, Vol. 14, No. 85, 259-264.

Benegora Leefmilieu, Redt de Voorkempen, Belt, Internationale Scheldewerkgroep (1985); *Smeerpijp is rampzalig voor kwaliteit Schelde*, Antwerpen.

Benson, J.K. (1975); The Interorganizational Network as a Political Economy. *Administrative Science Quarterly*, Vol. 20, No. 2, 229-249.

Berends, P., A. van Broekhoven, E. Jagtman, H. Peters, P. van Rooy, E. Turkstra, S. de Wit, K. Wulffraat (1995); *Ruimte voor water*, Visienotitie als aanzet voor discussie. Projectteam NW 4.

Bernauer, T., P. Moser (1996); *Reducing Pollution of the Rhine River: The Influence of International Cooperation*, Working Paper 96-7, IIASA. Laxenburg (Austria)

Bestuurlijk Overleg Westerschelde (1991a); *Beleidsplan Westerschelde*. Middelburg.

Bestuurlijk Overleg Westerschelde (1991b); *Western Scheldt Policy Plan*. Middelburg.

Bestuurlijk Overleg Westerschelde (1993); *Beleidsplan Westerschelde, Voortgangs- en evaluatierapportage periode 1991-1992*. Middelburg.

Bestuurlijk Overleg Westerschelde (1995); *Beleidsplan Westerschelde, Voortgangs- en evaluatierapportage, periode 1993-1994*. Middelburg.

Betlem, I. (1997); Gewässerbewirtschaftung auf der Grundlage von Fluszeinzugsgebieten. In: Correia, F.N., R.A. Kraemer (ed.); *Eurowater 2, Dimensions Europäischer Wasserpolitik, Themenberichte*. Berlin, 381-429.

Bijlsma, L. (1990); Schelde, wereldrivier, wereldvoorbeeld. In: Verslag *Symposium een Schelde zonder grenzen, Naar een duurzame ontwikkeling van een stroomgebied*, Neeltje Jans, 26 April 1990. Ministerie van Verkeer en Waterstaat, Rijkswaterstaat, Directie Zeeland, Middelburg, 8-13.

BIM (Brussels Instituut voor Milieubeheer) (1995); *Verslag over de staat van het leefmilieu in het Brussels Hoofdstedelijk Gewest*.

Bimberichten (1994); *Brussels Instituut voor Milieubeheer*, 1989-1994. BIM berichten, Nummer 11, Speciale uitgave "5 jaar BIM", December 1994.

Biswas, A.K. (ed.) (1994); *International waters of the middle east, From Euphrates-Tigris to Nile*. Oxford University Press, Bombay.

Bond Beter Leefmilieu, Stichting Reinwater (1989); *De Schelde, Vlaamse delta ecologisch rampgebied*. juli 1989.

Bouchez, L.J. (1978); The Netherlands and the Law of International Rivers. In: Panhuys, H.F. van, W.P. Heere, J.W.J. Jitta, K.S. Sik, A.M. Stuyt (eds.); *International Law in the Netherlands*. Volume I. Alphen aan den Rijn.

Bouman, N. (1996); A New Regime for the Meuse. *Review of European Community & International Environmental Law*, Vol. 5, No. 2, 161-168.

Braam, A. van (1989); *Filosofie van de bestuurswetenschappen*. Serie wetenschapsfilosofie. Leiden.

Brabander, K. De, K. De Greeve (1988); Waterkwaliteit van de Schelde. *Water*, Vol. 7, No. 43/2, 223-227.

Bradach, J.L., R.G. Eccles (1991); Price, authority and trust: from ideal types to plural forms. In: Thompson, G., J. Frances, R. Levacic, J. Mitchell (ed.); *Markets, Hierarchies and Networks, The Coordination of Social Life*. London.

Brassinne, J. (1994); *La Belgique fédérale*, Bruxelles.

Bremen, W.J. van den (1992); Seaport Development and State Boundaries: The Ems-Dollart Region and the Scheldt-Antwerp Region on Dutch Frontiers. *Ocean & Coastal Management*, 1992, Nr. 18, 197-213.

Bressers, J.Th.A. (1989); Beleidsevaluatie en beleidseffecten. In: Hoogerwerf, A. (red.); *Overheidsbeleid*. Alphen aan den Rijn.

Bressers, H., L.J. O' Toole, J. Richardson (1995); Networks as Models of Analysis: Water Policy in Comparative Perspective. In: Bressers, H., L.J. O' Toole, J. Richardson (red.); *Networks for Water Policy, A comparative perspective*. London.

Bruijn, J.A. de, E.F. ten Heuvelhof (1991); *Sturingsinstrumenten voor de overheid, over complexe netwerken en een tweede generatie sturingsinstrumenten*. Leiden.

Bruijn, J.A. de, W.J.M. Kickert, J.F.M. Koppenjan (1993); Inleiding: beleidsnetwerken en overheidssturing. In: Bruijn, J.A. de, W.J.M. Kickert, J.F.M. Koppenjan (red.) (1993); *Netwerkmanagement in het openbaar bestuur, over de mogelijkheden van overheidssturing in beleidsnetwerken*. Rotterdam/Delft.

Bruijn, J.A. de, A.B. Ringeling (1993); Normatieve kanttekeningen bij het denken over netwerken. In: Bruijn, J.A. de, W.J.M. Kickert, J.F.M. Koppenjan (red.) (1993); *Netwerkmanagement in het openbaar bestuur, over de mogelijkheden van overheidssturing in beleidsnetwerken.* Rotterdam/Delft.

Buuren, P.J.J. van (1997); Vergunningenwet Westerschelde niet voor herhaling vatbaar. *Milieu & recht,* 1997, Nr. 6, 122-124.

Caldwell, L.K. (1988); Beyond environmental diplomacy: the changing institutional structure of international cooperation. In: Carroll, J.E. (red.) (1988); *International environmental diplomacy, The management and resolution of transfrontier environmental problems.* Cambridge.

Caldwell, L.K. (1990); *International Environmental Policy, Emergence and dimensions.* Durham and London.

Camps, H. (1990); Jagen op cowboys. *Elsevier,* 17-2-1990.

Caponera, D.A. (1985); Patterns of cooperation in international water law: Principles and institutions. *Natural Resources Journal,* Vol. 25, July 1985, 563-587.

Caponera, D.A. (1992); *Principles of water law and administration, National and International.* Rotterdam.

Cappaert, I. (1988); Het waterzuiveringsprogramma voor het Scheldebekken. *Water,* Vol. 7, No. 43/2, 228-236.

Cappaert, I. (1990); Oppervlaktewaterkwaliteitsbeheer. *Water,* Vol. 9, Nr. 50, 65-68.

Carroll, J.E. (1986); Water Resources Management as an Issue in Environmental Diplomacy. *Natural Resources Journal,* Vol. 26, No. 1, 207-220.

CEDRE (1991); Centre d'étude du droit de l'environnement. *De toepassing door het Waalse gewest van de internationale verdragen inzake de bescherming van de oppervlaktewateren.* N. de Sadeleer, Faculté de Droit des Facultés universitaires Saint-Louis. Studie uitgevoerd in opdracht van Greenpeace Belgium vzw. Report number N/91/12.

Champion, D.J., S.B. Kurth, D.W. Hastings, D.K. Harris (1984); *Sociology.* New York.

Claessens, J., J. van Hoof, J.H.M. De Ruig (1991); Interactie morfologie en baggerwerken. *Water,* Vol. 10, No. 60, 182-189.

COIB (1994); *Bilaterale betrekkingen van V&W met België.*

Colijn, C.J. (1990); Water zonder paspoort. In: Ministerie van Verkeer en Waterstaat, Directie Zeeland, *Symposium een Schelde zonder grenzen, Naar een duurzame ontwikkeling van een stroomgebied.* Neeltje Jans, 26 April 1990, 28-33.

Collection environnement de l' Université de Montréal (1997); *Water Resources Outlook for the 21th Century: Conflicts and Opportunities,* Extended abstracts , No. 9, Volumes I and II.

Commissie Westerschelde (1997); *Advies Commissie Westerschelde over natuurcompensatie maatregelen in het kader van de verruiming van de vaarweg in de Westerschelde.* Uitgebracht aan de minister van Verkeer en Waterstaat, augustus 1997.

Couwenberg, S.W. (1997); De Vlaamse toekomst ligt in Nederland. *Intermediair,* Vol. 33, No. 8, 19-21.

Crozier, M., E. Friedberg (1980); *Actors & Systems, The Politics of Collective Action.* Chicago, London.

Dam, J.C. van (1992); Transboundary river basin management and sustainable development. In: Dam, J.C. van and J. Wessel (ed.); *Transboundary river basin management and sustainable development.* Proceedings Volume I Lustrum Symposium Delft University of Technology, Delft, The Netherlands, 18-22 May 1992. Technical Documents in Hydrology, International Hydrological Programme, UNESCO, Paris, 11-17.

Delmartino, F., J.M.L.M. Soeters (1994); Ambtelijke cultuur in Vlaanderen en Nederland, Een verkenning. *Bestuurskunde,* Vol. 3, No.6, 246-252.

Deschouwer, K., L. De Winter, D. Della Porta (1996); Partitocracies between crises and reforms: The cases of Italy and Belgium. *Res Publica*, No. 38, 215-235.

Dieperink, C. (1997); *Tussen zout en zalm, Lessen uit de ontwikkeling van het regime inzake de Rijnvervuiling*. Amsterdam.

Downs, P.W., K.J. Gregory (1991); How Integrated is River Basin Management. *Environmental Management*, Vol. 15, No. 3, 299-309.

Driessen, P. (1995); Activating a policy network, The case of the Mainport Schiphol. In: Glasbergen, P. (red.) (1995); *Managing Environmental Disputes, Network Management as an Alternative*. Dordrecht.

Driessen, P., W. Vermeulen (1995); Network management in perspective, Concluding remarks on network management as an innovative form of environmental management. In: Glasbergen, P. (red.) (1995); *Managing Environmental Disputes, Network Management as an Alternative*. Dordrecht.

Dupont , C. (1993); Switzerland, France, Germany, the Netherlands: The Rhine. In: Faure, G.O., J.Z. Rubin (ed.); *Culture and Negotiation*. Newbury Park.

Durth, R. (1996); *Grenzüberschreitende Umweltprobleme and regionale Integration, Zur Politischen Ökonomie von Oberlauf-Unterlauf-problemen an internationalen Flüssen*. Baden-Baden.

Eck, G.T.M. van, N. De Pauw, M. van den Langenbergh, G. Verreet (1991a); Emissies, gehalten, gedrag en effecten van (micro)verontreinigingen in het stroomgebied van de Schelde en Schelde-estuarium. *Water*, Vol. 10, No. 60, 164-181.

Eck, G.T.M. van (1991b); De ontwikkeling van een waterkwaliteitsmodel van het Schelde-estuarium. *Water*, Vol. 10, No. 61, 215-218.

Edelenbos, J., M. van Twist (1997); *Beeldbepalende bestuurskundigen, Een kennismaking met kernfiguren uit de bestuurskunde*. Alphen aan den Rijn.

Elek, A. (1996); *Institutions of the North Sea Co-operation and their Relation to the International Rhine Commission. RBA Series on River Basin Administration*, Working Paper Nr. 21, Delft University of Technology, Delft.

Ellemers, J.E. (1987); Veel kunnen verklaren of iets kunnen veranderen: krachtige versus manipuleerbare variabelen. In: Lehning, P.B., J.B.D. Simonis (red.) (1987); *Handboek beleidswetenschap*. Meppel.

Engel, H. (1990); Een deltaplan voor de Schelde. In: Ministerie van Verkeer en Waterstaat, Directie Zeeland, *Symposium een Schelde zonder grenzen, Naar een duurzame ontwikkeling van een stroomgebied*. Neeltje Jans, 26 April 1990, 34-40.

Faure, G.O., J.Z. Rubin (1993); Lessons for Theory and Research. In: Faure, G.O., J.Z. Rubin (ed.); *Culture and Negotiation*, Newbury Park.

Fisher, R., W. Ury, B. Patton (1993); Excellent onderhandelen, een praktische gids voor het best mogelijke resultaat in iedere onderhandeling. Amsterdam. Translation from 'Getting to Yes: Negotiating agreement without giving in' (1981). Boston.

Frances, J., R. Levacic, J. Mitchell, G. Thompson (1991); Introduction. In: Thompson, G., J. Frances, R. Levacic, J. Mitchell (ed.), *Markets, Hierarchies and Networks, The Coordination of Social Life*. London.

Franssen, G.W.T.M., A.J. Schuurman; *De ontwikkeling van de Scheldeloop, van de Romeinse tijd tot heden*. Vlissingen.

Frey, F.W. (1993); The Political Context of Conflict and Cooperatioon Over International River Basins. *Water International*, Vol. 18, No. 1, 54-68.

Glasbergen, P. (1990); Besturingsproblemen rond watersystemen, Ervaringen met de beleidsnetwerkbenadering. *Bestuurswetenschappen*, Nr. 1, 41-52.

Glasbergen, P. (1995); Environmental dispute resolution as a management issue. Towards new forms of decision making. In: Glasbergen, P. (red.); *Managing Environmental Disputes, Network Management as an Alternative.* Dordrecht.

Godfroij, A.J.A. (1981); *Netwerken van organisaties, strategieën, spelen, structuren.* 's-Gravenhage.

Godfroij, A.J.A. (1992); Dynamische netwerken. *M&O*, 1992/2, 365-375.

Gosseries, A. (1995); The Scheldt and Meuse rivers, The 1994 Agreements Concerning the Protection of the Scheldt and Meuse Rivers. *European Environmental Law Review*, January 1995, 9-14.

Gouvernement wallon (1995); *Plan d'Environnement pour le Développement durable en Région wallonne.*

Grijns, L.C., J. Wisserhof (1992); *Ontwikkelingen in integraal waterbeheer, Verkenning van beleid, beheer en onderzoek.* Delft Studies in Integrated Water Management. Delft.

Gustafsson, J.- E. (1990); The management of river basins in France. *Proceedings Nordic Hydrological Conference*, Kalmar, NHP-report, Nr. 26, 337-346.

Haas, E.B. (1980); Why collaborate? Issue-Linkage and International Regimes. *World Politics*, Vol. 32 (1980), 357-405.

Haas, P.M., R.O. Keohane, M.A. Levy (ed.) (1995); *Institutions for the Earth, Sources of Effective International Environmental Protection.* Cambridge.

Hanf, K., L.J. O 'Toole (1992); Revisiting old friends: networks, implementation structures and the management of inter-organizational relations. *European Journal of Political Research*, Vol. 21 (1992), 163-180.

Hardin, G. (1968); The Tragedy of the Commons. Science,. Vol. 162, 1243-1248.

Heady, F. (1991); *Public Administration, A comparative perspective.* New York.

Heip, C. (1995); Mens en maatschappij als ecologische factoren in de toekomst van de Schelde. *Referatenboek 3de Internationaal Schelde-symposium, Integraal waterbeheer Schelde-estuarium*, woensdag 6- donderdag 7 december 1995, Provinciehuis Antwerpen.

Hendriks, C.A.J., S.V. Meijerink (1996); Lobbyen: communicatie als pressiemiddel. In: Groen, M. (red.); *Handboek Milieucommunicatie.* Samsom. Alphen aan den Rijn.

Hendriksen, J. (1995); Schone Schelde, schone taak. *de Water, nieuwsbrief over integraal waterbeheer.* No. 27, July 1995, 3-4.

Herweijer, M. (1985); *Evaluaties van Beleidsevaluatie en de Arbeidsongeschiktheidsverzekering.* Deventer.

Heuvel, J.H.J. van den, D.M. Ligtermoet (1989); Rivierbeheer en buitenlandse politiek: de complexiteit van het Westerscheldebeleid. *Bestuurswetenschappen*, 1989, No. 4, 216-228.

Heylen, J. (1995); Een instrument voor een integraal waterbeheer in de Belgisch-Nederlandse grensstreek: de grensoverschrijdende stroomgebiedcomités. *Benelux Dossier Waterbeleid*, No. 1, 70-77.

Hofstede, G.H. (1980); *Culture's Consequences, International Differences in Work-Related Values.* Beverly Hills.

Hofstede, G.H. (1991); *Cultures and Organizations, Software of the mind.* London.

Hofstede, G. (1994); Images of Europe. *Netherlands journal of social sciences*, Vol. 30, 63-82.

Honigh, M. (1985); *Doeltreffend beleid.* Assen/Maastricht.

Hoogerwerf, A. (1989); Beleid, processen en effecten. In: Hoogerwerf, A. (red.); *Overheidsbeleid.* Alphen aan den Rijn.

Hoogland, J.R., M. Bruyneel (1995); De Internationale Commissie voor de Bescherming van de Schelde (ICBS): Schone Schelde in de schijnwerper. *Referatenboek 3de Internationale Schelde-symposium*, 6-7 december 1995, Antwerpen, 128-131.

Hoogweg, P.H.A., F. Colijn (1992); Management of Dutch estuaries, The Ems-Dollard and the Western Scheldt. *Water Science and Technology*, Vol. 26, No. 7-8. 1887-1896.

Hoogweg, P.H.A, C. J. Van Westen (1988); De ecologische waarde van de Westerschelde voor de Noordzee. *Water*, Vol. 7, No. 43/1, 205-210.

Hoppe, M.H. (1990); *A Comparative Study of Country Elites: International Differences in Work-related Values and Learning and their Implications for Management Training and Development*. Chapel Hill.

Horst, H. van der (1996); *The Low Sky, Understanding the Dutch*. Schiedam.

Huisman, P., J. De Jong (1995); Adapting capacity of institutions, a precondition for sustainable water management. In: *Netherlands experiences with Integrated Water Management*. Proceedings of a meeting in Delft, The Netherlands, 5 October 1995, 13-31.

Huisman, P. (1998); From degradation towards sustainable development in the Rhine basin. Unpublished paper. Delft University of Technology, the Netherlands.

ICBS/CIPE (Internationale Commissie voor de Bescherming van de Schelde, Commission Internationale pour la Protection de l' Escaut) (1997); *Rapport: la qualité de l' Escaut 1994, Rapport: de kwaliteit van de Schelde 1994*. Antwerpen.

Ingen, F. van, M. de Ruiter (1997); Oorlog is duurder dan vrede, Een interview met SER-voorzitter Klaas de Vries. *Interdisciplinair*, Vol. 8, December 1997, 4-7.

IPR (Interparlementaire Beneluxraad) (1991a); *De Schelde, Een evaluatie van het beleid, de functies en de waterkwaliteit*. Verslag in opdracht van de commissies Leefmilieu en Havenvraagstukken van de Raadgevende Interparlementaire Beneluxraad. Delta Instituut voor Hydrobiologisch Onderzoek, Vincent Klap en Carlo Heip, 22 oktober 1991.

IPR (Interparlementaire Beneluxraad) (1991b); *Stenografisch verslag van de Scheldeconferentie van 29 november 1991 te Antwerpen*. Raadgevende Interparlementaire Beneluxraad, 29 november 1991.

IPR (Interparlementaire Beneluxraad) (1992); *Report of the meetings of 13 and 14 March 1992. Evaluation of the Scheldt and Meuse Conferences*. N. 186-187.

Iribarne, P. d' (1989); *La logique de l'honneur, Gestion des entreprises et traditions nationales*. Paris.

ISG (International Scheldt Group) (1994a); *Water quality management in the Scheldt basin, interim progress report 1993*. Report AX/94/013.

ISG (International Scheldt Group) (1994b); *Water quality management in the Scheldt basin, appendices 1993*. Report AX/94/014.

ISG (International Scheldt Group) (1994c); *Waterkwaliteitsbeheer in het Scheldestroomgebied* (Also available in French).

Istendael, G. van (1989); *Het Belgisch labyrint of De schoonheid der wanstaltigheid*, Amsterdam.

Jansen, P.Ph. van, L. van den Bendegom, J. Berg, A. Zanen, M. De Vries (1979); *Principles of River Engineering, the nontidal alluvial river*. Pitman Publish. Limite.

Jenkins-Smith, H.C., P.A. Sabatier (1993a); The Study of Public Policy Processes. In: Sabatier, P.A., H.C. Jenkins-Smith (ed.); *Policy Change and Learning, An Advocacy Coalition Approach*. Boulder.

Jenkins-Smith, H.C., P.A. Sabatier (1993b); The Dynamics of Policy-Oriented Learning. In: Sabatier, P.A., H.C. Jenkins-Smith (ed.); *Policy Change and Learning, An Advocacy Coalition Approach*. Boulder.

Journet, J.M., D. Duhem (1996); Le SDAGE du bassin Artois-Picardie. In *Proceedings Hydrotop 96*, 16-18 Avril 1996, Marseille, France.

Kensen, S., T. Abma (1994); *Beleidsevaluatie...een uitnodiging tot een reflectie op de wereld(en) van onderzoek*. Rotterdam, Erasmus Universiteit. Instituut voor Beleid en Management in de Gezondheidszorg/Vakgroep Bestuurskunde.

Keohane, R.O. (1988); International Institutions: Two approaches. *International Studies Quarterly*, Vol. 32, December 1988, 379-396.

Keohane, R.O., E. Ostrom (ed.) (1995); *Local Commons and Global Interdependence, Heterogeneity and Cooperation in Two Domains*. London.

Kickert, W.J.M. (1991); *Complexiteit, zelfsturing en dynamiek, over management van complexe netwerken bij de overheid*. Alphen aan den Rijn.

Kickert, W.J.M., E.-H. Klijn and J.F.M. Koppenjan (1997); Introduction: A Management Perspective on Policy Networks. In: Kickert, J.M., E.-H. Klijn, and J.F.M. Koppenjan (ed.); *Managing Complex Networks, Strategies for the Public Sector*. London.

Kingdon, J.W. (1984); *Agendas, Alternatives and Public Policies*, Little, Brown, Boston.

Klijn, E.-H., J.F.M. Koppenjan (1997); Beleidsnetwerken als theoretische benadering: Een tussenbalans. *Beleidswetenschap*, Vol. 11, Nr. 2, 143-167.

Klijn, E.-H., J.F.M. Koppenjan (1997a); De netwerkbenadering van beleid: Een dupliek. *Beleidswetenschap*, Vol. 11, Nr. 2, 179-181.

Klijn, E-.H., G. Teisman (1992); Besluitvorming in beleidsnetwerken: Een theoretische beschouwing over het analyseren en verbeteren van beleidsprocessen in complexe beleidsstelsels. *Beleidswetenschap*, Vol. 6, Nr. 1, 32-51.

Klijn, E.-H., G.R. Teisman (1997); Strategies and Games in Networks. In: Kickert, J.M., E.-H. Klijn, and J.F.M. Koppenjan (ed.); *Managing Complex Networks, Strategies for the Public Sector*. London.

Knoppers, R., W. van Hulst (1995); *De keerzijde van de dam*. Amsterdam.

Kuijcken, E., P. Meire (1988); Het land van Saeftinghe, slikken en schorren: Ecologische betekenis van getijdengebieden langs de Schelde. *Water*, Vol. 7, No. 43/2, 214-222.

Kuijs, J. (1990); Duizenden actievoerders bij burgemeesters op de stoep, Schone Schelde moet politiek issue zijn. *NG*, 1990, No. 19, 6-7.

Kwarten, L. (1997); Cultuurkloven houden Europa blijvend verdeeld. *Intermediair*, Vol. 33, No. 23, 23-24.

Lammers, J.G. (1984); *Pollution of international watercourses*. The Hague/Dordrecht.

Lefébure, A. (1997); Ook Frankrijk ratificeert Verdrag. *Schelde nieuwsbrief*, Vol. 4, Nr. 13, p. 8.

LeMarquand, D.G. (1977); *International rivers, the politics of cooperation*. Westwater Research Centre, University of British Columbia.

Lemstra, W. (1990); Rotterdam en Antwerpen: één Euregio? Samenwerken van complementaire havens ligt voor de hand. *ROM*, Vol. 8, No. 10, 4-8.

Levacic, R. (1991); Markets and government: an overview. In: Thompson, G., J. Frances, R. Levacic, J. Mitchell (ed.); *Markets, Hierarchies and Networks, The Coordination of Social Life*. London.

Linnerooth, J. (1990); The Danube River Basin: negotiation Settlements to Transboundary Environmental issues. *Natural Resources Journal*, Vol. 30, 629-660.

Lutgen, G. (1996); *Annuaire de l'Environnement en Region wallonne*. Bruxelles.

Maes, F. (1990); *Bilateral Consultation between The Netherlands and Belgium regarding the Scheldt River and the Meuse River*. Case Study Course on International Water Law and Management, IHE Delft, 23 November 1990.

Maes, F. (1996); De verdragen ter bescherming van de Maas en de Schelde in een diplomatieke en internationaal milieurechtelijke context. *Tijdschrift voor Milieurecht*, 1996, 329-344.

Martin, L.L. (1995); Heterogeneity, Linkage and Commons problems. In: Keohane, R.O., E. Ostrom (Ed.); *Local Commons and Global Interdependence, Heterogeneity and Cooperation in Two Domains*. London.

Marty, F. (1997); *International River Management, The political determinants of Success and Failure, Introduction and Research Design*. Institut für Politikwissenschaft, Zürich.

Meijerink, S.V., C.A.J. Hendriks (1996); Grensoverschrijdende milieuproblemen en interculturele communicatie. In: Groen, M. (red.); *Handboek Milieucommunicatie*. Samsom. Alphen aan den Rijn.

Meire, P., M. Hoffman, T. Ysebaert (red.) (1995); *De Schelde: een stroom natuurtalent*. Instituut voor Natuurbehoud, Hasselt. Rapport 95.10.

Meire, P. (1996); Integraal waterbeheer in het Schelde-estuarium. Een uitdaging. *Nieuwsbrief Buren*, Vol. 17, No. 1, 13-16.

Merckx, G. (1995); De verdeling van het Maasdebiet, van struikelblok naar pragmatisme. *Water*, Vol. 14, No. 85, 265-268.

Mijs, A.A. (1992); Genesis and viability of inter-organizations: an institutional approach. *Netherlands Journal of Social Sciences*. Vol. 28, No. 2, 155-169.

Mingst, K.A. (1981); The functionalist and Regime Perspectives: The case of Rhine River Cooperation. *Journal of Common Market Studies*. Vol. XX, 161-173.

Ministère de l' Environnement; *Water, a common heritage, integrated development and management of river basins, the French approach*.

Ministerie van de Vlaamse Gemeenschap, Departement Leefmilieu en Infrastructuur (1991); *De baggerwerken in de Schelde en de kwaliteit van water en bodem, Stand van zaken*. Februari 1991.

Ministerie van de Vlaamse Gemeenschap, Departement Leefmilieu en Infrastructuur (1995); *Beleidsplan sanering waterbodem Beneden-Zeeschelde*.

Ministerie van Verkeer en Waterstaat (1994); *Besluitvorming over grote infrastructuurprojecten in een aantal Europese landen*, Project Integrale Prioriteitstelling infrastructuurprojecten in een aantal Europese landen. Rotterdam.

Ministerie van Verkeer en Waterstaat, Directoraat-Generaal Rijkswaterstaat, Directie Zeeland, Bestuurlijk Overleg Westerschelde, Heidemij Advies, Rijksinstituut voor Kust en Zee/RIKZ, Resource Analysis (1996a); *Herstel Natuur Westerschelde, Alternatieven*.

Ministerie van Verkeer en Waterstaat, Directoraat-Generaal Rijkswaterstaat, Directie Zeeland, Bestuurlijk Overleg Westerschelde, Heidemij Advies, Rijksinstituut voor Kust en Zee/RIKZ, Resource Analysis (1996b); *Herstel Natuur Westerschelde, Projectenbundel*.

Ministerie van Verkeer en Waterstaat, Directoraat-Generaal Rijkswaterstaat (1995); *Deltaplan Grote Rivieren*.

Ministerie van Verkeer en Waterstaat, Directoraat-Generaal Rijkswaterstaat, RIKZ (1996); *Ontwikkelingen in de Westerschelde, prognose voor de komende 25 jaar*.

Ministerie van Verkeer en Waterstaat, Directoraat-Generaal Rijkswaterstaat, RIKZ (1997); *Westerschelde stram of struis? Eindrapport van het Project OostWest, een studie naar de beïnvloeding van fysische en verwante biologische patronen in een estuarium*, Rapport RIKZ-97.023.

Ministerie van Verkeer en Waterstaat, Directoraat-Generaal Rijkswaterstaat, Directie Zeeland (1997); *De toekomst van de Westerschelde, Beschouwingen vanaf de dijk*.

Ministerie van Verkeer en Waterstaat, Directoraat-Generaal van de Rijkswaterstaat (1997); *Ontwerp Beheersplan voor de Rijkswateren, programma voor het beheer in de periode 1997 t/m 2000*.

Ministry of the Environment and Infrastructure of Flanders (1990); *Flemish Environmental Policy and Nature Development Plan*, Propositions for 1990-1995, Brussels (in Dutch).

Mintzberg, H. (1987); Crafting strategy. *Harvard Business Review*, Vol. 65 (1987), 66-75.

Mitchell, R.B. (1995); Heterogeneities at Two Levels: States, Non-state Actors and Intentional Oil Pollution. In: Keohane, R.O., E. Ostrom (Ed.); *Local Commons and Global Interdependence, Heterogeneity and Cooperation in Two Domains*. London.

Newson, M. (1992); *Land, Water and Development, River basin systems and their sustainable management*. London.

Nicolazo, J.L.; *Les Agences de l'eau*. Paris.

Niesing, H., L.L.P.A. Santbergen, S.A. de Jong (1996); *Marsh Amelioration along the River Schelde (MARS). Dutch projectplan, Phase I (1995&1996), The restoration of estuarine habitats in the Westerschelde*. Ministerie van de Vlaamse gemeenschap, Departement Leefmilieu en Infrastructuur, AMINAL, Afdeling Natuur, Ministerie van Verkeer en Waterstaat, Directie Zeeland, Report AX-95.072.

Nollkaemper, A. (1993a); Progressie en stagnatie in het internationale regime voor grensoverschrijdende waterverontreiniging. *Milieu en recht*, January 1993/1, 11-23.

Nollkaemper, P.A. (1993b); *The legal regime for transboundary water pollution: Between discretion and constraint*. Dordrecht.

Olson, M. (1965); *The logic of Collective Action, Public Goods and the Theory of Groups*. Harvard Economic Studies. Cambridge.

Olsthoorn, H., H. Hegeman (1991); Prosa, Voorkomen is beter dan zuiveren, maar er wordt al gezuiverd. *ROM*, No. 10, 17-20.

O'Toole, L. (1988); Strategies for intergovernmental management: implementing programs in interorganizational networks. *International Journal of Public Administration*, Vol. 4, 417-441.

Ovaa, B.P.S.A. (1991a); *Scheldestroomgebied, Naar een samenhangend beheer van het riviersysteem van de Schelde in het perspectief van duurzame onwikkeling*. Wageningen.

Ovaa, B.P.S.A. (1991b); *Scheldestroomgebied, Naar een samenhangend beheer van het riviersysteem van de Schelde in het perspectief van duurzame onwikkeling, Bijlagen*. Wageningen.

O-RSD (Onderzoekers Rijn-Schelde Delta) (1994); *Rijn-Schelde Delta, een ruimtelijk economische verkenning*.

Paepe, R. (1991); Implementatie in het beleid: Vlaamse visie. *Water*, Vol. 10, No. 60, 210- 212.

Pallemaerts, M. (1991); De toepassing van het Europees milieurecht in België: praktische uitvoering versus formele omzetting. In: Bronders, B., E. Goethals, L. Lavrysen (red.); *Rechtspraktijk en milieubescherming, uitgave ter ere van het eerste Antwerpse juristen congres*. Antwerpen.

Perdok, P.J. (1995); *Institutional framework for water management in the Netherlands*, Eurowater vertical report, RBA Series on River Basin Administration, Delft University of Technology, Delft.

Perdok, P.J. (1997); Niederlande. In: Correia, F.N., R.A. Kraemer (ed.); *Eurowater 1, Institutionen der Wasserwirtschaft in Europa, Länderberichte*. Berlin, 329-477.

Pieters, T., C. Storm, T. Walhout, T. Ysebaert (red.) (1991); *Het Schelde estuarium, Méér dan een vaarweg*. Ministerie van Verkeer en Waterstaat, Directoraat-Generaal Rijkswaterstaat, Dienst Getijdewateren, Directie Zeeland. Middelburg. Report GWWS-91.081.

Planchar, R.-A. (1993); *L'Escaut et la Meuse, Fleuves Hollandais? Fleuves Siamois?*

Plugge, M. (1993); *Taakgroep Westerschelde*. Description of activities of 12-10-1993.

Praag, C.S. van (1991); Het Bijzondere van Nederland: de Landenvergelijking in het Sociaal en Cultureel Rapport, *Mens en Maatschappij*, Vol. 66, 343-363.

Prescott, J.R.V. (1987); *Political Frontiers and Boundaries*. London.

Pröpper, I.M.A.M. (1996); Succes en falen van sturing in beleidsnetwerken: Enkele lessen ten behoeve van een theoretisch model. *Beleidswetenschap*, 1996, Nr. 4, 345-365.

Pröpper, I.M.A.M. (1997); Succes- en faalfactoren voor repliek: De discussie over beleidsnetwerken. *Beleidswetenschap*, Vol. 11, Nr. 2, 174-178.

Provincie Zeeland (1993); *De zin van de Westerschelde*.

Provincie Noord-Brabant (1993); Verslag "Water kent geen grenzen" Naar een pan-Europees waterbeleid, Samenwerkingsverband van de provincies Noord-Brabant, Limburg en Zeeland, 's-Hertogenbosch, maart 1993.

Reinwater (1992a); *Afvalwater en beleid België, Onderzoek naar de haalbaarheid van de internationale afspraken rond de Noordzee*.

Reinwater (1992b); *Afvalwater en beleid in Frankrijk, Onderzoek naar de haalbaarheid van de internationale afspraken rond de Noordzee.*

Reinwater (1992c); *Reinwater in 1992.* Jaarverslag.

Reitsma, D. (1995); Major public works: cultural differences and decision-making procedures. *Tijdschrift voor Economische en Sociale Geografie*, Vol. 86, No. 2, 186-190.

Retkowsky, Y. (1995); *La planification des ressources en eau: les apports de la Loi sur l'eau.* Ministère de l' Environnement.

Ringeling, A.B. (1997); *Stromingen en consensus, Een bestuurskundige analyse van de besluitvorming over de compensatie van natuurwaarden in de Westerschelde*, Advies voor de commissie Westerschelde.

Roovers, P. (1988); Verdiepingsprogramma van de maritieme toegang tot de Haven van Antwerpen. *Water*, Vol. 7, No. 43/1, 16-183.

Rooy, M. de *et al.* (1993); *De Schelde, ecologie, watervervuiling en waterkwaliteitsbeleid in het Scheldestroomgebied*, Grenzeloze Schelde, Escaut sans Frontières.

Rooy, M. de, B. Wijffels (1993); Grenzeloze Schelde. *Reinwater*, 1993, 14-18.

Rosillon, F., P. Van der Borght, C. Dasnoy, B. Tricot (1996); Les contrats de rivière en région Wallonne. *Hydrotop 1996, Colloque scientifique et technique, Textes des conférences*, Marseille, France, 16-18 Avril 1996.

Ruben, B.D. (1977); Guidelines for cross-cultural communication effectiveness. *Group and Organization Studies*, No. 2, 470-479.

RWS-DGW, SLD (Rijkswaterstaat, Dienst Getijdewateren, Directie Zeeland) (1993); *Projectplan OostWest, fase 1993-1994.* Werkdocument GWWS-93.813X. Middelburg.

RWS-Zld (Rijkswaterstaat, Directie Zeeland) (1994); *Een internationale blik op het Scheldeestuarium, Verslag van de studie van een reviewteam, bestaande uit 7 deskundigen uit België en Nederland, betreffende studies van Rijkswaterstaat naar nieuwe beheerswijzen voor het Schelde-estuarium.*

Sabatier (1993); Policy Change over a Decade or More. In: Sabatier, P.A., H.C. Jenkins-Smith (ed.); *Policy Change and Learning, An Advocacy Coalition Approach.* Boulder.

Saeijs, H.L.F., E. Turkstra (1991); Naar een Europese watersysteem benadering. Naar een duurzame ontwikkeling van het Scheldestroomgebied vanuit een Nederlandse invalshoek. *Water*, Vol. 10, No. 60, 204-209.

Saeijs, H.L.F., E. Turkstra (1994); Towards a Pan-European integrated river basin approach: plea for sustainable development of European river basins. *European Water Pollution Control*, Vol. 4, No. 3, 16-28.

Saeijs, H.L.F., L.L.P.A. Santbergen, S.A. de Jong, W. van der Hoofd (1995); *Natuurtalent in ontwikkeling! Op weg naar een herstelplan voor de Schelde.* Lecture at the occasion of the presentation of the report "De Schelde, een stroom natuurtalent".

Saeijs, H.L.F., M.J. van Berkel (1997); The Global Water Crisis: The Major Issue of the Twenty-first Century, a Growing and Explosive Problem. In: Brans, E.H.P., J. de Haan, A. Nollkaemper, J. Rinzema (red.) (1997); *The Scarcity of Water, Emerging Legal and Policy Responses.* London.

Salet, W.G.M. (1994); *Gegrond Bestuur, Een internationale ijking van bestuurlijke betrekkingen*, Intreerede. Delft.

Scharpf, F.W. (1978); Interorganizational policy studies: issues, concepts and perspectives. In: Hanf, K., F.W. Scharpf (1978); *Interorganizational Policy making; limits to coordination and central control.*

Scharpf, F.W., B. Reissert, F. Schnabel (1978); Policy effectiveness and conflict avoidance in intergovernmental policy formulation. In: Hanf, K., F.W. Scharpf (1978); *Interorganizational Policy making; limits to coordination and central control.*

Scheele, R.J., C.J. Colijn, C.J. van Westen (1987); Beleid en onderzoek ten behoeve van het watersysteem van de Westerschelde. *Milieu*, 1987, No. 3, 101-106.

Sebenius, J.K. (1992a); Negotiation analysis: A characterization and review. *Management science*, Vol. 38, No. 1, 18-38.

Sebenius, J.K. (1992b); Challenging conventional explanations of international cooperation: negotiation analysis and the case of epestemic communities. *International Organization*, Vol. 46, No. 1, 323-365.

Sironneau, J. (1990); *Administration et droit des eaux aux Pays-Bas*. Neuilly.

Sluijs, D., E. Vermeerbergen, J. Baccaert (1992); Prosa: Preventieonderzoek in het Scheldebekken gericht op afval- en emissiereductie, Exploratief onderzoek naar de mogelijkheden tot het introduceren van een afval- en emissiepreventiebeleid bij bedrijven uit de kanaalzone Gent-Terneuzen. *Water*, Vol. 11, No. 62, 3-6.

Smit, C. (1966); *De Scheldekwestie*. Uitgave der stichting Nederlands vervoerswetenschappelijk instituut.

Smit, H., R. Zijlmans, E. Turkstra (1995); Verdrag inzake de bescherming van de Schelde. *Zoutkrant*, Vol. 9, Nr. 3, 1-2.

Smitz, J.S. (1995); De instelling en de werkwijze van de internationale commissie voor de bescherming van de Maas. *Water*, Vol. 14, No. 85, 257-258.

Société belge de droit international, Association belge pour le droit de l'Environnement, T.M.C. Asser Instituut (1997); *The Charleville-Mézières agreements on the Meuse and the Scheldt: New trends in the law of international rivers*. Brussels.

Soeters, J.L. (1993); Managing Euregional Networks. *Organization Studies*, Vol. 14, No. 5, 639-656.

Soeters, J. (1991); Management van Eurogionale netwerken. *Openbaar bestuur*, No. 10, 21-27.

Soeters, J., G. Hofstede, M. van Twuyver (1995); Culture's consequences and the police: cross-border cooperation between police forces in Germany, Belgium, and the Netherlands. *Policing and Society*, Vol. 5, 1-14.

Soeters, J.L. (1995); Cultures gouvernementales et administratives en Belgique et aux Pays-Bas: de la divergence à la convergence? *RISA*, No. 2, 299-314.

Somers, E. (1992); *Internationaal publiekrechtelijke aspecten verbonden aan de maritieme samenwerking in de euregio Scheldemond*. Onderzoeksproject Euregio Scheldemond.

Stroink, F.A.M. (1998); Complexe besluitvorming: een opgave voor burger, bestuur en rechter. In: Heide, H.P., F.A.M. Stroink, P.C.E. van Wijmen (1998); *Complexe besluitvorming*. VAR-reeks 120. Alphen aan den Rijn.

Strubbe, J. (1988a); De Waterverdragen. *Water*, Vol. 7, No. 42, 148-153.

Strubbe, J. (1988b); De politieke geschiedenis en het internationale statuut van de Westerschelde. *Water*, Vol. 7, No. 43/2, 237-241.

Strubbe, J. (1995); Het verdrag inzake de verruiming van de Westerschelde in historisch perspectief. *Water*, Vol. 14, No. 85, 233-236.

Suykens, F. (1995); De historiek van het totstandkomen van de Vlaams-Nederlandse waterverdragen. *Water*, Vol. 14, No. 85, 227-232.

Swanborn, P.,G. (1987); *Methoden van sociaal-wetenschappelijk onderzoek*. Amsterdam.

Tatenhove, J., P. Leroy (1995); Beleidsnetwerken: Een kritische analyse. *Beleidswetenschap*, Vol. 9, No. 2, 128-145.

Teisman, G.R. (1992); *Complexe besluitvorming, Een pluricentrisch perspectief op besluitvorming over ruimtelijke investeringen*. 's-Gravenhage.

Termeer, C.J.A.M. (1993a); Een methode voor het managen van veranderingsprocessen in netwerken. In: Bruijn, J.A. de, W.J.M. Kickert, J.F.M. Koppenjan (red.); *Netwerkmanagement in het openbaar bestuur, over de mogelijkheden van overheidssturing in beleidsnetwerken*. Rotterdam, Delft.

Termeer, C.J.A.M. (1993b); *Dynamiek en inertie rondom mestbeleid: een studie naar veranderingsprocessen in het varkenshouderijnetwerk*. 's-Gravenhage.

Tombeur, H. (1995); De volkenrechtelijke aspecten van de exclusieve verdragen inzake de bescherming van Maas en Schelde. *Water*, Vol. 14, No. 85, 237-241.

Toonen, T.A.J. (1990); *Internationalisering en het openbaar bestuur als institutioneel ensemble, Naar een zelfbestuurskunde*, Intreerede. Leiden.

Tromp, D., J.P. Swart (1988); Consequenties van het op diepte houden van de vaargeul in de Westerschelde. *Water*, Vol. 7, No. 43/1, 195-199.

TSC (Technische Scheldecommissie) (1984a); *Nota verdieping Westerschelde, programma 48'/43'*. Middelburg, 15 Juni 1984.

TSC (Technische Scheldecommissie) (1984b); *Verdieping Westerschelde, studierapport, programma 48'/43'*. Middelburg-Antwerpen, Juni 1984.

Veen, R. van der (1989); Het riool van België. *Intermediair*, Vol. 25, No. 48, 1 december 1989, 54-55.

Veld, R.J. in 't (1995); *Spelen met vuur, over hybride organisaties*, Intreerede. 's-Gravenhage.

Veldkamp, C. (1994); Schuivende bestuurlijke verhoudingen in België en de relatie met Nederland. *Openbaar bestuur*, No. 3, 2-9.

Verheyen, R., P. Meire, J.A.W. De Wit, A. Schneiders, C. Wils, T. Ysebaert (1991); Naar een ecologisch herstelplan voor de Schelde. *Water*, Vol. 10, No. 60, 195-203.

Verlaan, P.A.J., S.V. Meijerink, V.J. Maartense, M. Donze (1997); Slibtransport in de Schelde over de Belgisch-Nederlandse grens. *H₂O*, Vol. 30, Nr. 8, 255-260.

Villeneuve, C.H.V. de (1996a); Western Europe's Artery: The Rhine. Natural resources journal, Vol. 36, Summer 1996, 475-488.

Villeneuve, C.H.V. de (1996b); Consistentie, transparantie en subsidiariteit - naar samenhang in het internationaal waterbeheer? Milieu & Recht, Nr. 12, 247-251.

VNR (Vereniging Nederlandse Riviergemeenten) (1990); *Beleidsplan 1991-1995*. Ingenieurs- en Adviesbureau Kobessen, Arnhem.

V&W (Ministerie van Verkeer en Waterstaat) (1985); *Omgaan met Water: Naar een integraal waterbeleid*.

V&W (Ministerie van Verkeer en Waterstaat) (1997); *Vierde Nota waterhuishouding, Regeringsvoornemen*. Den Haag.

V&W (Ministerie van Verkeer en Waterstaat) (1989); *Derde Nota Waterhuishouding*. Tweede Kamer, 21 250, No. 1-2. [summarized in English as : Ministry of Transport and Public Works, *Water in The Netherlands: A time for Action*.].

V&W (Ministerie van Verkeer en Waterstaat) (1994); Directoraat-Generaal Rijkswaterstaat, Directie Zeeland, *Grensoverschrijdende samenwerking Vlaanderen-Nederland op het gebied van de waterhuishouding*. Geïnventariseerd per januari 1994, AXW 94.012.

V&W, RWS-Zld (Ministerie van Verkeer en Waterstaat, Rijkswaterstaat, Directie Zeeland (1994a); *Beheer en onderzoek van Rijkswaterstaat in de Westerschelde, Aanbevelingen van een reviewteam en conclusies van Rijkswaterstaat*.

V&W, RWS-Zld (Ministerie van Verkeer en Waterstaat, Rijkswaterstaat, Directie Zeeland (1994b); *Het Schelde Estuarium voor verbetering vatbaar, Bijstelling van het beheer van de Westerschelde, Conclusie van Rijkswaterstaat naar aanleiding van een review*.

Vugt, G.W.M. van (1994); Bestuursculturen van België en Duitsland. *Openbaar bestuur*, Vol. 4, No. 12, 10-15.

Wallensteen, P., A. Swain (1997); *Comprehensive assessment of the freshwater resources of the world, International fresh water resources: conflict or cooperation?* Stockholm.

Waterloopkundig laboratorium (1996); Van zorgen om de Maas naar zorgen voor de Maas, Nota over de afvoerproblematiek van de Maas. Opdrachtgever: Raadgevende Interparlementaire Beneluxraad. Delft.

Weiss, C. (1977); Research for Policy's Sake: The enlightenment Function of Social Research. *Policy Analysis.* Nr. 3, 531-545.
Wel, H. De (1994); *Organisatie van de rivierbekkenwerking in Vlaanderen.* AMINAL, Bestuur Algemeen Milieubeleid.
Wessel, J. (1988); Makelaars en middelaars in het waterbeheer. *De Ingenieur*, No. 12, 36-37.
Wessel, J. (1989); Development in and tasks for decision making in river basin management. *Aqua Fennica*, Vol. 19, No. 2, 163-168.
Wessel, J. (1991); The feasability of a river basin management approach to the sustainable development of watercourses. In: *Proceedings of the International symposium on "Impacts of watercourse Improvements: Assessment, Methodology, Management Assistance"*, 10-12 September 1991. Wépion, Belgium, 9-19.
Wessel, J. (1992a); Institutional and administrative aspects of international river basins. In: Dam, J.C. van and J. Wessel (ed.); *Transboundary river basin management and sustainable development.* Proceedings Volume I Lustrum Symposium Delft University of Technology, Delft, The Netherlands, 18-22 May 1992. Technical Documents in Hydrology, International Hydrological Programme, UNESCO, Paris, 23-28.
Wessel, J. (1992b); De Franse waterbeheerwetgeving. *De Ingenieur*, No. 4, 34-38.
Wessel, J. (1994a); Bestuurlijk-juridische verantwoordelijkheden bij rivierbeheer. Maas 1993, Wat ging er mis? *De Ingenieur*, No. 4, 32-35.
Wessel, J. (1994b); Hoe hanteren we de Maas na Kerstmis 1993? *De Ingenieur*, No. 19, 30-31.
Wessel, J. (1996); *Dimensies van waterbeheer*, afscheidsrede. TU Delft, Delft.
Wisserhof, J. (1994); Matching research and policy in integrated water management. Delft Studies in Integrated Water Management, Delft University of Techology, Delft.
Wit, J.A.W. De, W. Admiraal, P. Meire (1991); Ecologisch herstel van stroomgebieden: voorbeelden en aanbevelingen. *Water*, Vol. 10, No. 60, 190-194.
Wolf, A.T. (1997); International Water Conflict Resolution: Lessons from Comparative Analysis. *Water Resources Development*, Vol. 13, No. 3, 333-365.
World Bank (1993); *Water Resources Management, A world bank policy paper.* Washington D.C.
World Commission on Environment and Development (WCED) (1987); *Our common future (Brundtland-report).*

Yin, R.K. (1994); *Case study research, Design and Methods.* Aplied Social Research Methods Series, Volume 5. Thousand Oaks.
Young, O.R. (1984); Regime dynamics: The rise and fall of international regimes. In: Krasner, S.D. (ed.); *International Regimes.* Cornell University Press.

Zahn, E. (1989); *Regenten, rebellen en reformatoren, Een visie op Nederland en de Nederlanders.* Amsterdam.
Zijlstra, K., J.J. Lilipaly (1995); Conferenties inzake de Maas en de Schelde. *Dossier Waterbeleid*, Benelux, 1995, Nr. 1, 26-32.

B. Official documents

Advice of the Dutch advisory body, the 'Raad van de Waterstaat', concerning the implementation of the 48'/43'/38' deepening program, 30-10-1985.
Bestuurlijk Overleg Westerschelde: Covenant, Middelburg, 22-10-1992.
Commission proposal for a Council directive establishing a framework for community action in the field of water policy, 26.02.1997.
Convention on the Protection and Use of Transboundary Watercourses and International Lakes, United Nations Economic Commission for Europe, Geneva, E/ECE/1267, 1992.

Covenant between the Flemish Administration Water Infrastructure and Marine Affairs and Flemish environmental NGOs concerning the implementation of the deepening program and works for nature restoration, 17-7-1996.

Covenant signed by the chairman of the BOW, and the Directors-General of the Flemish Administration Water Infrastructure and Marine Affairs, and the Administration Environment, Nature- Land- and Water management, Temse, 15-5-1997.

Declaration of Arles by the Ministers for the Environment of France, Germany, Belgium, Luxembourg and the Netherlands on tackling the problems caused by the high water level of the Rhine and the Meuse, Arles, 4-2-1995.

Draft Articles on the Law of the Non-Navigational Uses of International Watercourses, Draft Report of the International Law Commission, U.N., GAOR, 43rd Sess., at 1, U.N. Doc. A/CN/.4/L.463/Add.4 (1991).

EK 1995-1996, Report of the 36th meeting, Tuesday, 25-6-1996.

European Parliament, 1992-1993, Resolutie over de waterkwaliteit van de Maas (Resolution concerning the water quality of the Meuse), 9-4-1992.

IPR recommendation concerning the Meuse and the Scheldt, 12-6-1993.

List of joint decisions approved by representatives of the Main Directorate of the Ministry of Transport, Public Works, and Water Management, the regional Directorate Zeeland of the same Ministry, the ZMF, the 'Stichting het Zeeuwse Landschap', and the 'Vereniging Natuurmonumenten', Goes, 15-11-1995.

Ministerie van Verkeer en Waterstaat, Hoofddirectie van de Waterstaat, Hoofdafdeling Bestuurlijke en Juridische Zaken (1988); WVO-vergunning voor België, Nr. RJW 51525.

Ministerie van Verkeer en Waterstaat, Hoofddirectie van de Waterstaat, Hoofdafdeling Bestuurlijke en Juridische Zaken (1988); WVO-vergunning voor België, Nr. RJW 51526.

Ministerie van Verkeer en Waterstaat (1991); WVO-vergunning baggerspecie voor België, 28-11-1991.

Resolution of the Interparliamentary Meuse conferences, Approved by the Interparliamentary Council of the Benelux on December 1990.

TK (Reports of discussions in the Second Chamber of the Parliament of the Netherlands) 1977-1978, Aanhangsel van de Handelingen, 854.

TK 1985-1986, 19 200 No. 12.

TK 1985-1986, Aanhangsel van de Handelingen, 122.

TK 1985-1986, Aanhangsel van de Handelingen, 909.

TK 1986-1987, Aanhangsel van de Handelingen, 597.

TK 1988-1989, Aanhangsel van de Handelingen, 68.

TK 1990-1991, Aanhangsel van de Handelingen, 356.

TK 1990-1991, Aanhangsel van de Handelingen, 679

TK 1990-1991, 21 800 XII No. 61.

TK 1991-1992, Aanhangsel van de Handelingen, 686.

TK 1991-1992, Aanhangsel van de Handelingen, 853.

TK 1992-1993, 23 075 No. 1.

TK 1992-1993, 23 075 No. 1, Annex.

TK 1992-1993, 23 075 No. 2.

TK 1992-1993, Aanhangsel van de Handelingen, 621.

TK 1993-1994, 23 075 No. 3.

TK 1993-1994, 23 075 No. 4.

TK 1993-1994, 23 530 No. 6.

TK 1993-1994, 23 536, No. 1: *De nieuwe bestuurlijke verhoudingen in België en de betrekkingen met Nederland.*

TK 1994-1995, 22 026.

TK 1994-1995, 23 075 No. 5.

TK 1994-1995, 23 075 No. 10.

TK 1994-1995, 24 041 No. 3.

TK 1994-1995, 24 041 No. 5.

TK 1994-1995, Report of the 97th meeting, 31-8-1995, pp. 5919-5939.

TK 1995-1996, 24 041 No. 33a.

TK 1995-1996, Report of the 56th meeting, 15-2-1996, pp. 4074-4114.

Verdrag inzake de verbetering van het kanaal van Gent naar Terneuzen (Treaty concerning the improvement of the Canal from Ghent to Terneuzen), Trb. 1960, No. 105.

Verdrag inzake de verbinding tussen de Schelde en de Rijn (Treaty concerning the connection between the Scheldt and the Rhine), Trb. (Treaty Series of the Netherlands) 1963 No. 78.

Verdrag inzake de bescherming van de Maas tegen verontreiniging (Convention on the protection of the Meuse against pollution), Trb. 1994 No. 149.

Verdrag inzake de bescherming van de Schelde tegen verontreiniging (Convention on the protection of the Scheldt against pollution), Trb. 1994 No. 150.

Verdrag tot herziening van het Reglement ter uitvoering van artikel IX van het Tractaat van 19 april 1839 en van het hoofdstuk II, afdelingen 1 en 2 van het Tractaat van 5 november 1842, zoals gewijzigd, voor wat betreft het loodswezen en het gemeenschappelijk toezicht daarop (Scheldereglement) (Scheldt regulations) , Trb. 1995 No. 48.

Verdrag inzake de afvoer van het water van de Maas (Convention on the flow of the river Meuse), Trb. 1995 No. 50.

Verdrag inzake de verruiming van de vaarweg in de Westerschelde (Convention on the deepening of the navigation channel in the Western Scheldt), Trb. 1995 No. 51.

C. Files

Dutch draft convention concerning the cooperation on the management of the rivers Scheldt and Meuse, 12-4-1991, including explanatory report from the Dutch negotiation delegation for the Dutch Council of Ministers (Files Ministry of Transport, Public Works, and Water Management, Directorate-General Rijkswaterstaat, The Hague).

Dutch draft convention concerning the cooperation on the management of the Scheldt and the Meuse, 14-4-1992 (Files Ministry of Transport, Public Works, and Water Management, Directorate-General Rijkswaterstaat, The Hague).

Dutch-Flemish communiqué concerning the water conventions and the HSL Antwerp/Rotterdam, 1-12-1994.

Dutch report of meeting between the Dutch Minister of Transport, Public Works, and Water Management, Smit-Kroes, and the Belgian Minister of Public Works, Olivier, 18-7-1984, Brussels. (Files TSC Ministry of Transport, Public Works, and Water Management, Directorate-General Rijkswaterstaat, The Hague).

Internal Dutch strategy paper (Files TSC-meeting 28, Ministry of Transport, Public Works, and Water Management, Directorate-General Rijkswaterstaat, The Hague).

ISG/DWS/6: Minutes of ISG-meeting 1, 29-4-1992, Yerseke (Files Ministry of Transport, Public Works, and Water Management, Directorate-General Rijkswaterstaat, Regional Directorate Zeeland, Middelburg).

ISG/DWS/7: Paper describing objectives of the ISG (Files Ministry of Transport, Public Works, and Water Management, Directorate-General Rijkswaterstaat, Regional Directorate Zeeland, Middelburg).

ISG/DWS/9: Proposal for Scheldt GIS (Files Ministry of Transport, Public Works, and Water Management, Directorate-General Rijkswaterstaat, Regional Directorate Zeeland, Middelburg).

ISG/DWS/10: Proposal for organization of an ecological workshop (Files Ministry of Transport, Public Works, and Water Management, Directorate-General Rijkswaterstaat, Regional Directorate Zeeland, Middelburg).

ISG/DWS/20: Minutes of ISG-meeting 2, 10-9-1992, Aalst (Files Ministry of Transport, Public Works, and Water Management, Directorate-General Rijkswaterstaat, Regional Directorate Zeeland, Middelburg).

ISG/DWS/29: Minutes of ISG-meeting 3, 3-12-1992, Douai (Files Ministry of Transport, Public Works, and Water Management, Directorate-General Rijkswaterstaat, Regional Directorate Zeeland, Middelburg).

ISG/DWS/33: Minutes of ISG-meeting 2, 26-3-1993, Aalst (Files Ministry of Transport, Public Works, and Water Management, Directorate-General Rijkswaterstaat, Regional Directorate Zeeland, Middelburg).

ISG/DWS/40: Minutes of the workshop ISG-DWS, 2/3-6 1993, Namur (Files Ministry of Transport, Public Works, and Water Management, Directorate-General Rijkswaterstaat, Regional Directorate Zeeland, Middelburg).

ISG/50: Minutes of ISG-meeting 6 (Files Ministry of Transport, Public Works, and Water Management, Directorate-General Rijkswaterstaat, Regional Directorate Zeeland, Middelburg).

Joint communiqué of the Netherlands, France, the Walloon region, and the Brussels Capital region, 26-4-1994.

Letter from the Dutch Minister of Transport, Public Works, and Water Management, Westerterp, to the Chairman of the Second Chamber of the Dutch Parliament, 1-10-1975 (Files TSC Ministry of Transport, Public Works, and Water Management, Directorate-General Rijkswaterstaat, The Hague).

Letter from the Dutch Minister of Transport, Public Works, and Water Management, Smit-Kroes, to the Belgian Minister of Public Works, Olivier, 28-10-1983. (Files TSC Ministry of Transport, Public Works, and Water Management, Directorate-General Rijkswaterstaat, The Hague).

Letter from the Dutch Minister of Transport, Public Works, and Water Management to the chairman of the 'Raad van de Waterstaat', 2-11-1984 (Files TSC Ministry of Transport, Public Works, and Water Management, Directorate-General Rijkswaterstaat, The Hague).

Letter from the Dutch Minister of Transport, Public Works, and Water Management, Smit-Kroes, to the Belgian Minister of Public Works, Olivier, 7-6-1985 (Files TSC Ministry of Transport, Public Works, and Water Management, Directorate-General Rijkswaterstaat, The Hague).

Letter from the Dutch Minister of Transport, Public Works, and Water Management, Smit-Kroes, to the Belgian Minister of Public Works, Olivier, 12-7-1985 (Files TSC Ministry of Transport, Public Works, and Water Management, Directorate-General Rijkswaterstaat, The Hague).

Letter from the Dutch Minister of Foreign Affairs, Van den Broek, to the Belgian Minister of Foreign Affairs, Tindemans, 3-12-1986 (Files Ministry of Transport, Public Works, and Water Management, Directorate-General Rijkswaterstaat, The Hague).

Letter from the Belgian Minister of Foreign Affairs, Tindemans, to the Dutch Minister of Foreign Affairs, Van den Broek, 26-1-1987 (Files Ministry of Transport, Public Works, and Water Management, Directorate-General Rijkswaterstaat, The Hague).

Letter from the Dutch Minister of Transport, Public Works, and Water Management to the Belgian Minister of Public Works, 26 June 1986 (Files SWC-meetings, Ministry of Transport, Public Works, and Water Management, Directorate-General Rijkswaterstaat, The Hague).

Letter from the Province of Zeeland to the Dutch Minister of Transport, Public Works, and Water Management, 12-3-1986 (Files SWC-meetings, Ministry of Transport, Public Works, and Water Management, Directorate-General Rijkswaterstaat, The Hague).

Letter from the Dutch Minister of Transport, Public Works, and Water Management, Smit-Kroes, to her Flemish colleague, Dupré, 22-9-1988 (File TSC-meeting 32, Ministry of Transport, Public Works, and Water Management, Directorate-General Rijkswaterstaat, The Hague).

Letter from the chairman of the Belgian delegation, Poppe, to the chairman of the Dutch delegation, Biesheuvel, 27-11-1990 (Files Ministry of Transport, Public Works, and Water Management, Directorate-General Rijkswaterstaat, The Hague)

Letter from the chairman of the Belgian delegation, Poppe, to the chairman of the Dutch delegation, Biesheuvel, 10-12-1990 (Files Ministry of Transport, Public Works, and Water Management, Directorate-General Rijkswaterstaat, The Hague)

Letter from the Belgian Minister of Foreign Affairs, Eyskens, to the chairman of the Core Working Group for the Western Scheldt, De Vries-Hommes, 12 January 1990 (Files Ministry of Transport, Public Works, and Water Management, Directorate-General Rijkswaterstaat, The Hague)

Letter from the chairman of the Dutch delegation, Biesheuvel, to the chairman of the Belgian delegation, Poppe, 19-11-1991 (Files Ministry of Transport, Public Works, and Water Management, Directorate-General Rijkswaterstaat, The Hague).

Letter from the Belgian Director-General of Water infrastructure, Demoen, to the Director-General of Rijkswaterstaat, Blom, 9-8-1991 (Files Ministry of Transport, Public Works, and Water Management, Directorate-General Rijkswaterstaat, The Hague).

Letter from the Regional Directorate Zeeland of Rijkswaterstaat to the Director-General of Rijkswaterstaat, 20-6-1990, Notitie Nr. AX 90.092 (Files Ministry of Transport, Public Works, and Water Management, Regional Directorate Zeeland, Middelburg).

Letter from the Flemish Minister of Public Works and Transport, Sauwens, to the Dutch Minister of Transport, Public Works, and Water Management, 18-12-1990 (Files Ministry of Transport, Public Works, and Water Management, Regional Directorate Zeeland, Middelburg).

Letter from the chairman of the Dutch delegation, Biesheuvel, to Janssens of the General Directorate of Politics of the Belgian Ministry of Foreign Affairs (Files Ministry of Transport, Public Works, and Water Management, Directorate-General Rijkswaterstaat, The Hague).

Letter from the Dutch Minister of Transport, Public Works, and Water Management to the Second Chamber of the Dutch Parliament, 1-9-1992 (Files Ministry of Transport, Public Works, and Water Management, Directorate-General Rijkswaterstaat, The Hague).

Letter from the Task Group for the Western Scheldt to the Dutch Minister of Transport, Public Works, and Water Management, 15-10-1992 (Files Ministry of Transport, Public Works, and Water Management, Directorate-General Rijkswaterstaat, The Hague).

Letter from the Dutch Minister of Transport, Public Works, and Water Management, Jorritsma-Lebbink, to the chairman of the Commission for Infrastructure and Water Management of the Dutch Parliament, 20-1-1995.

Letter from the Secretary of the 'Stichting het Zeeuwse Landschap, Prof.dr. Nienhuis, to the Head-Engineer-Director of the Regional Directorate Zeeland of Rijkswaterstaat, Saeijs, 27-10-1993 (Files Ministry of Transport, Public Works, and Water Management, Directorate-General Rijkswaterstaat, Regional Directorate Zeeland, Middelburg).

Letter from the Municipal Executive of Oostburg to the Regional Directorate Zeeland of Rijkswaterstaat, 5-1-1994 (Files Ministry of Transport, Public Works, and Water Management, Directorate-General Rijkswaterstaat, Regional Directorate Zeeland, Middelburg).

Letter from the Provincial Executives of Zeeland, Northern-Brabant, and Limburg to the Dutch Minister of Transport, Public Works, and Water Management, 22-6-1993 (Files Ministry of Transport, Public Works, and Water Management, Directorate-General Rijkswaterstaat, The Hague).

Letter from the Dutch Minister of Transport, Public Works, and Water Management to the Provincial Executives of Zeeland, Northern-Brabant and Limburg, 8-10-1993 (Files Ministry of Transport, Public Works, and Water Management, Directorate-General Rijkswaterstaat, The Hague).

Letter from the Dutch Minister of Transport, Public Works, and Water Management, Jorritsma, to the chairman of the Commission for Transport and Water Management of the Dutch parliament, 20-1-1995.

Letter from the Municipality of Flushing to the Regional Directorate Zeeland of Rijkswaterstaat, 29-6-1994 (Files Ministry of Transport, Public Works, and Water Management, Directorate-General Rijkswaterstaat, Regional Directorate Zeeland, Middelburg).

Letter from the Municipality of Reimerswaal to the Ministry of Transport, Public Works, and Water Management, 9-1-1995 (Files Ministry of Transport, Public Works, and Water Management, Directorate-General Rijkswaterstaat, Regional Directorate Zeeland, Middelburg).

Letter from Daamen and van Heteren, Regional Directorate Zeeland of Rijkswaterstaat, to the Director-General of Rijkswaterstaat, 22-3-1995 (Files Ministry of Transport, Public Works, and Water Management, Directorate-General Rijkswaterstaat, Regional Directorate Zeeland, Middelburg).

Letter from the Flemish Prime Minister, Van den Brande, and the Flemish Minister of Public Works, Transport, and Physical Planning, Baldewijns, to the Dutch Prime Minister, Kok, 13-12-1995 (Files Ministry of Transport, Public Works, and Water Management, Directorate-General Rijkswaterstaat, Regional Directorate Zeeland, Middelburg).

Letter from the Dutch Prime Minister, Kok, to his Flemish colleague, Van den Brande, 22-1-1996 (Files Ministry of Transport, Public Works, and Water Management, Directorate-General Rijkswaterstaat, Regional Directorate Zeeland, Middelburg).

Letter from the BBL to the Dutch Council of State, 22-1-1996 (Files Ministry of Transport, Public Works, and Water Management, Directorate-General Rijkswaterstaat, Regional Directorate Zeeland, Middelburg).

Letter from Strubbe to Zijlmans, 24-1-1996 (Files Ministry of Transport, Public Works, and Water Management, Directorate-General Rijkswaterstaat, The Hague).

Letter from the chairman of the project group Nature Restoration Western Scheldt, Kop, to the members of the BOW, 26-6-1995 (Files Ministry of Transport, Public Works, and Water Management, Directorate-General Rijkswaterstaat, Regional Directorate Zeeland, Middelburg).

Letter from the executive committee of the BOW to the Minister of Transport, Public Works, and Water Management, 8-5-1996 (Files Ministry of Transport, Public Works, and Water Management, Directorate-General Rijkswaterstaat, Regional Directorate Zeeland, Middelburg).

Letter from the chairman of the Commission Western Scheldt, Hendrikx, to the Minister of Transport, Public Works, and Water Management, Jorritsma-Lebbink, 29-8-1997.

Ministerie van Verkeer en Waterstaat, Rijkswaterstaat, Directie Zeeland, Notitie Nr. RFO 85.078. Proposal to formulate a policy plan for the Western Scheldt, 23-12-1985 (Files Province of Zeeland).

Minutes meeting of representatives of the DGRNE, the University of Liège, RIKZ, and RWS-Zld on the exchange of data between the Scheldt-GIS and the LIFE-2 project, 6-7-1994 (Files Ministry of Transport, Public Works, and Water Management, Directorate-General Rijkswaterstaat, Regional Directorate Zeeland, Middelburg).

Minutes of a preparatory meeting for a third Scheldt symposium, 17-2-1995 (Files Ministry of Transport, Public Works, and Water Management, Directorate-General Rijkswaterstaat, Regional Directorate Zeeland, Middelburg).

Minutes meeting of the excutive committee of the BOW, 6 July 1995, Middelburg.

Minutes meeting of the executive committee of the BOW, 10-10-1995, Middelburg.

Minutes meeting of the excutive committee of the BOW, 18-12-1995, Middelburg.

Minutes meeting of the excutive committee of the BOW, 31-1-1996, Middelburg .

Minutes meeting of the excutive committee of the BOW, 8-5-1996, Middelburg.

Minutes SWC-meeting 2, 16-1-1986 (File Ministry of Transport, Public Works, and Water Management, Directorate-General Rijkswaterstaat, The Hague).

Minutes SWC-meeting 3, 23-10-1986, Renesse (File Ministry of Transport, Public Works, and Water Management, Directorate-General Rijkswaterstaat, The Hague).

Minutes TSC-meeting, 10-3-1970, The Hague (File Ministry of Transport, Public Works, and Water Management, Directorate-General Rijkswaterstaat, The Hague).

Minutes TSC-meeting, 2-3-1971, Antwerp (File Ministry of Transport, Public Works, and Water Management, Directorate-General Rijkswaterstaat, The Hague).

Minutes TSC-meeting 22, 3-10-1978, Brussels (File Ministry of Transport, Public Works, and Water Management, Directorate-General Rijkswaterstaat, The Hague).

Minutes TSC-meeting 24, 20-6-1980, Bokrijk, Belgian Limburg (File Ministry of Transport, Public Works, and Water Management, Directorate-General Rijkswaterstaat, The Hague).

Minutes TSC-meeting, 24-6-1983 (File Ministry of Transport, Public Works, and Water Management, Directorate-General Rijkswaterstaat, The Hague).

Minutes TSC-meeting 32, 14 October 1988, Beveren (File Ministry of Transport, Public Works, and Water Management, Directorate-General Rijkswaterstaat, The Hague).

Minutes TSC-meeting 33, 23-11-1989, Middelburg (File Ministry of Transport, Public Works, and Water Management, Directorate-General Rijkswaterstaat, The Hague).

Minutes TSC-meeting 34, 2-1-1991 (File Ministry of Transport, Public Works, and Water Management, Directorate-General Rijkswaterstaat, The Hague).

Minutes TSC-meeting 35, 3-12-1991, Terneuzen (File Ministry of Transport, Public Works, and Water Management, Directorate-General Rijkswaterstaat, The Hague).

Minutes TSC-meeting 36, 10-12-1992, Antwerp (File Ministry of Transport, Public Works, and Water Management, Directorate-General Rijkswaterstaat, The Hague).

Minutes TSC-meeting 39, 28-3-1995, Schuddebeurs (File Ministry of Transport, Public Works, and Water Management, Directorate-General Rijkswaterstaat, The Hague).

Minutes TSC-meeting 40, 27-9-1995, Cleydael (File Ministry of Transport, Public Works, and Water Management, Directorate-General Rijkswaterstaat, The Hague).

Minutes TSC-meeting 41, 15-2-1996, Slot Moermont (File Ministry of Transport, Public Works, and Water Management, Directorate-General Rijkswaterstaat, The Hague).

Minutes TSC-meeting 42, 1-10-1996, Ghent (File Ministry of Transport, Public Works, and Water Management, Directorate-General Rijkswaterstaat, The Hague).

Minutes TSC-meeting 43, 19-6-1997, Flushing (File Ministry of Transport, Public Works, and Water Management, Directorate-General Rijkswaterstaat, The Hague).

Press release Ministry of Transport, Public Works, and Water Management, No. 3702, 10 March 1987.

Press release Ministery of Transport, Public Works, and Water Management, No. 4466, 9 January 1991.

Press release ICPS, 11-5-1995, Antwerp.

Press release ICPS, 5-12-1995, Antwerp.

Press release ICPS, 16-4-1996, Ghent.

Proposal LIFE programme, Development of a computer decision support framework for the assessment of reduction of specific waste water discharge in the river Scheldt basin (Agence de l'Eau Artois-Picardie, Ministère de la Région wallone, VMM) (Files Ministry of Transport, Public Works, and Water Management, Directorate-General Rijkswaterstaat, Regional Directorate Zeeland, Middelburg).

Proposals made by the Dutch delegation in the Commission-Biesheuvel-Poppe concerning international Scheldt and Meuse issues, numbered A to E, including additional notes made by the Dutch delegation (Files Ministry of Transport, Public Works, and Water Management, Directorate-General Rijkswaterstaat, The Hague).

Regional Directorate Zeeland of Rijkswaterstaat, Memo RVO-94.081, 18-11-1994.

Report informal meeting to discuss the issue of the pipeline Tessenderlo, 28-6-1988 (File TSC-meeting 32, Ministry of Transport, Public Works, and Water Management, Directorate-General Rijkswaterstaat, The Hague).

Report of a meeting between representatives of Flemish and Dutch national government agencies to discuss the implementation of the 48'/43'/38' deepening program, Middelburg, 18-5-1995 (Files Ministry of Transport, Public Works, and Water Management, Directorate-General Rijkswaterstaat, Regional Directorate Zeeland, Middelburg).

Report of deliberations between Belgian and Dutch civil servants of 15-4-1987 (Files SWC, Ministry of Transport, Public Works, and Water Management, Directorate-General Rijkswaterstaat, The Hague).

Reports of participation rounds organized by the Dutch advisory council, the 'Raad van de Waterstaat'. (Files TSC-meeting 29, 12-7-1985, Delft. Ministry of Transport, Public Works, and Water Management, Directorate-General Rijkswaterstaat, The Hague).

Report of Dutch preliminary talks first SWC-meeting, 13-5-1985, The Hague (Files SWC-meetings, Ministry of Transport, Public Works, and Water Management, Directorate-General Rijkswaterstaat, The Hague).

Report of meeting between representatives of the Ministry of Transport, Public Works, and Water Management, the Ministry of the Flemish Community, and the ZMF concerning the content and the conditions formulated in the WVO Permit for the Flemish region, 8-9-1994 (Files Ministry of Transport, Public Works, and Water Management, Directorate-General Rijkswaterstaat, Regional Directorate Zeeland, Middelburg).

Report of the Dutch Embassy concerning the Tessenderlo pipeline (Files Ministry of Transport, Public Works, and Water Management TSC-meeting 30, 7 March 1986, Ghent)

Report of the Flemish region for a discussion in the meeting of the SWC, 16-1-1986. (Files SWC-meetings, Ministry of Transport, Public Works, and Water Management, Directorate-General Rijkswaterstaat, The Hague).

Stenographic report meeting BOW, 8-5-1996, Middelburg

Traité entre le Royaume des Pays-Bas et le Royaume de Belgique au sujet du partage et de la qualité des eaux de la Meuse (In: Proceedings of *"The Charleville-Mézières agreements on the Meuse and the Scheldt: New trends in the law of international rivers*. Tuesday 6 May 1997, Association belge pour le droit de l' Environnement, T.M.C. Asser Instituut. Palais des Académies, Brussels, Belgium).

Traité entre le Royaume des Pays-Bas et le Royaume de Belgique au sujet de la construction du canal de Baalhoek (In: Proceedings of *"The Charleville-Mézières agreements on the Meuse and the Scheldt: New trends in the law of international rivers*. Tuesday 6 May 1997, Association belge pour le droit de l' Environnement, T.M.C. Asser Instituut. Palais des Académies, Brussels, Belgium).

Traité entre le Royaume des Pays-Bas et le Royaume de Belgique au sujet de l' amélioration de la voie navigable dans l' Escaut près du goulet de Bath (In: Proceedings of *"The Charleville-Mézières agreements on the Meuse and the Scheldt: New trends in the law of international rivers*. Tuesday 6 May 1997, Association belge pour le droit de l' Environnement, T.M.C. Asser Instituut. Palais des Académies, Brussels, Belgium).

D. Newspaper articles

Anonymous, 12-2-1979: "Antwerpen tast milieu aan in België en Nederland."

PZC, 13-11-1984.

De Standaard, 28-8-1985: "Smeerpijp Albertkanaal werkt uitdiepen van Schelde tegen."

De Standaard, 17/18-10-1987.

Gazet van Antwerpen, 16-10-1987: "Waterverdragen geblokkeerd."

NRC/Handelsblad, 27-10-1987.

De Stem, 9-3-1991: "Soap opera rond Maas en Schelde."

De Volkskrant, 28-4-1990: "Schoonmaak van de Schelde wordt een wereldklus".

PZC, 20-11-1990: "Zieke Schelde en oeverloos gepraat".

Le Soir, 8-4-1992: "Louables intentions mais moyens limités. Fleuve et canal: poumons Liégeois".

PZC, 12-5-1993: "Nederland vertraagde verdragen, Senatoren Zijlstra (PvdA) en Eversdijk (CDA) kritisch over Maij."

NRC, 9-6-1993:"Maij-Weggen schopte Walen nodeloos tegen de schenen, Nieuw rivierenverdrag is urgent." K. Zijlstra

PZC, 12-6-1993: "Taakgroep zoekt over de grens steun voor schone Westerschelde."

PZC, 14-6-1993: "Belgen frustreren studies voor delta van Rijn-Schelde."

De Volkskrant, 17-6-1993: "Lubbers en Van den Brande willen nog dit jaar akkoord, Nederland en Vlaanderen koppelen TGV en Schelde."

Trouw, 17-6-1993: "Lubbers sluit 'deal' met Vlaanderen over TGV-tracé."

Het financieele dagblad, 22-6-1993: "Nederlands kapitaal voor Vlaming tweede keus. Zelfbewuste minister-president Van den Brande schetst Lubbers nog eens het nieuwe België."

PZC, 18-6-1993: "Langgekoesterde wens België voor Nederland niet langer onbespreekbaar. Baalhoekkanaal niet in verdrag."

PZC, 4-8-1993: "Belgische gemeenten werken samen met Schelde-taakgroep."

PZC, 6-10-1993: "Maij offert schone Schelde op aan komst van flitstrein."

PZC, 20-10-1993: "Provincies eisen handhaving van waterverdragen."

PZC, 25-10-1993: "Minister laat Zeeland in de steek."

Trouw, 28-10-1993: "Vlaanderen woedend over uitspraken minister Maij-Weggen."

PZC, 28-10-1993: "Toch overleg over Westerschelde."

NRC, 6-11-1993: "Maas is vooral in België een open riool."

PZC, 24-11-1993: "Belgen dringend nodig in Delta."

NRC, 30-11-1993: "Nederlanders vervuilen de Maas even zeer als de Belgen."

NRC, 1-12-1993a: "Vervuiler aansprakelijk voor lozingen."

NRC, 1-12-1993b: "Maij niet zelf naar de rechter om lozen Maas."

NRC, 2-12-1993: "Maas toch schoner dan Rijn."

NRC, 3-12-1993: "Maij is nodeloos agressief over vervuiling van de Maas, Minister uit Wallonië wil wel praten."

NRC, 28-1-1994: "Zicht op schonere Maas nu partijen aan één tafel zitten."

PZC, 5-2-1994: "Optimisme over diepere Schelde."

PZC, 26-2-1994: "Vlaanderen wil snel overleg uitdiepen van Westerschelde."

PZC, 26-3-1994: "Antwerpen dreigde bezegeling reglement Schelde te bederven."

PZC, 1-4-1994: "Vlaanderen zuivert de Schelde."

PZC, 21-4-1994: "Vlaams kabinet zet Nederland mes op de keel over de Schelde."

NRC, 21-4-1994: "Vlaamse premier voert druk op Nederland rond waterverdrag op."

PZC, 26-4-1994: "Vlaamse regering niet bij tekenen waterverdragen."

NRC, 27-4-1994: "Maij ligt niet wakker van Vlaamse weigering waterverdrag te tekenen."

PZC, 1-6-1994: "Stuurgroep tegen koppeling HSL en uitdiepen Schelde."

PZC, 5-7-1994: "Nederlands beleid rond verdieping op de korrel, Antwerpen hekelt 'obstructie'."

PZC, 27-8-1994: "Diepere Schelde is nodig voor groei economie Delta."

PZC, 7-9-1994: "Nut verdieping Schelde betwist."

PZC, 21-9-1994: "Voorwaarden aan diepe Schelde."

PZC, 30-9-1994: "Snel akkoord over Schelde nodig."

Cobouw, 3-10-1994: "Jorritsma gaat met Belgen praten over Westerschelde."

PZC, 4-10-1994: "Vispopulatie oostelijke Schelde telt 32 soorten."

De Volkskrant, 29-11-1994: "Uitdieping van Westerschelde is verliesgevend."

Haagse Courant, 29-11-1994: "Vlaamse milieubond: 'Uitdiepen Schelde onrendabel'."

PZC, 21-12-1994: "GS trekken bezwaar tegen verdiepen van de Westerschelde in."

PZC, 22-12-1994: "Zeeuwse gemeenten krijgen stem in Scheldecommissie."

De Stem, 11-1-1995: "Zeeland krijgt zitting in Scheldecommissie."

PZC, 17-1-1995: "Waterverdragen geen garantie voor schonere rivieren."
Gazet van Antwerpen, 18-1-1995: "Milieuprotest buiten de muren."
De Stem, 18-1-1995: "Jorritsma: Snel plan reiniging Maas en Schelde."
NRC Handelsblad, 25-2-1995: "Romeins rijk werpt nog steeds zijn schaduw over Europa."
PZC, 2-3-1996: "Van Gelder: ontpolderen, hoe dan ook."
PZC, 5-3-1996: "Saeijs haalt uit naar landbouw."

E. Respondents/ Informants[1]

France

M. Grandmougin   Water Agency Artois-Picardie, Vice-Director responsible for scientific and international affairs

J.-M. Journet   Water Agency Artois-Picardie, Chief of the Department of technical data

D. Martin   Water Agency Artois-Picardie, Water quality Engineer

J. Prygiel   Water Agency Artois-Picardie, Chief of the Department of Ecology

Walloon Region

B. de Kerckhove   Ministry if the Walloon Region, Directorate-General for Natural Resources and the Environment, Correspondent for the Belgian Coordination Commission for Environmental affairs and International Relations

F. Paulus   Ministry of the Walloon Region, DGRNE, Attaché

B. Tricot   Ministy of the Walloon Region, DGRNE, Agricultural Engineer

J.-M. Wauthier   Ministry of the Walloon Region, DGRNE, Director

Brussels Capital Region

A. Thirion   Brussels Institute for Environmental Management, Strategy Department

T. Varet   Ministry of the Brussels Capital Region, Administration of Equipment and Transport

---

1. The positions mentioned are the positions at the time of the interview.

Flemish Region

K. De Brabander — Flemish Environment Agency, Director International Environmental policy

M. Bruyneel — Flemish Environment Agency, Chief Department of Information

H. Maeckelberghe — Flemish Environment Agency, Director-Engineer

P. Meire — Ministry of the Flemish Community, Department of the Environment and Infrastructure, Administration for the Environment, Nature, Land and Water Management, Institute for Nature Conservation, Scientific Attaché

F. De Mulder — Province of Eastern Flanders, Chief Department of Environmental planning and Nature conservation

M. Verdievel — Flemish Environment Agency, Project-engineer

M.-P. De Vroede — Ministry of the Flemish Community, Department of the Environment and Infrastructure, AMINAL, Coordinator basin policies

The Netherlands

L. Adriaanse — Ministry of Transport, Public Works, and Water Management, Directorate-General for Public Works and Water Management, Regional Directorate Zeeland, Chief Department of Regional development, Environment, and Strategy

Blondeel — Province of Zeeland, Coordinator of Euregio projects

F. de Bruijckere — Ministry of Transport, Public Works, and Water Management, Directorate-General for Public Works and Water Management, Regional Directorate Zeeland, Department of policy preparation and planning Integrated Water management

J.J. Cappon — Ministry of Transport, Public Works, and Water Management, Directorate-General for Public Works, and Water Management, Institute for Inland Water Management and Waste Water Treatment, Head Department of Land-affairs

B. van Eck — Ministry of Transport, Public Works, and Water Management, Directorate-General for Public Works and Water Management, Research Institute for Coastal and Marine Management, Main Department of Advice and Policy, Senior project manager

C. Heip	Netherlands Institute for Ecological Research-Centre for Marine and Estuarine Ecology, Director
P. Herman	Netherlands Institute for Ecological Research-Centre for Marine and Estuarine Ecology, Researcher
J. Hollears	Ministry of Transport, Public Works, and Water Management, Directorate-General for Public Works and Water Management, Regional Directorate Zeeland, Main Department of Infrastructure and Navigation
B. de Hoop	Ministry of Transport, Public Works, and Water Management, Directorate-General for Public Works and Water Management, Regional Directorate Zeeland, Chief Department of Planning and management of navigation channels
M. Meulblok	Ministry of Transport, Public Works, and Water Management, Directorate-General for Public Works and Water Management, Regional Directorate Zeeland, Chief Department of River engineering
F. van Pelt	Province of Zeeland, Direction of Economy, Physical Planning, and Welfare, Coordinator Delta waters
M. de Rooy	Reinwater Foundation
H. Saeijs	Ministry of Transport, Public Works, and Water Management, Directorate-General for Public Works and Water Management, Regional Directorate Zeeland, Head-Engineer-Director
L. Santbergen	Ministry of Transport, Public Works, and Water Management, Directorate-General for Public Works and Water Management, Regional Directorate Zeeland, Employee International River Scheldt Commission and International Commission on Large Dams (ICOLD)
H. Smit	Ministry of Transport, Public Works, and Water Management, Directorate-General for Public Works and Water Management, Chief Main Department of Advice and Policy
E. Turkstra	Ministry of Transport, Public Works, and Water Management, Directorate-General for Public Works and Water Management, Regional Directorate Zeeland, Chief Department of Integrated Water management
S. Vereeke	Ministry of Transport, Public Works, and Water Management, Directorate-General for Public Works and Water Management, Regional Directorate Zeeland, Employee Department of Integrated Water management)

C. de Villeneuve	Ministry of Transport, Public Works, and Water Management, Directorate-General for Public Works and Water Management, General Directorate, Deputy Chief International Water Policy Division
J. Vreeke	Ministry of Transport, Public Works, and Water Management, Directorate-General for Public Works and Water Management, RIKZ, Hydrographic researcher
J. Vroon	Ministry of Transport, Public Works, and Water Management, Directorate-General for Public Works and Water Management, RIKZ, Project manager
J. de Wit	Ministry of Transport, Public Works, and Water Management, Directorate-General for Public Works and Water Management, RIZA, Chief Emission Department
R. Zijlmans	Ministry of Transport, Public Works, and Water Management, Directorate-General for Public Works and Water Management, General Directorate, Deputy Chief International Water Policy Division

Benelux

Abts	Benelux Economic Union

Appendices

Appendix 1: List of Figures

Appendix 2: List of Tables

Appendix 3: Glossary

Activated policy network	See *Policy arena*
Actor	Unit that by a certain unity of action acts as an influencing party: State, (department of) organization, individual
Arrangement	(implicit) agreement
Boundary river	River that forms the border between two or more states (= contiguous river)
Cognitive learning	Gaining knowledge on the intellectual relationships in an issue-area, i.e on cause-effect relationships, policy alternatives, impacts of policy alternatives, or on the intellectual relationships between issues
Communication	Dimension of interaction. Exchange of information, demands, announcements, claims and opinions
Conflict	Game type. Whenever an actor attempts to exert power over another to overcome that actor's perceived blockade of the first actor's goal and faces significant resistance
Contiguous river	See *Boundary river*
Cooperation	Game type. Working together for common benefits
Crucial decision	Decision that turns out to be a point of reference for the actors involved in succeeding decision making games
Decision making	A complex of series of decisions taken by different actors (As a dimension of interaction: type of communication in which the actors agree implicitly or explicitly on an exchange formula)
Decision making game	See *Decision making round*
Decision making round	Game that is played between two crucial decisions

Exchange	Dimension of interaction. Implementation of an exchange agreement that is made between actors
Game	The whole of interactions, incentives and interventions that develops in a policy arena
Game type	Characterization of the game that is played between the actors. See *Conflict* and *Cooperation*
Implementation gap	Discrepancy between a formulated policy, and the implementation of this policy
Incentive position	Position in which an actor is able to influence the behaviour of other actors by creating (dis)incentives
Individualism/Collectivism	Cultural dimension that indicates the relationship between the individual and the collectivity which prevails in a given society
Institutionalization	Development or change of structural or cultural network characteristics
Institutional arrangement	Organizational arrangement regulating the interactions between actors
Interaction	Communication, Negotiation, Decision making, Exchange
Interaction position	Position in which an actor is able to interact with other actors to achieve his objectives
Interactive learning	See *Strategic learning*
Interactive strategy	Type of strategy in which an actor tries to acquire indispensable resources for achieving his objectives in interaction with other actors (either offensive or reactive)
Interdependence	Situation in which actors are at least partly dependent on each other for the solution of their problems or the achievement of their objectives (= Mutual dependence)

Intervention position	Position in which an actor is able to influence patterns of interaction between other actors by intervention
Issue	An issue exists whenever an actor perceives that some of his goals are being blocked (frustrated, denied) by one or more other actors
Issue-area	A recognized cluster of concerns involving interdependence not only among the actors, but also among the issues themselves
Lower level governments	Councils, executives, and government agencies on the regional and local level (provinces, departments, municipalities, waterboards etc.)
Maintaining autonomy	(*Strategy aimed at-*): Type of strategy in which an actor tries to solve his problems or to achieve his objectives without interaction with other actors
Masculinity/ Femininity	Cultural dimension that indicates whether the roles of men and women are clearly separated or overlapping in a given society
Multi-level game	Indication of the linkages between games that are played on two or more administrative levels, such as between a game that is played on the international level and an *intra*national decision making game
Mutual dependence	See *Interdependence*
Negotiation	Dimension of interaction. A type of communication in which the actors try to find exchange formulas that are attractive to the parties involved
Offensive strategy	Type of strategy in which an actor takes the initiative to achieve his objective (either interactive or aimed at maintaining autonomy)

Perception	Generally: Subjective observation. In this study: broader concept than the concept problem perception, which relates to: • the interdependencies between the actors arising from the division of the resources needed to fulfil their ambitions; • the ambitions, and 'stakes in the game' of the actors; • the policy game and the importance of the policy problem being addressed in the game, compared with the policy problems being addressed in other games.
Policy arena	That part of the policy network where the interactions take place (the game is played) (= Activated policy network)
Policy network	Multi-actor structure of interdependence
Policy window	Opportunity to get political attention for a problem or policy alternative, i.e. to place that problem or policy alternative on the political agenda (= Window of opportunity)
Position	Indication of the type of resources an actor possesses, and the possible strategies he may use (See *Incentive*, *Interaction*, and *Intervention position*)
Power Distance	Cultural dimension that indicates the extent to which the less powerful members of institutions and organizations within a country expect and accept that power is distributed unequally
Problem	A discrepancy between an actual and a desired situation
Problem perception	A discrepancy between a *perceived* actual situation and a desired situation
Reactive strategy	Type of strategy in which an actor reacts on initiatives taken by other actors (either interactive or aimed at maintaining autonomy)
Reputation	A conviction of others that an actor will use a specific strategy

Resource	All things that may help an actor to solve his problems or to achieve his objectives
River basin approach	Steering and research concept emphasizing that river basins should be managed and analyzed as a coherent unity
Shadow of the future	The likelihood that actors will need each other in the future to solve their problems or to achieve their objectives
Strategic learning	Gaining information on (inter) dependencies and problem perceptions of other actors involved in a decision making game, and to discover score possibilities, and possible win-win situations (= Interactive learning)
Strategy	Pattern of behaviour aiming at the realization of individual or common objectives
Substantive issue-linkage	Linkage between issues because of their intellectual coherence. See *Issue-area* and *Cognitive learning*
Successive river	See *Transboundary river*
Tactical issue-linkage	Linkage between issues for tactical or strategic reasons
Transboundary river	River that crosses the border between two or more states (= Successive river)
Trust	A type of expectation that alleviates the fear that one's exchange partner will act opportunistically
Uncertainty Avoidance	Cultural dimension that indicates the tolerance for uncertainty in a given society
Upstream-downstream issue	Water management issue characterized by an asymmetric distribution of interests and resources among the actors involved in decision making on that issue, which is caused by upstream-downstream relationships (mostly river issues).
Window of opportunity	See *Policy window*

Win-win situation   A situation in which the actors involved perceive that they are better off with an agreement than without one

Appendix 4: Abbreviations

AED	Administration de l'Equipement et des Déplacements (=BUV) (Administration of Equipment and Transport, Brussels Capital Region)
AGHA	Antwerpse Gemeenschap voor de Haven (Antwerp Port Association)
AMINAL	Administratie Milieu, Natuur-, Land- en Waterbeheer (Administration for the Environment, Nature, Land and Water Management, Flemish Region)
AMINAL-IN	Instituut voor Natuurbehoud (Institute for Nature Conservation, Flemish Region)
ARNE	Administration des Ressources Naturelles et de l'Environnement (Administration for Natural Resources and the Environment, Brussels Capital Region)
Awbm	Algemene wet bepalingen milieuhygiëne (General Environmental Act, the Netherlands)
AWP	Algemeen Waterzuiveringsprogramma (General Waste Water Treatment Programme, Flemish Region)
AWZ	Administratie Waterwegen en Zeewezen (Administration Water Infrastructure and Marine Affairs, Flemish Region)
BBL	Bond Beter Leefmilieu (Association for a Better Environment)
BIM	Brussels Instituut voor Milieubeheer (Brussels Institute for Environmental Management) (=IBGE)
BIWM	Brusselse Intercommunale Watermaatschappij (Brussels Intermunicipal Water Company)
BUV	Bestuur van Uitrusting en van Vervoerbeleid (=AED)
CCIM	Coördinatiecommissie voor het Internationaal Milieubeleid (Coordination Commission for International Environmental Policies, Belgium)
CEMO	See NIOO-CEMO
CIPE	Commission Internationale pour la Protection de l'Escaut (=ICBS/ICPS)
CIPM	Commission Internationale pour la Protection de la Meuse(=ICBM/ICPM)
COIB	Coördinatie Overleg Internationale Betrekkingen (Coordination Commission for International Relations, Ministry of Transport, Public Works, and Water Management, the Netherlands)
DGRNE	Direction Générale des Ressources Naturelles et de l' Environnement (Directorate General for Natural Resources and the Environment, Walloon Region)
DGW	Dienst Getijdewateren (Tidal Waters Divion, Rijkswaterstaat) [Presently: RIKZ]

DIREN	Direction régionale de l'environnement et de la nature (Regional direction for environmental and nature management, France)
DSS	Decision Support System
EIA	Environmental Impact Assessment
EK	Eerste Kamer (First Chamber of the Parliament of the Netherlands)
EU	European Union
HD	Hoofd-Directie (Main Directorate)
HID	Hoofd-Ingenieur-Directeur (Head-Engineer-Director)
HSL	Hoge Snelheidslijn (High Speed Alignment)
IBGE	Institut Bruxellois pour la Gestion de l'Environnement (= BIM)
ICBM	Internationale Commissie voor de Bescherming van de Maas (= ICPM/CIPM)
ICBS	Internationale Commissie voor de Bescherming van de Schelde (=ICPS/CIPE)
ICPM	International Commission for the Protection of the Meuse (=ICBM/CIPM)
ICPS	International Commission for the Protection of the Scheldt (= ICBS/CIPE)
ICPR	International Commission for the Protection of the Rhine
IDV	Individualism
i.e.	Inhabitant Equivalent
IMP	Indicatief Meerjarenprogramma (Long term water quality policy programme, The Netherlands)
IN	See AMINAL-IN
IPR	Interparlementaire Beneluxraad (Interparliamentary Council of the Benelux)
ISG-DWS	International Study Group-Description of the Water quality of the Scheldt (Later ISG)
ISG	International Scheldt Group
ISG-DES	International Scheldt Group-Description of the Ecology of the Scheldt
MARS	Marsh Amelioration along the River Schelde
MAS	Masculinity
MINA	Milieu- en Natuurbeleidsplan voor Vlaanderen (Flemish Policy Document for Nature and Environmental management)
MP	Member of Parliament
NAP	Noordzee Aktie Programma (North Sea Action Programme)
NIMBY	Not In My Back-yard (Name of Dutch Act)

NIOO-CEMO	Nederlands Instituut voor Oecologisch Onderzoek-Centrum voor Mariene en Estuariene Ecologie (Netherlands Institute for Ecological Research-Centre for Marine and Estuarine Ecology)
NW3	Derde Nota Waterhuishouding (Third Policy Document on Water Management)
PAWN	Policy Analysis of Water management for the Netherlands
PC	Permanente Commissie van Toezicht op de Scheldevaart (Permanent Commission for Supervising navigation on the Scheldt)
PDI	Power Distance
POM	Project Onderzoek Maas (Project Research Meuse)
POR	Project Onderzoek Rijn (Project Research Rhine)
RAP	Rhine Action Programme
RIKZ	Rijksinstituut voor Kust en Zee (Research Institute for Coastal and Marine Management, Rijkswaterstaat)
RIZA	Rijksinstituut voor Integraal Zoetwaterbeheer en Afvalwaterbehandeling (Institute for Inland Water Management and Waste Water Treatment, Rijkswaterstaat)
RSD	Platform Rijn Schelde Delta (Rhine-Scheldt Delta)
RWS	Rijkswaterstaat (National Institute for Public Works and Water Management, The Netherlands)
SAGE	Schéma d'Aménagement et de Gestion des Eaux (French water management plan)
SAP	Schelde Aktie Programma (Scheldt Action Programme)
SAWES	Systeem Analyse Westerschelde (Systems Analysis of the Western Scheldt)
SDAGE	Schéma Directeur d'Aménagement et de Gestion des Eaux (French strategic water management plan)
Stb.	Staatsblad (Official Law Journal of the Netherlands)
SWC	Scheldewater Commissie (Scheldt Water Commission)
TK	Tweede Kamer (Second Chamber of the Parliament of the Netherlands)
TMC	Technische Maascommissie (Technical Meuse Commission)
Trb.	Tractatenblad (Treaty Series of the Netherlands)
TSC	Technische Scheldecommissie (Technical Scheldt Commission)
UAI	Uncertainty Avoidance
UN-ECE	United Nations-Economic Commission for Europe
UN-ILC	United Nations-International Law Commission
VLAREM	Vlaams Reglement voor de Milieuvergunning (Flemish Regulation on Environmental Permits)

VMM	Vlaamse Milieumaatschappij (Flemish Environment Agency)
VMZ	Vlaamse Maatschappij voor Waterzuivering (Flemish Agency for Waste water treatment) [Presently: VMM]
VNR	Vereniging Nederlandse Riviergemeenten (Association of Dutch river municipalities)
VROM	Ministerie van Volkshuisvesting, Ruimtelijke Ordening en Milieubeheer (Ministry of Housing, Land-use Planning and Environmental Management)
V&W	Ministerie van Verkeer en Waterstaat (Ministry of Transport, Public Works and Water Management)
WVO	Wet Verontreiniging Oppervlaktewateren (Dutch Surface Water Pollution Control Act)
Zld.	Zeeland (Sealand)
ZMF	Zeeuwse Milieufederatie (Zeeland Environment Federation)

Appendix 5: Hydrographic description of the Scheldt basin[1]

Table A1: Hydrographic basins with surface area, population density (1993), main tributaries and connected waters, and most important towns

Basin	Surface area (km²)	Population density (km⁻²)	Tributaries and connected waters	Most important towns
Upper-Scheldt (*Haut-Escaut*)	6,088	313	Scarpe Haine Espierre	Tourcoing Denain Valenciennes Douai Roubaix
Leie (*Lys*)	4,305	746	Deûle	Lille Lens Béthune
Dender (*Dendre*)	1,386	281		Dendermonde Aalst
Zenne (*Senne*)	1,171	1291		Mechelen Brussel
Demer	2,188	326		Hasselt Aarschot
Nete	1,560	342	Grote Nete Kleine Nete	
Dijle (*Dyle*)	1.265	397	Demer	Leuven
Upper Sea Scheldt	1,007	472	Upper Scheldt Dender Leie	Ghent
Lower Sea Scheldt	1,854	520	Rupel	Antwerp
Canal Ghent-Terneuzen	668	526		Zelzate
Western Scheldt	2,841	204	Canal Ghent-Terneuzen	Flushing Middelburg Terneuzen

1. This Annex is made by P.A.J. Verlaan.

**Table A2: Inflow and outflow canals and the connected catchment areas:
1. Meuse/Rhine catchment area, 2. Somme/Seine catchment area, and 3. North Sea.
The notes in the first and third columns refer to the Map of the Scheldt basin presented
in this appendix**

Outflow canals		Inflow canals	
Canal de St.Quentin[1]	2	Canal du Centre[11]	1
Canal du Nord[4]	2	Albertkanaal[20]	1
Canal de Neufossé[9]	3	Schelde-Rijn connection[23]	1
Canal Ghent-Oostende[15]	3	Bathse spuikanaal[24]	1
Afleidingskanaal van de Leie[14]	3	Kanaal door Zuid Beveland[25]	1
Rigole D'oise et du Norrieu[2]	1	Kanaal door Walcheren[26]	1
		Zuid-Willemsvaart[22]	1

1. Upper Scheldt (Haut Escaut) basin

The Upper Scheldt basin contains the main course of the River Scheldt. It covers almost 30% of the Scheldt catchment area. Most important tributaries are the Haine, Scarpe, and Espierre. The Haine, which has its origin in Wallonia, flows out in the Upper Scheldt in France. The Scarpe flows through Arras and Douai, and joins the Upper Scheldt at the Belgian-French border. Most water from Lille, Roubaix and Tourcoing (in the Leie basin) is discharged to the Upper Scheldt via the Espierre. The Upper Scheldt is connected to the Meuse catchment area by the Canal de St.Quentin[1] and the Rigole d'Oise et du Norrieux[2], and to the Somme catchment area by the Canal du Nord[4]. Other canals that connect the Upper Scheldt basin with adjacent basins are described hereafter.

2. Leie (Lys) basin

The Leie rises in the north of France. Its main tributary is the Deûle. The Leie and Deûle have their source from an area with a high population density including the cities of Lille and Béthune. The Leie follows its natural course until Aire-sur-La Lys where it crosses the Canal à Grand Gabarit. This canal consists of several sub-canals that connect the Upper Scheldt with the Scarpe (Canal de la Sensee[3]), the Scarpe to the Deûle (Canal de la Deûlc[6]), the Deûle to the Leie (Canal d'Aire[8]) and the Leie to the Aa drainage basin and the North Sea (Canal de la Neufossé[9]). Further downstream, the Leie forms the border between France and Wallonia over a distance of 15 km. Here the Leie is connected with the Upper Scheldt via the Deûle and the Canal de Roubaix[7]. The Leie joins the Upper Scheldt at Ghent. Near to Ghent, most Leie water is diverted and thus will not reach the Upper Sea Scheldt. First, Leie water is diverted to the North Sea by the Afleidingskanaal van de Leie[14] and the Canal Ghent-Oostende[15]. Secondly, water of the Leie is diverted to the Western Scheldt by the Gentse ringvaart[13] and the Canal Ghent-Terneuzen[16]. Upstream of Ghent, small amounts of water from the Upper Scheldt are discharged to the Leie via the Canal Bossuit-Kortrijk[12].

3. Dender (Dendre) basin

The Dender has her source in Wallonia and joins the Upper Sea Scheldt at Dendermonde. The Dender is connected with the Upper Scheldt by the Canal Nimy-Blaton-Péronnes[11] and to the Haine by the Canal Pommeroel-Condé[5].

4. Zenne (Senne) basin

The Zenne rises in Soignies, flows through Brussels and debouches in the Dijle at Mechelen. The two major canals in this basin are the:
1. Willebroekkanaal[18] connecting Brussels directly with the Rupel.
2. Canal de Charleroi à Brussel[19], connecting the Zenne basin with the Meuse catchment area.

The Canal du Centre[11] and the Canal Nimy-Blaton-Peronnes[11] form a further connection with the Upper Scheldt and Dender and thus allow these basins to have a navigation function.

5. Demer basin

The Demer has its source at Tongeren. It collects the water of many small tributaries and brooks such as the Grote Gete, Kleine Gete. The Demer debouches in the Dijle at Werchter. North of Hasselt, the Albertkanaal[20] crosses several tributaries in the Demer basin, and connects the basin of the Lower Sea Scheldt with the Meuse catchment area.

6. Nete basin

The Nete basin has two tributaries: the Grote Nete and the Kleine Nete. North of Mechelen, the Nete flows out on the Dijle. The Nete basin is, like the Demer basin crossed by the Albertkanaal[20] and its cross-canals. Furthermore, the Netekanaal[21] directly connects the lower Nete with the Albertkanaal[20].

7. Dijle (Dyle) basin

The Dijle has her source about 15 km north of Charleroi. The Dijle debouches in the Rupel, along with the Zenne and the Nete.

8. Upper Sea Scheldt basin

This basin is part of the estuarine zone of the Scheldt catchment area. Since the water is fresh in this area, it is also called the freshwater estuary of the Scheldt. The Upper Sea Scheldt receives her water from the Upper Scheldt, the Dender and, during high discharge conditions, also from the Leie. The Durme also discharges water to the Upper Sea Scheldt. The Rupel joins the Upper Sea Scheldt at Rupelmonde.

9. Lower Sea Scheldt basin

This basin is situated between Rupelmonde and the Dutch-Belgian border. As the tide dominates the water movement it is also called the upper estuary. The Lower Sea Scheldt receives her water from the Rupel (and her tributaries: Zenne, Dijle, Demer and Nete) and the Upper Sea Scheldt (and her tributaries: Dender and Leie). Both rivers discharge about the same amount of water on the Lower Sea Scheldt. The agglomeration Antwerp is situated in this basin. The Albertkanaal[20] connects the Antwerp harbour basins via the Nete and Demer basin with the Meuse catchment area. A part of the water from the Albertkanaal[20] is used to feed the Schelde-Rijn conncection[23]. Another part of this water reaches the Lower Sea Scheldt via the Zandvliet/Berendrecht and the Boudewijn/Cauwelaert sluices.

10. Canal Ghent-Terneuzen basin

The Canal Ghent-Terneuzen[16] receives its water mainly from the Leie and to a minor extent from the Durme and the Upper Scheldt. Water from the Leie basin and the Upper

Scheldt reach the Canal Ghent-Terneuzen[16] via the Gentse ringvaart[13]. The Durme is enclosed by a transverse dam at Lokeren. Water upstream of Lokeren feeds the Canal Ghent-Terneuzen[16] via the Moervaart[17] whereas downstream of Lokeren, Durme water feeds the Upper Sea Scheldt.

11. Western Scheldt basin

The Western Scheldt is the only basin entirely situated in the Netherlands. It receives water from the Lower Sea Scheldt and sea water from the North Sea. Also a number of man-made canals supply water to the Western Scheldt :

1. Since 1988 superfluous water in the freshwater Zoom lake is discharged to the Western Scheldt by the Bathse spuisluis[24]. The Zoom lake receives its water from the Rivers Rhine and Meuse.

2. The Western Scheldt is connected with the saline Eastern Scheldt by the Kanaal door Zuid-Beveland[25].

3. A part of the water from the Leie, Durme, and Upper Scheldt reaches the Western Scheldt by the Canal Ghent-Terneuzen[16].

4. Small amounts of water are discharged to the Western Scheldt by the Kanaal door Walcheren[26].

Finally, the Western Scheldt debouches in the North Sea at Flushing.

The Scheldt Basin

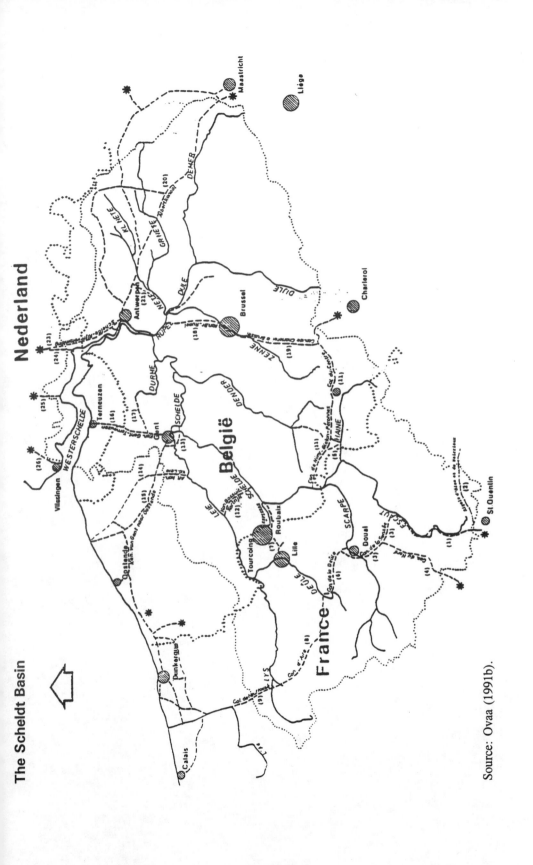

Source: Ovaa (1991b).

Appendix 6: Administrative organization of water management in the Scheldt basin[1]

France

France is a unitary and centralized state in which the central government is considered generally to be very influential, and public administration to be organized rather hierarchically. Although this largely is true, efforts at decentralization have given more power to local government authorities in the past twenty years.[2] Administrative levels are the State or Central government (*Etat*), Regions (*Régions*), Departments (*Départements*), and the Municipalities (*Communes*). Finally, in the Water Act 1964 six basin agencies (*Agences de Bassin*) were created, five of them corresponding with the river basins Rhône, Loire, Garonne, Seine and Rhine. The sixth agency, the *Agence de l'Eau Artois-Picardie*, was set up in the northwestern part of France, and comprises, among other things, the French part of the Scheldt basin.[3] Structurally, the central government is divided into ministries, ranging in recent years from eleven to twenty-one, each in turn subdivided into several *directions* as the principle operating units, plus a ministerial *cabinet* or secretariat with staff assisting the minister, and various consultative and control organs.[4] Traditionally, apart from the state the more than 36,000 French municipalities

---

1. The descriptions of the administrative organization of water management in the Scheldt basin states that can be found in Ovaa (1991a and b), and ISG (1994a and b) helped the author a lot to find his way in French and Belgian water management.

2. Heady (1991, p. 193).

3. Loi sur l'Eau 1964. Although the hydrographic basins certainly were an important criterion, it is known that there were other reasons as well. Nicolazo (p.10) quotes the first director of the Seine-Normandie agency, who states that: "*A first problem has been how many agencies would have to be installed. In first instance, one had though not to install six of them, but eight: obviously there were the big rivers, the Seine, Garonne, Loire, and the Rhône, which according to some would have to be divided into two parts: the Rhône from its source to Avignon, and for the downstream part, the whole mediterranean coast, and furthermore the two agencies for the north, Artois-Picardie, and the partly French rivers, the Meuse and the Rhine, and one was thinking of making a last one comprising the coastal rivers between the Seine and the Loire. The Permanent Secretariat had worked a bit in this way. And finally, when they had made a map of this situation, there turned out to be a very serious problem! Who would have to direct the Agencies? Because there were three corps of engineers, each willing their people to direct each agency (Mining, Civil Engineering, and Agricultural Engineering), and eight could not be divided by three, one had decided to reduce the number of agencies to six!. In this way it was decided to reduce the number of agencies to six!*" ("*Une première hésitation a éte de savoir combien d'Agences. A l'orgine, on avait pensé qu'il en fallait non pas six, mais huit: il y avait bien sûr les grands fleuves, la Seine, la Garonne, la Loire, le Rhône, dont certains pensaient qu'il fallait le scinder en deux: prendre le Rhône jusqu'à Avignon et puis, pour la partie basse, prendre toute la côte méditerranéenne et en plus des deux Agences concernant le Nord (Artois-Picardie) et les fleuves partiellement francais que sont la Meuse et le Rhin, on avait pensé faire un tout qui aurait comporté les fleuves côtiers entre la Seine et la Loire. Le Sécrétariat permanent avait travaillé un peu dans cette voie. Et finalement. si l'on avait écarté cela, c'est qu'il avait un très grave problème! Qui allait prendre la direction des Agences? Comme il y avait trois corps techniques qui briguaient d'avoir des gens à eux dans chacune des Agences (les Mines, les Ponts et le Génie rural) et huit n'étaient pas divisible par trois, il a bien fallu arriver à un découpage par six! Le découpage en six s'est donc fait de cette facon!*")

4. Heady (1991, p. 193).

are very powerful.[5] The 95 departments do each have a prefect as the principle representative of the central government (*Préfet du Département*). In the decentralization process 22 regions have been interposed between the central government and the departments, each with a regional prefect (*Préfet de la Région*) for coordinating the field services of most but not all ministries.[6] For two reasons French administrative relations are less hierarchical than one may expect on the basis of the number of different administrative units. First, since the administrative reforms between 1982 and 1986 the relation between the *collectivités locales* (region, department, and municipality) is not hierarchical any more. Each of them only has a hierarchical relation with the state level.[7] Secondly, because of the so-called *cumul des mandats* many local governments are represented in the central government.[8] In the following, the present administrative organization is described.

On the central level, several ministries do have competencies related to water management, such as the Ministry of the Environment (*Ministère de l'Environnement*), the Ministry for Public Health (*Santé*) responsible for the control of drinking water, the Ministry of Public Works (*Ministère de l'Equipement*) responsible for navigation, and the Ministry of Industry (*Industrie*) responsible for underground waters and pollution control. The Ministry of the Environment is responsible for the coordination of these water policies. The *Direction de l'Eau* of this ministry is the secretary of the *Commission Interministérielle de l'Eau* and of the *Comité National de l'Eau*, the two organizations where national water policy is discussed.[9] The former consists of representatives of the sectoral ministries, whereas in the latter, water users, local governments and ministries are represented. Most ministries do have deconcentrated services in the regions and departments. The prefects, being the state representatives at the regional and departmental levels, take care for the local coordination of the various ministerial policies.[10]

With the exception of the activities of the regional services of the Ministry of the Environment, the DIRENs (*Directions régionales de l'environnement et de la nature*), the regions do not play an important role in French water management.[11] In each of the six hydrographic basins mentioned above, the prefect of the region in which the main office of the basin agency is situated, is the *Préfet coordinateur de bassin*. This prefect is responsible for the coordination of water policies that go beyond the boundaries of the regions. The Scheldt basin is situated in the regions *Nord-Pas de Calais* and *Picardie*.

The Departmental prefects are responsible for the implementation and coordination of national water policies in the 95 departments. Since the completion of the deconcentration rounds in France, the parliament of the department, the *Conseil Général* has the possibility to support water management initiatives financially.[12] The Scheldt basin is

---

5. Because of this huge number of mostly small municipalities France has more local administrative units than all other EU members states together (Toonen, 1990, p. 52).

6. Heady (1991, p. 193).

7. Salet (1994, p. 20).

8. Toonen (1990, p. 58).

9. Barraqué *et al.* (1997, p. 253).

10. Ibid., p. 201.

11. Ibid., p. 202.

12. Ibid.

situated in the departments *Nord, Pas de Calais* and *Aisne.*

The municipalities are responsible for the construction and exploitation of sewage systems and waste water treatment plants. In addition, they traditionally are in charge of delivering water services. These tasks often are delegated to privatized water industries.[13]

The basin level probably is the most important administrative level in French water management. In each of the six hydrographic basins there is a Basin Agency (*Agence de l'eau*) and a Basin Committee (*Comité de bassin*). The basin agencies work with a fund that is supplied with charges on the extraction and pollution of water. The fund is used to subsidize waste water treatment by industries and municipalities. The Basin Committee is a *de facto* "regional water parliament", consisting of representatives of users, associations and local authorities, who form the majority, and state representatives. It pronounces on the fixing of charges and on investment programmes. According to the Water Act of 1992 each of the six basin committees has to make an integrated water management plan[14], the SDAGE (*Schéma Directeur de l'Aménagement et de Gestion des Eaux*). On the subbasin level the SDAGE is elaborated in the SAGE-plans (*Schéma de l'Aménagement et de Gestion des Eaux*), which are developed by the *Comités locales de l'Eau*. These committees are composed of local authorities, user groups, and riparian owners.[15] The French part of the Scheldt basin is entirely situated in the Artois-Picardie basin.

Because of the decentralization of French administration, the *conseils généraux* and *conseils régionaux* received competencies to develop environmental departments and make their own water management plans. This has caused a competition between these *collectivités locales* and the basin committees, and raised the question of the legitimacy of the plans developed by the basin committees, which are not elected directly.[16]

Apart from the *comités de bassin* and the *comités locales de l'eau*, there is one more type of basin commission, the *comités de rivière*. These committees, in which both local governments and local interest groups are cooperating, prepare so-called *contrats de rivière*, which contain action programmes. In one respect the *comités locales de l'eau* are similar to the *comités de rivière*. Both committees are initiated by so-called bottom up processes, and therefore voluntary. There are some major differences as well. First, the *contrats de rivière* are valid for a period of five years, whereas there is no period of validity defined for the SAGE plans. Secondly, the SAGE plans have to be in conformity with the SDAGE plans, and therefore are, by definition integrated plans. A *contrat de rivière* might deal with only some aspects of water management, such as the ground water quality. Finally, the SAGE-plans are legally binding, whereas the *contrats* are not.[17]

Belgium

During the past decades, the successive state reforms have changed the Belgian institutional landscape considerably. Since the constitutional amendment of 1988 Belgium

---

13. Ibid.

14. Loi N° 92-3 du 3 janvier 1992 sur l'eau. Journal Officiel de la République francaise.

15. For a discussion of French basin planning, see for example Betlem (1997, pp. 389-396), Retkowsy (1995), or Wessel (1992b). For more information about the SDAGE of the Artois-Picardie basin, see Journet *et al.* (1996).

16. Barraqué *et al.* (1997, pp. 209 and 255).

17. Information on these differences was provided by J. -M. Journet.

is a full federal state.

Nowadays, Belgium consists of communities and regions, a relatively small national parliament, and directly elected representatives in the parliaments of the regions and communities. The three regions, the Flemish, Walloon, and Brussels Capital Regions, have received competencies for territory-related issues, such as land-use planning, environmental, water, and nature management. The regions do also have the competence to develop economic policies. The three communities, the Dutch-, French-, and German-speaking Communities, have received competencies for person-related issues, such as social welfare, family-, and immigration policies.[18] Since 1993, the regions and communities do also have competencies to conclude international conventions concerning the issues they are authorized to.[19] Because in the Flemish region, the government and parliament of the region and the Dutch-speaking community are unified, there exists only one Flemish parliament, government and one ministry, the Ministry of the Flemish Community. Like in France Belgian ministers, either on the federal, regional, or community level, do have personal cabinets with main advisory competencies. In the following paragraphs, the main actors on the federal state level, and the regional actors will be introduced. The Scheldt basin is situated in each Belgian region. For the international Scheldt issues discussed in this thesis, the Belgian communities are less relevant.

Federal state
Because of the constitutional amendments of 1980, 1988, and 1993 most water management competencies are attributed to the three regions. During the last stage of the Belgian state reforms the regions did even receive competencies to conclude international agreements. The federal state, however, still plays a role in some aspects of water policy. First, the federal state is responsible for nuclear policies.[20] Secondly, the federal state still plays a role in the conclusion of some international treaties. For those issues, where the federal state still has competencies, both the federal state and the region concerned have to approve an international agreement. As regards issues where only the regions have competencies, the federal state only has to be informed about the international agreement that the region intends to conclude.[21]

Although Belgium is a federal state, and the regions and communities do even have competencies to conclude international conventions, international organizations, such as the European Union (EU) and the United Nations Economic Commission for Europe (UN-ECE), do only recognize the state of Belgium as a member state. For the Belgian representation in these organizations the regions and/or communities have to agree on a joint point of view. To reach such agreements on environmental issues the federal state of Belgium and the regions have concluded a cooperation agreement (*Samenwerkingsakkoord*), which entailed the establishment of the coordination commission CCIM (*Coördinatiecommissie voor het Internationaal Milieubeleid*). In this commission, where regional representatives meet once every week, Belgian international environmental

---

18. A detailed description of the distribution of competencies in Belgium can be found in Brassinne (1994, pp. 14-25).

19. Veldkamp (1994, p. 3).

20. BIM (1995, p. 278).

21. Ibid.

policies are prepared.[22]

The Flemish Region
By far the biggest part of the Belgian part of the Scheldt basin is situated in the Flemish Region. One Department of the Ministry of the Flemish Community is especially of importance for this study, namely the Department of the Environment and Infrastructure (*Departement voor Leefmilieu en Infrastructuur*). This Department plays an important role in Flemish water management, since within the separate regions government is very centralized. For administrative purposes Flemish watercourses are divided in navigable and non-navigable watercourses, the latter being subdivided into three categories: from the sources (category three), via the middle parts (category two), to the mouth (category one) of a watercourse. National water and environmental policies are formulated in the MINA-plan (*Milieu- en Natuurbeleidsplan Vlaanderen*).

The Administration Environment, Nature-, Land-, and Water Management (*Administratie Milieu, Natuur-, Land- en Waterbeheer*, AMINAL) of the Department is responsible for strategic water management, the operational water quality management of the navigable water courses, and the operational water quality and quantity management of the non-navigable watercourses of the first category. The Water Division of this administration, among other things, is responsible for the coordination of the basin policy (*Stroombekkenbeleid*) in the Flemish region, which will be discussed below.

The Administration Water Infrastructure and Marine Affairs (*Administratie Waterwegen en Zeewezen*, AWZ) of the Department is responsible for the operational water quantity management of the navigable watercourses. In addition, it is responsible for the infrastructure and maintenance dredging works in the important navigation channels, and the maritime access to the port of Antwerp. The responsibilities of the former Antwerp Seaport Service (*Antwerpse Zeehavendienst*) are transferred to three divisions.[23] The Division Maritime Scheldt (*Afdeling Maritieme Schelde*), among other things, is responsible for the maritime access to the ports of Ghent, Antwerp and Brussels, the maintenance and infrastructure dredging works, and the salvage of wrecks. The Division Scheldt Maritime Affairs (*Afdeling Zeewezen Schelde*), among other things, is responsible for the navigation on the Western Scheldt and the Lower Sea Scheldt. Finally, the Division Sea Scheldt (*Afdeling Zeeschelde*), among other things, is responsible for the infrastructure works in the port of Antwerp.

Besides the departments of the Ministry of the Flemish Community, pararegional institutions (*pararegionalen*) play an important role in Flemish water management. These organizations are not part of the Ministry, but are directly accountable to the Minister. They do have a contract with a minister that contains a description of their tasks. The Flemish Environment Agency (*Vlaamse Milieu Maatschappij*, VMM) is one of the pararegional institutions, and is responsible for the water quality policy in the Flemish region. Its main task is the development of General Waste Water Treatment Programmes (AWPs[24]). The private organization AQUAFIN implements these programmes, and is responsible for the construction, improvement, and exploitation of waste water treatment

---

22. Baakman *et al.* (1994, p. 245), Interview K. De Brabander.

23. Letter from the head of the department Maritime Scheldt, Belmans, to the Secretary of the executive committee for the Western Scheldt, van Pelt, of 12 January 1995.

24. AWP =Algemeen Waterzuiveringsprogramma.

plants. These works are financed by the MINA-fund, which is supplied with revenues of taxes on waste water discharges.

The Belgian Provinces are responsible for the management of the non-navigable watercourses of the second category. The Scheldt basin is situated in the Flemish Provinces *Antwerpen*, *Limburg*, *Oost-Vlaanderen*, *Vlaams-Brabant*, and *West-Vlaanderen*.

The Municipalities are responsible for the construction and exploitation of sewage systems, and for the operational water quality management of the non-navigable watercourses of the third category.

Finally, the polderboards (*Polders/Wateringen*) are responsible for the water quantity management of the watercourses of the second and third category. The *Polders* take care of the maintenance of the dikes and the regulation of the inland water levels. Unlike polders, which indeed can only be found in the Flemish polders, *Wateringen* can be found throughout Flanders. They are responsible for the regulation and protection of suitable conditions for agriculture and hygiene.[25]

A relatively new phenomenon in Flemish water management are the basin committees (*Bekkencomités*). The Flemish region is divided into 10 hydrographic subbasins, of which eight are part of the Scheldt basin. In these committees the Flemish ministries, provincial and local government agencies, and NGOs jointly prepare basin reports (*Bekkenrapporten*) yearly, in which the situation of the subbasin, the main bottlenecks, and policy options are discussed. The basin committees do have an advisory function as regards the development and implementation of integrated water management in the Flemish region. The eight basin committees in the Scheldt basin are the committees for the Upper Scheldt, Leie, Dender, Dijle, Demer, Nete, Lower-Scheldt, and the Polder and Ghent Canal basin.[26]

The Walloon Region[27]

Only a relatively small part of the Scheldt basin is situated in the Walloon region. The Walloon region uses the same subdivision of watercourses as the Flemish region.[28] The Directorate-General for Environment and Natural Resources (*Direction Générale des Ressources Naturelles et de l'Environnement* (DGRNE) of the Ministry of the Walloon region (*Ministère de la Région wallone*) is responsible for the water quality management of all watercourses, and the water quantity management of the non-navigable watercourses of the first category. The Ministry of Equipment and Transport (*Ministère wallon de l'Equipement et des Transports*) is responsible for the water quantity management of the navigable watercourses. In 1995, the Environmental management plan for sustainable development (*Plan d'Environnement pour le Développement durable*) was approved by the Walloon government.[29] Among other things, this plan deals with water quality management in the Walloon region. Furthermore, the DGRNE produces a yearly report

---

25. For more information on the role of the Polders and Wateringen in Flemish water management, see Anonymous (1990, pp. 39-46).

26. For more information on the Flemish *Bekkencomités*, see De Wel (1994).

27. This section is partly based on an actualization of the ISG-report (ISG, 1994a and b), made by the Walloon government in 1996 (Letter from J.-M. Wauthier (DGRNE) to H. Smit (Ministry of Transport, Public Works, and Water Management, the Netherlands) of 13-5-1996.

28. See above.

29. Gouvernement wallon (1995).

describing the state of the art of the Walloon environment (*Etat de l'Environnement wallon*). The 1995 report focused on the water quality, and water quality management in the Walloon region.[30] The Walloon provinces are responsible for the water quantity management of the non-navigable watercourses of the second category, unless these watercourses fall under the jurisdiction of the waterboards. The Scheldt is situated in the Walloon Provinces *Hainaut, Brabant, Namur,* and *Liège*.

The Municipalities manage the sewage systems and are responsible for the water quantity management of the non-navigable watercourses of the third category. The Ministry of the Walloon region has created communal district organizations, the so-called *intercommunales*, to construct and exploit the waste water treatment infrastructure. The eight Walloon *intercommunales* depend financially on the Walloon region. The *intercommunales* IDEA (Haine, Canal du Centre, upstream part of Senne), IPALLE (Lys, Espierre, Escaut, and Dendre), and IBW (Senne, Dyle, Gete) are situated in the Scheldt basin.

Finally, also the Walloon region has several *wateringues*, which are responsible for the water quantity management of the non-navigable watercourses of the third category.

The Walloon government, oriented toward and being inspired by the French, also introduced the management concept of the *contrats de rivière*.[31] In the river contracts governmental and non-governmental actors cooperate on the development of an integrated plan for a subbasin. River contracts are voluntary, but have to be in accordance with a ministerial decree.[32]

The Brussels Capital Region
Since July 1989 the Brussels Capital Region has its own parliament, government and civil service. The Brussels Capital Region comprises 19 Municipalities. Unlike the Walloon and Flemish regions, which received many competencies after the constitutional amendments of 1980 and 1988, the Brussels Capital Region has the competence and responsibility to develop environmental and water policies only since 1988. This explains the backlog the Brussels Capital Region has compared with the other Belgian regions.[33] Since the first administrations were formed in 1990, several reorganizations have taken place in the Brussels Capital Region. At present (1997), the Brussels Capitial Region has two Ministries, one of them being responsible for the general management of the region, and one for the construction of public works.[34] The BUV (*Bestuur van Uitrusting en van Vervoerbeleid*) is part of the latter Ministry. Two water management departments of the BUV, A5 and B4 deal with groundwater management and waste water treatment, and work under the leadership of the Brussels Minister for the Environment.[35] The BUV also is responsible for the construction of waste water treatment plants in the Brussels region.

---

30. Lutgen (1996).

31. For more information on the organization of these river contracts, and the first experiences with these contracts, see Rosillon *et al.* (1996).

32. Moniteur Belge (Belgisch staatsblad)- 26.05.1993, Ministeriële omzendbrief betreffende de toelaatbaarheidsvoorwaarden en de uitwerkingsmodaliteiten van de riviercontracten in het Waalse gewest.

33. BIM (1995, p. 9).

34. Interview A. Thirion.

35. Before the reorganization, the Administrations B4 and A5 were part of the administration ARNE (*Administration des Ressources Naturelles et de l'Environnement*)

Next to the administrations, the Brussels Capital Region has an important pararegional institution, the *Brussels Instituut voor Milieubeheer* (BIM), which is responsible for the development of environmental policies, and, among other things, is responsible for the issuing of environmental permits.[36] The BIM is not part of the ministry, but depends directly on the Minister of the Environment.

Within the Brussels Capital Region the production and transport of drinking water is the responsibility of the *Brusselse Intercommunale Watermaatschappij* (BIWM). Most water is diverted from the river Meuse, and transported to the Brussels Region by a pipeline. The distribution of drinking water, which is the competence of the municipalities, presently is delegated to the *Brusselse Intercommunale voor Watervoorziening*.[37]

The Netherlands[38]

Compared with Belgium the water management organization in the Netherlands has been rather constant during the past decades. The Netherlands is a decentralized unitarian state with three administrative levels: The Central government, Provinces (*Provincies*), and on the local level the Municipalities (*Gemeenten*) and Water Boards (*Waterschappen*). For administrative purposes the Dutch water systems are divided into national and regional waters. The Western Scheldt, the Dutch parts of the Canal Ghent-Terneuzen, and the Scheldt-Rhine connection do all belong to the national waters.

On the national level the Directorate-General for Public Works and Water Management (*Rijkswaterstaat*) of the Ministry of Transport, Public Works, and Water Management (*Ministerie van Verkeer en Waterstaat*) is responsible for most water management tasks. The Directorate-General *Rijkswaterstaat* consists of a Central Directorate in the Hague, and several regional Directorates, and research institutes. For the Scheldt water management the Main Department Water of the Central Directorate, the regional Directorate Zeeland, and the research institute for Coastal and Marine Management (*Rijksinstituut voor Kust en Zee*) are most important. Besides the Ministry of Transport, Public Works, and Water Management, the Ministry of Housing, Physical Planning and the Environment (*Volkshuisvesting, Ruimtelijke Ordening, en Milieubeheer* (VROM) plays an important role in the development of national water policies, because it is responsible for the setting of water quality standards, drinking water, and sewerage. National water policies are formulated in the Third Policy Document on Water Management (*Derde Nota Waterhuishouding*)[39], which contains water quantity and quality policies for both surface waters and groundwaters. When this thesis is completed the draft version of the Fourth Policy Document on water Management will be issued. In addition to the formulation of strategic water policies, the central government is responsible for the operational management of the national waters, and the supervision of the management of the regional waters, and the groundwater systems.

The Dutch Provinces are the linking pin between on the one hand the national

---

36. Bimberichten (1994).

37. ICBS/CIPE (1997, p. 49).

38. More extensive overviews of the Dutch water management organization can be found in (Barraqué, 1995), Sironneau (1990), and Perdok (1995, 1997).

39. V&W (1989).

ministries and on the other hand the Municipalities and Water Boards. They have to translate national policies into regional policies, and have to communicate local experiences with national policies to the national level. The provinces are responsible for the management of the regional waters, and the groundwater systems. Their main tasks are the development of provincial water management plans, and the coordination of regional water policies with land-use and environmental policies. The Provinces have delegated the larger part of the implementation of the regional water policies to the Water Boards. The biggest part of the Scheldt basin on Dutch territory is situated in the Province of Zeeland, whilst a relative small part is situated in the Province of Northern-Brabant.

The two administrative unities on the local level, the Municipalities and Water Boards, do not have a hierarchical relation with each other, and are supervised by the Provincial and National government.

Whereas the central government, provinces and municipalities are so-called "general democracies", the water boards are "functional" governmental bodies, which do only have water management tasks. They cannot be compared with the *Polders* and *Wateringen/wateringues* in Flanders and Wallonia, because they are much more powerful. During the past decennia the scope of activities of the Dutch water boards has gradually extended from flood protection and water quantity management, to water quality management, and finally to ecological management. The increase in scale of the water boards, and the approval of the Water Board Act of 1991 (*Waterschapswet*), which gave the water boards almost the same status as the Provinces and Municipalities[40], strengthened the position of the Water Boards even more. The Water Boards develop policy plans which contain operational water policies. Water Boards situated in the Dutch part of the Scheldt basin are *Het Vrije van Sluis*, *De Drie Ambachten*, *Het Hulster Ambacht*, the *Zeeuwse Eilanden*, and the *Hoogheemraadschap West-Brabant*.

Dutch municipalities are responsible for the management of some harbours, canals, and all sewage systems. Municipalities and Water Boards along the Western Scheldt cooperate in the task group for the Western Scheldt (*Taakgroep Westerschelde*) of the Dutch Association of River Municipalities (*Vereniging Nederlandse Riviergemeenten*, VNR).

Since 1992 regional Departments of several Ministries, the Province of Zeeland, Water Boards and Municipalities along the Western Scheldt cooperate in a joint platform, the *Bestuurlijk Overleg Westerschelde*. In this platform they have jointly prepared the Policy Plan for the Western Scheldt (*Beleidsplan Westerschelde*), which contains a concrete Action plan. In 1996 the platform has discussed whether it would be useful to extend the platform with representatives of NGOs. The result of this discussion was that it would not be wise to increase the number of participants.

International institutional arrangements

Since 1839 Belgium and the Netherlands jointly address issues related to pilotage and marking in the so-called Permanent Commission for supervising navigation on the Scheldt (PC). Presently, this commission consists of two representatives of the Flemish government and two representatives of the Dutch government.

---

40. The Provincial Council does have the competence to create or abolish water boards.

Since 1948 Belgium and the Netherlands discuss issues concerning the improvement and maintenance of the navigation channel in the Sea Scheldt and Western Scheldt in the Technical Scheldt Commission (TSC). The TSC has various subcommissions, which in turn may consist of several working groups.

Since 1995 the Scheldt basin states and regions address Scheldt water quality issues in the provisionally installed International Commission for the Protection of the Scheldt against pollution (ICPS). The ICPS has two permanent and one temporary working group. The permanent working groups are the working group 1 for the water quality, and the working group 2 for emission policies. Working group 3 studied the coordination of the ICPS and other transboundary platforms, and presently is in charge of the coordination of the activities of the ICPS and other international organizations, such as the EU.

Regional transboundary issues between Belgium and the Netherlands are discussed in several transboundary basin committees. In these committees the water quantity and quality management of transboundary water courses and groundwater systems are discussed. The transboundary basin committee *Kreken en Polders* is entirely situated in the Scheldt basin, whereas the basin committee *Molenbeek/Kleine AA-Mark* is partly situated in the Scheldt basin.[41]

---

41. Heylen (1995, p. 73).

Appendix 7: The Rhine, Meuse and Scheldt basins

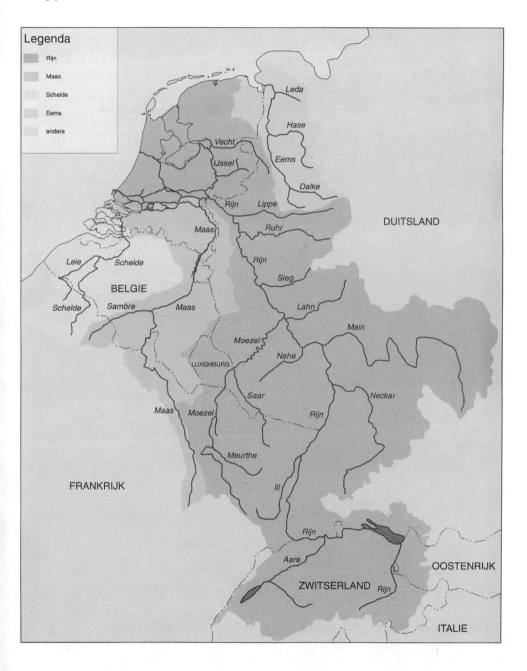

Source: Rijkswaterstaat, Meetkundige dienst. In: Ministerie van Verkeer en
Waterstaat (1997, p.11).

Appendix 8: The Scheldt basin

Source: ICBS/CIPE (1997, p. 11).

Appendix 9: The Scheldt estuary

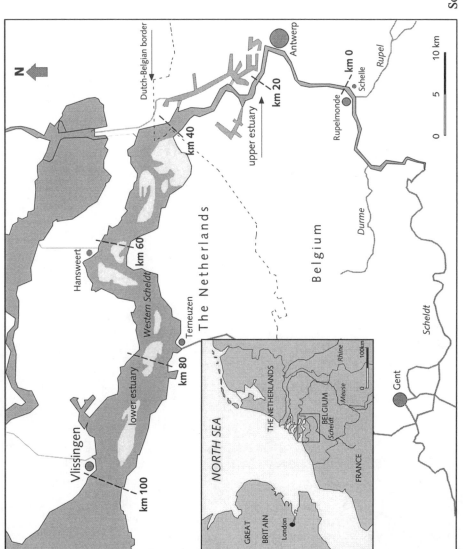

Source: Verlaan (1998).

Appendix 10: The alignment of the Baalhoek canal

Source: TSC (1984a, p. 28).

Appendix 11: The alignment of the Baalhoek and Bath canals

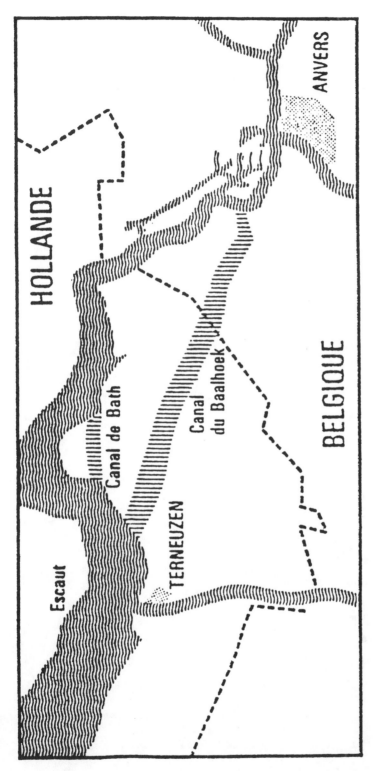

Source: Société belge de droit international et al. (1997).

Appendix 12: Series of decisions in decision making rounds I to XIIb

Round I (1967-1968)

9-1967   The Belgian government informs the Dutch government that it would like to start negotiations on two infrastructure projects: the construction of the Baalhoek Canal and the straightening of the bend near Bath in the Western Scheldt

1967   The Dutch government links the two infrastructure issues to the issues of the water quantity and quality of the Meuse, and the water quality of the Scheldt

1968   The Belgian and Dutch governments decide to install a negotiation commission, and to start negotiations on:
1. the construction of the Baalhoek Canal
2. the straightening of the bend near Bath
3. the water quality of the Scheldt
4. the water quality of the Meuse
5. the water quantity of the Meuse

Round II (1969-1975)

14-5-1969   The Belgian and Dutch governments start negotiations on the five issues

19-6-1975   The Belgian and Dutch delegations conclude the negotiations with three draft conventions dealing with the five issues:
1. a convention on the distribution and the quality of the water of the river Meuse
2. a convention on the construction of the Baalhoek canal
3. a convention aiming at an improvement of the navigation channel in the Scheldt near Bath, which contains regulations concerning the water quality of the Scheldt

Round III (1975-1983)

8-9-1975   The regional council of Wallonia complains that the region was not involved in the negotiations on the water conventions

16-9-1975   The regional council of Wallonia rejects the three draft conventions, because they would not be beneficial for the Walloon region. The region does not want to construct storage reservoirs on its territory, and to pay for a clean-up of the rivers Scheldt and Meuse

8-1977   Belgian government agencies issue a report with the proposal of the 48'/43'/38' deepening programme

1-1978   The Belgian government installs a commission with representatives of several ministerial cabinets to discuss a Belgian reaction on the draft conventions of 1975

1979   The commission stops its activities, because it is unable to reach a Flemish-Walloon consensus

20-11-1979   Belgian and Dutch Ministers decide to jointly carry out a study on the impacts of the 48'/43'/38' deepening programme on the water system of the Western Scheldt

7-7-1980   The Belgian government installs a working group with representatives of several cabinets that has to reach a Flemish-Walloon consensus on the water conventions

23-7-1981   The working group issues a study report

1981   The Walloon government does not accept the results of the study report, mainly because the report contains a proposal to ensure a minimum discharge of the river Meuse

1982   The Belgian government installs an *ad hoc* commission of ministers including representatives of the Belgian regions Wallonia and Flanders to discuss the study report issued in 1981

Round IV (1983-1985)

20-10-1983 Belgium and the Netherlands install the Technical Meuse Commission

28-10-1983 The Dutch government informs the Belgian government that a deepening of the navigation channel in the Western Scheldt requires a convention to be approved by both the Dutch and Belgian parliaments, and links the approval of this convention to the approval of the other water conventions

15-6-1984 Belgian and Dutch experts finish the study on the 48'/43'/38' deepening programme, and conclude that implementation of the deepening programme is acceptable

4-7-1984 Decision of the Belgian Ministerial Committee for the Water Conventions to propose the Netherlands:
1. to restart negotiations on the water conventions
2. to place the issue of the deepening of the navigation channel in the Western Scheldt on the negotiation agenda
3. to skip the issue of the straightening of the bend of Bath from the negotiation agenda

12-7-1984 The Dutch government proposes to start Belgian-Dutch negotiations on a convention concerning the protection of the banks along the Western Scheldt against erosion

13-7-1984 The Belgian Minister of Foreign Affairs informs the Dutch government on the decision of the Belgian Ministerial Committee of 4-7-1984

18-7-1984 The Dutch government confirms the linkage it had made on 28-10-1983, in particular the linkage with the issue of the distribution of the water of the Meuse
Belgian-Dutch decision to install a Scheldt Water Commission (SWC)

7-6-1985 The Dutch government informs the Belgian government that Belgium needs a WVO permit for dumping dredged material in the Western Scheldt

13-6-1985 First meeting of the SWC. Discussion on the Tessenderlo pipeline

7-10-1985 Declaration of intent of the Belgian and Dutch Ministers of Foreign Affairs. They express their willingness to solve the following issues as soon as possible in a way that is acceptable for both parties:
1. the distribution of the water of the Meuse during low flows
2. the water quality of the river Meuse
3. the 48'/43'/38' deepening programme
4. the construction of the Baalhoek canal
The ministers reach an agreement on a linkage between the two infrastructure issues and the two Meuse issues

19-11-1985 The Belgian government applies for two WVO permits for dumping dredged material in the Western Scheldt: a first one for dumping dredged material from dredging works on the bar near Zandvliet, and a second one for dumping dredged material from dredging works on the bars in the Western Scheldt

Round V (1985-1987)

30-10-1985 The Dutch advisory council, the 'Raad van de Waterstaat', formulates an advice concerning the implementation of the deepening programme. The main conclusion is that implementation of the deepening programme is acceptable

5-11-1985 The Dutch Embassy issues a report expressing its concerns about the Belgian plan to issue a permit for the discharge of waste water in the Western Scheldt via the pipeline Tessenderlo-Antwerp

29-1-1986 Dutch local, regional and national governments start the preparation of an integrated policy plan for the Western Scheldt

1986 A Belgian working group with representatives of national and regional governments rejects the declaration, mainly because the Walloon representatives do not want to construct one or two storage reservoirs in Wallonia

3-12-1986 The Dutch Minister of Foreign Affairs gives in to the Belgian (Walloon) objections, and proposes to start negotiations on the basis of the declaration of intent of 1985, except for the passages referring to the construction of storage reservoirs on Walloon territory

12-12-1986 The Belgian Council of Ministers approves the declaration of intent of 1985, except for the passages referring to the construction of storage reservoirs

26-1-1987 The Belgian government informs the Dutch government that it would like to restart negotiations on the water conventions on the basis of the modified declaration of intent

1987 The Belgian government withdraws its delegation from the SWC, because it does not want to continue formal deliberations concerning the water quality of the river Scheldt during the negotiations on the 'water conventions' in the commission Biesheuvel-Davignon

Round VI (1987-1989)

3-3-1987 The Walloon Prime Minister, Wathelet, raises objections to the composition of the Belgian delegation, since no representatives of the regions are included

1987 The Belgian ministerial commission for Foreign Affairs installs a ministerial Commission *ad hoc* for the Relations with the Netherlands, with representatives of the Flemish, Walloon, and Brussels Capital Regions

18-5-1988 The Dutch government issues two WVO permits for dumping dredged material in the Western Scheldt to the Belgian state. The permits contain the condition that the water quality of the Scheldt has to be improved

8-9-1988 The Belgian government informs the Dutch government that in spite of the WVO Act the freedom of navigation on the Western Scheldt should not be threatened

11-1988 The Dutch Province of Zeeland sends a motion to the European Commission in which it demands the commission to mediate in the negotiations on the water conventions

10-1989 The Belgian national government, and the governments of the regions conclude a cooperation agreement, and decide to restart international negotiations with the Netherlands, and to add representatives of the Walloon, Flemish, and Brussels Capital Regions to the Belgian delegation. The chairman of the Belgian delegation, Davignon, is succeeded by the former Secretary-General of the Belgian Ministry of Public Works, Poppe

Round VII (1989-1991)

13-10-1989 Belgian-Dutch negotiations are restarted

12-2-1990 The Belgian government applies for an extension of the permit for dumping dredged material from the bar near Zandvliet

1990 The Dutch government denies the Belgian request for a new WVO-permit for dumping dredged material from the bar near Zandvliet

1-12-1990 The Interparliamentary Council of the Benelux (IPR) approves a resolution, which states, among other things, that the negotiations on the water conventions have to be concluded in the near future

19-12-1990 Decision of the Belgian Ministerial Commission *ad hoc* that, among other things, the water quality objectives of EC-Directive 78/659, the water quality for cyprinids, will be applied to Scheldt and Meuse, and that these objectives only apply to the main course of the rivers

8-1-1991 The Dutch government does not accept the Belgian proposal, because Belgium seems to be unwilling to apply other EC-Directives, and the agreements made at the North Sea Ministerial Conferences to the Scheldt and Meuse. The Dutch delegation suspends the negotiations, and decides to prepare a draft convention itself

30-1-1991   The Flemish region applies for a new permit for dumping dredged material from the bars in the Western Scheldt

Round VIII (1991-1992)

15-4-1991   The Dutch delegation presents its unilaterally prepared draft convention to the Belgian delegation. This draft convention is an integrated convention dealing with all Scheldt and Meuse issues at stake

1991   The Benelux parliaments formulates a resolution demanding the conclusion of the water conventions

6-6-1991   Meeting of the plenary negotiation commission in Brussels, where the Belgian delegation is not able to give a formal reaction on the Dutch draft convention, because Wallonia and Flanders have not reached agreement on a joint standpoint

21-10-1991   The Belgian delegation presents three draft conventions, which are not yet approved by the Belgian government, because no new national and regional governments are installed yet. The three conventions successively deal with the protection of the Meuse and the Scheldt against pollution, the discharge of the river Meuse, and the management of the navigation channel in the Western Scheldt

19-11-1991   The Dutch delegation formulates a reaction on the draft conventions presented by the Belgian delegation, and state that the Belgian proposal cannot be accepted, mainly because the proposal seems to be entirely based on the decision of the Belgian ministerial Commission ad hoc of 19-12-1990

28-11-1991   The Dutch grant another permit for dumping dredged material from the dredging works on the bars in the Western Scheldt. This permit contains several conditions, among which the condition to extract large amounts of polluted sediments from the Lower Sea Scheldt

12-1991   The Dutch Rijkswaterstaat issues the Pilot Study OostWest, which indicates, among other things, that implementation of the 48'/43'/38' deepening programme will have negative consequences on the ecology of the Western Scheldt. This report also introduces the policy alternative 'ontpolderen'

13-3-1992   The Benelux parliament formulates a resolution demanding, among other things, the implementation of the deepening programme, the installation of an International Scheldt Commission, and the preparation of a Scheldt Action Programme

Round IX (1992-1993)

13-3-1992   The Dutch minister of Transport, Public Works and Water Management meets her Flemish and Walloon colleagues. The ministers express their willingness to restart negotiations, and to aim at an 'umbrella convention' dealing with all Scheldt and Meuse issues simultaneously

18-3-1992   The member states of the UN-ECE conclude the Convention on the Protection and Use of Transboundary Watercourses and International Lakes making obligatory international cooperation on a basin-scale

1992   The Dutch delegation contacts the other Scheldt and Meuse basin states

9-4-1992   The European Parliament formulates a resolution demanding the basin states to comply with all EC guidelines, and the European Commission to install an International Meuse Commission

29-4-1992   Water quality management administrations of the Scheldt basin states establish the ISG, an informal study group for the water quality of the Scheldt

18-6-1992   Decision of Dutch and Flemish Ministers that the construction of a Baalhoek canal requires a transboundary EIA-study. Decision of the Dutch Minister to postpone this study until an agreement is reached on the other water conventions

6-1992   The Belgian and Dutch delegations agree that negotiations on the water conventions should aim at three separate conventions: (1) a multilateral convention on the water quality of Scheldt and Meuse, and Belgian-Dutch conventions on (2) the distribution of the water of the Meuse, and (3) the deepening of the navigation channel in the Western Scheldt

13-7-1992 Decision of Dutch and Belgian ministers to restart negotiations on the water conventions

1-9-1992   Dutch Minister of Transport, Public Works, and Water Management declares that no Environmental Impact Assessment (EIA) is necessary for the implementation of the deepening programme

22-10-1992 Dutch national, regional, and local government agencies along the Western Scheldt sign a covenant approving the Policy plan for the Western Scheldt

10-12-1992 Presentation of Pilot Study OostWest in the TSC. Start of Belgian-Dutch research on morphological and ecological impacts of dredging works in the Scheldt estuary

6-1-1993   Belgium and the Netherlands conclude the so-called 'small water conventions': (1) an Agreement for the regulation of navigation and recreation on the border Meuse, (2) a Convention concerning a correction of the Belgian-Dutch border in the Canal Ghent-Terneuzen, and (3) a Convention on the protection of the banks along the Western Scheldt

6-1993   Dutch decision to unlink the issues of the deepening of the Western Scheldt and the water quality of Scheldt and Meuse. The Commission Biesheuvel-Poppe is dissolved

1993   Negotiations on the water conventions are split up into multilateral negotiations between the Scheldt and Meuse riparian states, and bilateral Flemish-Dutch negotiations. The multilateral negotiations aim at the conclusion of conventions on the water quality of Scheldt and Meuse. Parallel to the multilateral negotiations Flanders and the Netherlands start negotiations on the following bilateral issues:

- the deepening of the navigation channel in the Western Scheldt according to the 48'/43'/38' deepening programme;
- the flow of the river Meuse

In addition, they discuss the issues of:

- the maritime access to the Waasland harbour;
- possible extra deepening programmes in the Western Scheldt;
- the bilateral aspects of the water quality of Scheldt and Meuse

Flanders and the Netherlands agree on a linkage between the 48'/43'/38' deepening programme and the flow of the river Meuse

Round Xa (1993-1994)

5-1993   Flemish-Dutch agreement on a linkage between the 48'/43'/38' deepening programme and the flow of the river Meuse

13-5-1993 Start of bilateral Flemish-Dutch deliberations concerning the deepening of the navigation channel in the Western Scheldt and the flow of the river Meuse

16-6-1993 The Belgian and Flemish Prime Ministers agree on parallel negotiations on the deepening of the navigation channel in the Western Scheldt and the alignment of the HSL Antwerp-Rotterdam

5-11-1993 Start of formal negotiations on the deepening of the Western Scheldt and the flow of the river Meuse

1993/1994 Municipalities along the Western Scheldt prepare decisions based on the Dutch Act on Land-Use Planning

3-1994   The Dutch government refuses to sign the convention on the deepening programme until Flanders and the Netherlands would have reached an agreement on the alignment of the HSL

Round Xb (1993-1994)

9-6-1993 First round of informal deliberations between all Scheldt and Meuse basin states and regions on the multilateral Scheldt and Meuse issues

12-6-1993 The Benelux parliament formulates a recommendation demanding, among other things, a rapid implementation of the deepening programme, a rapid start of formal multilateral negotiations on the installation of an International Commission for the Protection of the Scheldt and the formulation of a Scheldt Action Programme, and a joint solution of the financial problems related to the clean-up of the river Scheldt

20-9-1993 Second round of informal deliberations between all Scheldt and Meuse basin states and regions on multilateral conventions for the Scheldt and the Meuse. Decision to install a formal negotiation commission

12-1993 Start of formal negotiations between all Scheldt and Meuse basin states and regions on multilateral conventions for the Protection of the Scheldt and the Meuse

3-1994 The delegations reach an agreement on the texts for multilateral conventions for the Scheldt and the Meuse

21-4-1994 Flanders re-establishes the linkage between bilateral and multilateral issues, and refuses to sign any other convention on Scheldt and Meuse issues as long as the Dutch do not sign the Convention on the deepening of the navigation channel in the Western Scheldt

26-4-1994 France, the Walloon and Brussels Capital Regions, and the Netherlands sign the multilateral Conventions on the protection of the Meuse and the Scheldt

Round XI (1994-1995)

29-4-1994 The Flemish government applies for another WVO permit for dumping dredged material in the Western Scheldt with a duration of six years

1-7-1994 Flanders and the Netherlands reach an agreement on the final text of the Convention on the deepening of the navigation channel in the Western Scheldt

25-10-1994 Flanders and the Netherlands reach an agreement on the procedures that will be followed in decision making on the alignment of the HSL

4-11-1994 The Dutch Minister of Transport, Public Works, and Water Management issues a new WVO Permit for the Flemish region , which is valid until 31 December 2000

1-12-1994 Delegations of the Flemish and Dutch governments declare that they will sign the:
● Convention on the deepening of the navigation channel in the Western Scheldt
● Convention on the flow of the river Meuse
● Convention on the revision of the Scheldt regulations
The Flemish region will sign the:
● Convention on the protection of the Scheldt
● Convention on the protection of the Meuse
The Flemish and Dutch parties reach an agreement on a detailed planning of the joint decision making on the alignment of the HSL

11-1-1995 Flanders and the Netherlands sign the Convention on the revision of the Scheldt regulations

17-1-1995 Flanders and the Netherlands sign the Conventions on the deepening of the navigation channel in the Western Scheldt and the flow of the river Meuse, and Flanders signs the multilateral Conventions on the protection of the Scheldt and the Meuse

Round XIIa (1995-1997)

11-5-1995 Provisional installation of the ICPS in Antwerp. Installation of three working groups

5-12-1995 Second plenary meeting of the ICPS. Approval of mandates of the working groups

4-3-1996 Meeting between Walloon and Dutch delegations in the ICPS. Agreement on interpretation of the Scheldt and Meuse conventions

16-4-1996 Third plenary meeting of the ICPS. Decision on list of parameters
2-10-1996 Dutch and Walloon ministers responsible for water management decide to organize regular meetings at high administrative level
6-1997 ICPS issues its first official report describing the water quality of the river Scheldt in 1994
20-11-1997 France ratifies the Scheldt and Meuse conventions at last

Round XIIb (1995-1997)

Implementation of the 48'/43'/38' deepening programme

Round 1 (1/1995-11/1995)
17-1-1995 Flanders and the Netherlands sign the Convention on the deepening of the navigation channel in the Western Scheldt
1-1995 The Dutch national government starts deliberations with Dutch municipalities on the issuing of permits based on the Act on Land-use planning
15-11-1995 The Ministry and environmental NGOs conclude a covenant in which the NGOs state that they will no longer appeal to court as regards the issuing of the permits needed for the implementation of the deepening programme

Round 2 (11/1995-6/1996)
11-1995 Members of the ZMF decide to continue the appeal to court concerning the issuing of a WVO Permit
19-1-1996 The EC declares that according to EC Law an EIA of the deepening programme is not obligatory
14-6-1996 The Dutch Council of State cancels the WVO Permit for the Flemish region, which implies that both maintenance and infrastructure dredging works are no longer possible, and that an EIA on the deepening programme is obligatory

Round 3 (6/1996-6/1997)
12-7-1996 The Dutch Minister of Transport, Public Works, and Water Management decides that:
 • maintenance dredging works in the Western Scheldt will be tolerated;
 • special legislation will be drafted for the implementation of the deepening programme and the related maintenance dredging works
 • all regular procedures will be continued, for the case that the special legislation would not be accepted by the Dutch Parliament
17-7-1996 The Flemish Administration Water Infrastructure and Marine Affairs and Flemish environmental NGOs conclude a covenant in which the NGOs state that they will no longer oppose the issuing of permits needed for the implementation of the deepening programme
23-8-1996 The Dutch Cabinet approves the special legislation for the implementation of the deepening programme
17-6-1997 The First Chamber of the Dutch Parliament approves the special legislation
19-6-1997 The Permit Act for the Western Scheldt enters into force

Decision making on nature restoration

Round A (1/1995-1/1996)
17-1-1995 Flanders and the Netherlands sign the Convention on the deepening of the navigation channel in the Western Scheldt

3-1995   The Dutch Minister of Transport, Public Works, and Water Management demands the BOW to prepare a plan for nature restoration

27-3-1995 The BOW decides to organize an 'open planning process', and installs a working group that has to study alternatives of nature restoration

8-5-1995   The working groups starts and asks the RIKZ and the consultancy Heidemij to prepare a study report on possibilities to restore nature in the Scheldt estuary

1-1996   The working group issues a study report with alternative projects for nature restoration:

- Projects of 'ontpolderen'
- nature restoration projects situated on the landside of the dikes along the estuary
- nature restoration projects situated outside the dikes along the estuary

Round B (1/1996-5/1996)

2/3 1996   The BOW organizes participation meetings to discuss the alternatives presented by the working group

3/4 1996   The provincial and various local councils formulate their advices

8-5-1996   The BOW concludes that there is no regional support for a plan for nature restoration, and advises the Dutch Minister to carry out additional research on possibilities of nature restoration

Round C (6/1996-8/1997)

13-9-1996 The Minister installs a commission of wise persons that has to formulate an advice on possibilities to comply with the provisions concerning nature restoration in the Convention on the deepening of the navigation channel in the Western Scheldt

29-8-1997 The Commission Western Scheldt issues its advise for the Minister. According to the commission, in the short run decision making should focus on the compensation of nature losses caused by the deepening programme, and should not take into account the policy alternative 'ontpolderen'. In addition, a long term vision for the Scheldt estuary would have to be developed, which might, on strict conditions, include projects of 'ontpolderen'

Appendix 13: Commissions and working groups where international Scheldt issues were discussed

Timeline chart (years 70–97) showing the active periods of the following commissions and working groups:

- Negotiation commission on the water conventions
- Biesheuvel-Davignon
- Biesheuvel-Poppe
- Multilateral commission
- Scheldt water comm.
- Bilateral commission
- ICPS
- TSC
- TSC-subcomm. Western Scheldt
- Working group WVO
- OostWest
- ISG

SUMMARY

CONFLICT AND COOPERATION ON THE SCHELDT RIVER BASIN[1]

Introduction and problem statement

Conscious management of international river basins is of vital importance for the quality of the environment and economic welfare in the basin states. Many problems in international river basins are so-called upstream-downstream problems. Downstream basin states may depend on upstream basin states as regards the discharge of a sufficient amount of water of a reasonable quality. As regards navigation or fish migration, however, the upstream basin states may be dependent on the downstream basin states.

The objectives of this thesis are to acquire knowledge of decision making on international river issues, and to contribute to the solution of upstream-downstream problems in international river basins. The threefold problem statement of the research is:

- How does decision making on international river issues develop?
- How can this development be explained?
- Which strategies can be used to contribute to the solution of upstream-downstream problems in international river basins?

Decision making on international river issues

The theoretical part of the thesis consists of four parts. First, the network perspective on decision making is treated. Secondly, the characteristics of decision making on *international* issues are discussed. Thirdly, the characteristics of decision making on international *river* issues are addressed. Fourthly, a conceptual framework is presented that can be used for the analysis and explanation of decision making on international river issues.

In the network perspective decision making is conceived of as a strategic interaction process. The actors involved in a decision making process do have different perceptions of the issues at stake. Because of the distribution of the (control of) resources (problem solving capacity) among them, the actors mostly depend on other actors as regards the achievement of their objectives. Because of these (inter) dependencies the actors start

---

[1]Summary of: Meijerink, S.V. (1999); Conflict and cooperation on the Scheldt river basin, A case study of decision making on international Scheldt issues between 1967 and 1997. Dordrecht.

interacting with each other. These strategic interactions may be conceived of as a decision making game. If actors help each other achieving their objectives, the game may be characterized by cooperation. If they hinder each other achieving their objectives, the game may be characterized by conflict. In most interactor relations elements of conflict and cooperation are present simultaneously. During the decision making games played institutionalization processes develop. First, the actors may reach an agreement on joint policies, i.e. on joint policy objectives or joint policy programmes. Secondly, they may reach an agreement on institutional arrangements to regulate their cooperation or to prevent conflicts. Finally, the cultural characteristics of their relationship may change. For example, during the interactions a culture of trust or distrust may develop. Networks are defined as (changing) patterns of social relationships between interdependent actors which take shape around policy issues and/or policy programmes, and that are formed, reproduced and changed by series of games in which actors try to influence policy processes as much as possible by strategic behaviour. Whilst on the one hand during the games played structural and cultural network characteristics are influenced, on the other hand these network characteristics influence the perceptions of (inter)dependence of the actors, and consequently the games played between them. Structural and cultural network characteristics, in turn, should be related to the broader context of the network. In this research five interrelated dimensions of the context of decision making on international river issues are distinguished:

- the underlying hydrological structure of the issues at stake;
- the international context of the decision making process;
- the cultural context of the decision making process;
- the *intra*national developments within each of the basin states;
- the history of conflict and cooperation between the basin states.

In addition to institutionalization processes, during the interactions strategic and cognitive learning processes may develop. Strategic learning concerns learning on (inter) dependencies, learning on problems perceptions of other actors involved in the decision making game, and the discovery of score possibilities and possible win-win situations. Cognitive learning is learning on the intellectual relationships in an issue-area, i.e. on cause-effect relationships, policy alternatives, impacts of policy alternatives, or on relationships between issues.

Although there are no fundamental differences between *intra*national and international decision making processes, decision making on *international* issues does have three important characteristics, which should be taken into account in a study of these processes. First, international decision making mostly is characterized by the absence of central authority. Therefore, the dominant decision making rule is unanimity. In some cases the competencies of the EU are an exception to this general rule. Secondly, decision making on international issues may be conceived of as a multilevel game. The international decision making game is related to various *intra*national decision making games. Thirdly, there often are considerable structural and cultural differences between the states involved in an international decision making process.

Decision making on international *river* issues does have two additional characteristics. First, many international river issues are upstream-downstream issues. Decision making on upstream-downstream issues is characterized by an asymmetric distribution of interests and resources among the actors. Because of the asymmetries within single issue-areas, issue-linkage may be an important explanation for the development of international institutional arrangements and policies for solving upstream-downstream problems.

Secondly, river issues are natural resources issues. Consequently, the attitude toward the environment and cognitive learning do play a relatively important role in decision making on these issues, and natural or environmental disasters may have an important impact on the course of the decision making processes.

The conceptual framework for the analysis and explanation of decision making on international river issues focuses on the relationship between the (See Figure 0.1):

- perceptions and strategies of the actors involved in the decision making process (Relationship 1);
- strategic interactions and institutionalization processes (Relationship 2);
- strategic interaction and learning processes (Relationship 3);
- structural and cultural characteristics of the context and network characteristics, and between these network characteristics and the perceptions of the actors involved in the decision making process (Relationships 4 and 5).

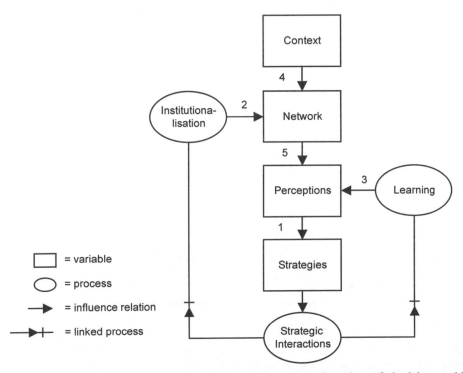

Figure 0.1 *Conceptual framework for the analysis and explanation of decision making on international river issues* The perceptions of the actors involved in decision making on an international river issue influence their strategic behaviour (Relationship 1). In the games that are played the structural and cultural network characteristics may change (Relationship 2). During the strategic interactions strategic and cognitive learning processes may develop as well (Relationship 3). Finally, the structural and cultural characteristics of the context influence the structural and cultural characteristics of the network (Relationship 4), which in turn influence the perceptions of the actors (Relationship 5).

Case study design

The empirical part of the research contains a single case study covering a long time span. It comprises a reconstruction, analysis and explanation of decision making on international Scheldt issues between 1967 and 1997, and covers the negotiations on the Belgian-Dutch water conventions, notorious for their length and deadlocks. For the description and analysis of this decision making process the rounds model of decision making is used. First, the decisions taken by the various actors involved in the decision making process are inventoried. Secondly, crucial decisions marking the beginning and/or end of decision making games or rounds are distinguished. Crucial decisions are decisions taken by one or more (representatives of the) governments of the Scheldt basin states and regions, which had an important impact on the composition of the international arena and/or the issues placed on the international agenda.

On the basis of the theory of decision making on international river issues the problem statement is elaborated, and seven research questions are formulated:

1. *How did decision making on international Scheldt issues develop?*
2. *What were the characteristics of the context of decision making on international Scheldt issues?*
3. *How did the perceptions of the actors involved in decision making on international Scheldt issues influence their strategies?*
4. *How did the strategic interactions between the actors involved in decision making on international Scheldt issues influence structural and cultural network characteristics?*
5. *Did learning processes develop?*
6. *How did the context of decision making on international Scheldt issues influence the perceptions and strategies of the actors involved in decision making on these issues?*
7. *Which strategies can be used to contribute to the solution of upstream-downstream problems in international river basins?*

For the reconstruction of the decision making process two main methods of data collection were used: the analysis of written material, such as files of international negotiations, minutes of meetings of international commissions, and newspaper articles, and a series of 40 interviews with water management experts in France, the Walloon, Flemish, and Brussels Capital regions, and the Netherlands.

Decision making on international Scheldt issues between 1967 and 1997: Fourteen decision making rounds

The Scheldt catchment area stretches out over France, Belgium, and the Netherlands. Because in the basin state Belgium water management competencies are distributed among the federal state and the Walloon, Brussels Capital, and Flemish regions, there actually are six main parties involved in decision making on international Scheldt issues. On the basis of the intellectual coherence between the issues two main international issue-areas are distinguished: the issue-area of the maritime access to the port of Antwerp, and the issue-area of water and sediment pollution. In total, fourteen decision making rounds can be distinguished. Two times two decision making rounds took place simultaneously. Below, a concise description of the decision making rounds is presented.

Round I (1967-1968) begins with the Belgian request to start negotiations on the construction of a canal connecting the Antwerp harbours situated on the left bank of the river Scheldt to the Western Scheldt, the Baalhoek canal, and the straightening of a sharp bend in the Western Scheldt, the bend near Bath. After the Dutch have decided to link these issues to three issues in which they perceive a main interest, namely the water quality and flow of the river Meuse, and the water quality of the Scheldt, it is decided to start bilateral negotiations on these five issues simultaneously in 1968.

Round II (1969-1975) starts with the installation of a Belgian-Dutch negotiation commission, and concludes with an agreement between the two negotiation delegations on three draft conventions dealing with the five issues introduced above, the so-called 'water conventions'.

Round III (1975-1983) addresses the intranational Belgian disagreement on the three draft conventions. According to Walloon politicians the proposed Belgian-Dutch package deal would be beneficial for both the Netherlands and the Flemish part of Belgium, but would not be beneficial at all for the Walloon region. On the contrary, this region would have to construct storage reservoirs on its territory to guarantee a minimum flow of the river Meuse, and would have to increase investments in waste water treatment infrastructure to clean-up the rivers Meuse and Scheldt.

Round IV (1983-1985) starts with the introduction of a new issue, namely the deepening of the navigation channel in the Western Scheldt, the so-called 48'/43'/38' deepening programme. The Belgian government would like to implement this programme to further improve the maritime access to the port of Antwerp. The Dutch government links the negotiations on the deepening programme to the negotiations on the water quality and flow of the river Meuse. This round is concluded with a declaration of intent signed by the Belgian and Dutch Ministers of Foreign Affairs, which contains statements that joint solutions will be searched for the water quality and flow of the river Meuse, the 48'/43'/38' deepening programme, and the construction of the Baalhoek canal. The issue of the straightening of the bend near Bath is skipped from the international agenda.

Round V (1985-1987) concerns the Belgian-Dutch discussions on a modification of the declaration of intent signed in 1985. Because Belgium, in particular the Walloon region, opposes passages in the declaration dealing with the construction of storage reservoirs, the Dutch do make a concession, and Belgium and the Netherlands reach an agreement to start negotiations on the basis of a modified declaration of intent.

Round VI (1987-1989) begins with the installation of the Belgian-Dutch negotiation commission Biesheuvel-Davignon. Shortly after this commission has been installed, the Walloon region opposes the composition of the negotiation commission, because of the absence of direct representatives of the Belgian regions in the commission. The decision making round is concluded with the decision of the Belgian government to admit representatives of the regions to the Belgian delegation, and to replace the chairman of the commission.

Round VII (1989-1991) comprises the negotiations in the Commission Biesheuvel-Poppe. In this commission the joint formulation of water quality policies for the Scheldt and Meuse is a major bottleneck. Because the Belgian delegation does not want to formulate water quality policies that go beyond the dated EC Directive 78/659 for carps, and with that proposes water quality objectives that are less ambitious than the objectives formulated at the North Sea Ministerial Conferences, the Dutch delegation decides to suspend the negotiations, and to draft a convention unilaterally.

Round VIII (1991-1992) concerns the discussions on the Dutch draft convention and the succeeding Belgian draft conventions. Like in the preceding decision making rounds, in particular the Walloon region strongly opposes the Dutch proposals. The region, among other things, is afraid of losing part of its sovereignty over the river Meuse, and would like to involve the upstream basin state France in the negotiations on the water quality of the Scheldt and the Meuse.

Round IX (1992-1993) starts with an agreement between the Belgian regions and the Netherlands to restart the negotiations on the 'water conventions'. In this decision making round two important developments take place. First, the UN-ECE Convention on the protection and use of transboundary watercourses and international lakes emphasizing the need for cooperation between all basin states of an international river is concluded. Secondly, because of the ongoing federalization process, it is decided that the Belgian regions will receive treaty-making competencies. Among other things, these two developments cause the Dutch government to unlink the bilateral and multilateral Scheldt issues. Shortly after the Dutch government has taken this decision, the Flemish region and the Netherlands reach an agreement to start bilateral negotiations on the deepening of the navigation channel in the Western Scheldt and the flow of the river Meuse. In addition, all Scheldt and Meuse basin states and regions agree to start multilateral deliberations concerning the management of the rivers Scheldt and Meuse.

Round Xa (1993-1994) starts with the installation of a Dutch-Flemish negotiation commission for the bilateral Scheldt and Meuse issues. In this decision making round Flemish and Dutch Ministers agree to link the negotiations on the 48'/43'/38' deepening programme to negotiations on the alignment of a high speed train from Antwerp to Amsterdam (the HSL). The Flemish and Dutch delegations are able to reach an agreement on draft conventions concerning the flow of the river Meuse and the deepening of the navigation channel relative easily. However, because negotiations on the alignment of the HSL appear to be rather laborious, the Dutch government refuses to sign the Convention on the deepening of the navigation channel in the Western Scheldt as long as no agreement is reached on the alignment of the HSL.

Round Xb (1993-1994) concerns the informal deliberations between all Scheldt and Meuse basin states and regions concerning the management of the rivers Scheldt and Meuse, and the following formal multilateral negotiations. After the delegations have reached an agreement on multilateral Scheldt and Meuse water quality conventions, on 26 April 1994 the governments of France, the Brussels Capital region, the Walloon region, and the Netherlands sign these conventions in Charleville-Mézières (France). The Flemish government re-establishes the linkages between the bilateral and multilateral Scheldt and Meuse issues, and refuses to sign the multilateral Scheldt and Meuse conventions as long as the Dutch government does not sign the Convention on the deepening of the navigation channel in the Western Scheldt.

Round XI (1994-1995) starts with Flemish-Dutch disagreement on the alignment of the HSL. At a meeting between the Dutch and Flemish Prime Ministers, they are able to reach an agreement in the procedures to be followed in decision making on the alignment of the HSL. From that moment the Dutch government is willing to sign the Convention on the deepening of the navigation channel in the Western Scheldt and, after more than 25 years of intermittent negotiations, Flanders and the Netherlands conclude the negotiations on the water conventions. On 17 January 1995 the Flemish region and the Netherlands sign the Conventions on the deepening of the navigation channel in the Western Scheldt and the flow of the river Meuse, and the Flemish region signs the multilateral Conventions on the protection of the Scheldt and the Meuse in Antwerp.

Round XIIa (1995-1997) begins with the provisional installation of the International Commission for the Protection of the Scheldt (ICPS), describes the multilateral negotiations on the first Scheldt Action Programme (SAP), and concludes with the French ratification of the Convention on the protection of the Scheldt, which enables the contracting parties to formally install the ICPS.

Round XIIb (1995-1997) concerns the implementation of the Convention on the deepening of the navigation channel in the Western Scheldt. In this decision making round several subrounds are distinguished concerning the implementation of the 48'/43'/38' deepening programme, and the development of a plan for compensating nature losses caused by the implementation of that programme. The case study period concludes with two main decisions. The first decision is the approval of special legislation enabling the Dutch government to set aside permitting procedures, and the Flemish region to implement the deepening programme. The second one is the advice of a commission of wise persons concerning the compensation of nature losses along the Western Scheldt.

Conclusions

The main conclusions of the case study concern the characteristics of the context of decision making on international Scheldt issues, the influence of the perceptions of the actors involved in the decision making process on their strategies, strategic interaction and institutionalization processes, learning processes, and the influence of the context of the decision making process on the perceptions and strategies of the actors involved.

Characteristics of the context
  Underlying hydrological structure of the issues at stake. Most international Scheldt issues are typical upstream-downstream issues, i.e. the underlying hydrological structure of these issues causes an asymmetric distribution of interests and resources among the actors involved in decision making on that issue. In the issue-area of the maritime access to the port of Antwerp the upstream basin state Belgium, and later the Flemish region, perceives a dependence on the downstream basin state the Netherlands. In the issue-area of water and sediment pollution dependencies are less unambiguous, because each basin state or region perceives a dependence on the basin states and regions situated upstream.
  International context. During the case study period two or more of the Scheldt basin states and regions were involved in decision making on international Rhine and Meuse issues. Furthermore, they were involved in decision making on more general water policies, such as EU-water policies, North Sea policies, and international river policies of the UN-ECE.
  Cultural context. There are considerable differences between the (decision making) cultures of the Scheldt basin states and regions. France and Belgium do have typical Latin cultures, which are characterized by relatively high scores on the cultural dimensions Power Distance and Uncertainty Avoidance. The Netherlands has a typical Nordic culture, which is characterized by relatively low scores on the cultural dimensions Power Distance and Uncertainty Avoidance, and a low score on the dimension Masculinity.
  Intranational developments. The most eye-catching *intra*national developments took place in Belgium. During the case study period the Belgian state was transformed from a unitarian state into a full federal state. Because of this federalization process, which comprised three stages of state reforms, the distribution of water management competencies among the federal state and the three Belgian regions changed fundamentally. Another main conclusion from the analysis of the intranational

developments concerns the different stages of development of water quality policies within the Scheldt basin states and regions. In the Netherlands water quality policies were developed first. Subsequently, such policies were developed in Flanders and France. Only recently water quality policies are developed in the Walloon and Brussels Capital regions. Supplementary explanations for these different stages of development are the upstream-downstream relations in the Scheldt and Meuse rivers, the Belgian federalization process, the different socio-economic situations in the upstream and downstream parts of the basin, and the different scores on the cultural dimensions Masculinity and Uncertainty Avoidance.

Historical context. The history of Belgian-Dutch relations concerning the management of the river Scheldt mainly relates to the management of the navigation channel in the Western Scheldt, and can be characterized by conflict better than by cooperation.

Influence of perceptions on strategies

The main actors involved in decision making on international Scheldt issues were the national and regional governments and their bureaucracies. The Belgian and later the Flemish government perceived a dependence on the Netherlands as regards the maintenance and improvement of the maritime access to the port of Antwerp, and used offensive interactive strategies to realize the infrastructure projects it had planned. Because of river bed contamination in the port of Rotterdam, and the use of Meuse water for drinking water production, the Dutch were mainly interested in the water quality and flow of the river Meuse. This perception explains the various tactical linkages the Dutch made between Scheldt infrastructure and Meuse water management issues. For a long time the Walloon and Brussels Capital regions did not perceive an interest in any of the international Scheldt or Meuse issues at stake, and mainly used reactive strategies aimed at maintaining autonomy. From 1992 these regions started to use more interactive strategies. The actors in the upstream basin state France had a different problem perception than the other upstream parties. They perceived the international environmental issues in the Scheldt basin as commons problems, and therefore were willing to cooperate on a clean-up of the river Scheldt if the other parties would do the same. The different preference intensities of the basin states and regions in different issue-areas explain the many tactical linkages made in the decision making process.

Beside national governments lower level governments and NGOs played a role in the decision making process. The different perceptions lower level governments had of the issues at stake may be explained partly by their different geographical scopes. For example, lower level governments in the Province of Zeeland were in favour of a tactical linkage between decision making on the deepening of the navigation channel in the Western Scheldt and the water quality of the Scheldt rather than a linkage between Scheldt and Meuse issues, or a linkage between decision making on the deepening of the navigation channel in the Western Scheldt and the negotiations on the alignment of a new high speed train. For the national Dutch government, however, the latter tactical linkages were rational. Environmental NGOs used various strategies to place and keep environmental issues on the international agenda, and to monitor the implementation of environmental policies. They, for example, organized protest marches, formed coalitions, carried out research, and started litigation.

Because the resources needed for the implementation of international policies were distributed among national governments, lower level governments, and NGOs, the latter parties were able to exert influence on decision making by national governments, and in some cases were able to make tactical linkages. After the conclusion of the Convention on the deepening of the navigation channel in the Western Scheldt in 1995 the national governments faced regional and local resistance against the implementation of the deepening programme, and plans for nature compensation. Decision making on the deepening of the navigation channel in the Western Scheldt is a perfect illustration of the risk of an implementation gap in decision making processes.

Strategic interaction and institutionalization
As regards the development of international institutional arrangements and policies two main conclusions are drawn. First, almost any agreement on the (further) development of international institutional arrangements or policies did involve tactical issue-linkage. The case study clearly demonstrates that if parties do have heterogeneous preference intensities in different issue-areas, tactical issue-linkage may be a useful means for the establishment of international cooperation. The second main conclusion of the research is that if tactical-linkages do not create a situation in which all parties perceive that they are better off with a package deal than without one, these linkages may cause lengthy deadlocks in the international negotiation process, and the erosion of existing institutional arrangements.

The history of conflicts over the management of the river Scheldt caused the growth of a culture of distrust between the parties. Only since the Dutch decision to unlink bilateral Scheldt infrastructure and multilateral Meuse water management issues in 1993, and the start of the bilateral and multilateral negotiations, trust between the basin states and regions is developing gradually.

Learning
Strategic learning processes played an important role in decision making in both issue-areas distinguished in this study, and enhanced consensus on the development of international institutional arrangements and policies. Beside strategic learning cognitive learning processes played an important role. In the issue-area of the maritime access to the port of Antwerp cognitive learning seems to have been more important than in the issue-area of water and sediment pollution. This may be explained by the complexity of the morphodynamics in the Scheldt estuary, and the uncertainties involved in the assessment of the consequences of policy alternatives to improve or maintain the maritime access to the port of Antwerp. There is, however, a broad consensus among experts on the types of policies that should be implemented to improve the water quality of the river Scheldt.

Influence of context on perceptions and strategies
Upstream-downstream relations seem to have strongly influenced the problem perceptions and strategies of the actors involved in decision making on international Scheldt issues. Because of the tactical linkages made between Scheldt and Meuse issues, decision making in the Meuse basin formed the most important international context. The example of Rhine river cooperation was frequently used in the international discussions concerning the development of international institutional arrangements and policies in the Scheldt basin. The development of general water policies by the EU, the UN-ECE and the North Sea co-states has influenced decision making on international Scheldt policies as well. The conclusion of the UN-ECE convention in 1992 was a main reason for the involvement of the upstream basin state France in the negotiations. Cultural differences between the Scheldt basin states and regions partly explain their different preferences concerning the

openness of decision making processes, and the form of institutional arrangements. In addition they may explain why Belgian and French civil servants generally do have less mandate than their Dutch colleagues.

Undoubtedly, the most influential intranational development was the Belgian federalization process. The combination of the Dutch linkage between Scheldt and Meuse issues, and Belgian internal disagreement on the strategies to be used in the negotiations with the Netherlands, caused lengthy deadlocks in decision making on these issues. The different stages of development of water (quality) policies in the Scheldt basin states and regions seem to have had a considerable influence on their perceptions of the international issues at stake, and their strategic behaviour. The general pattern that may be observed in the Scheldt and Meuse basins is that the basin states developed national water quality policies first, and subsequently were willing to reach an international agreement on the development of such policies.

Finally, the history of Belgian-Dutch conflicts concerning the management of the river Scheldt has shaped mutual reputations. The Belgians did distrust the Dutch as regards their willingness to maintain or improve the navigation channel to the port of Antwerp, whereas the Dutch did distrust the Belgians as regards their willingness to clean-up the rivers Scheldt and Meuse. Distrust between the basin states may explain partly their rigid mutual strategies, and why neither of them wanted to cooperate unconditionally and to build up a 'reservoir of goodwill'.

Additional conclusions

The case study made it possible to draw some additional conclusions, which do not relate directly to any of the research questions formulated beforehand.

The first additional conclusion concerns the effectiveness of international river commissions. In the last years before the conclusion of the Convention on the protection of the Scheldt against pollution, the Flemish government assigned large budgets for the construction of waste water treatment plants, and in the Walloon and Brussels Capital regions investments in waste water treatment infrastructure increased rapidly as well. Furthermore, in 1996 none of the respondents did expect the first Scheldt Action Programme (SAP) to go beyond existing national policies. These two observations may lead to the rather negative conclusion that the ICPS does hardly influence the development of water quality policies, and therefore is hardly effective. This hypothesis, however, may be challenged for several reasons. First, the Scheldt basin states and regions may have anticipated the conclusion of the Convention on the protection of the Scheldt, and the (provisional) installation of the river commission. Secondly, the interactions in the commission may induce cognitive learning processes. Thirdly, although the commission possibly does not have a large impact on the development of national water quality policies, it does enable the parties to better coordinate their policies. Fourthly, the basin states may generate economies of scale. Fifthly, the establishment of an international river commission may support national environmental ministries in the national competition on the division of government budgets. Sixthly, the interactions in the commission may induce strategic learning processes. Finally, international joint action theoretically may solve the classical commons problem. For the reasons discussed above, it is most likely that the benefits of the ICPS do exceed its costs, such as the costs of the secretariat, and the costs made by the individual basin states and regions.

The second additional conclusion concerns the geographical scales for dealing with international river issues. The case study has shown that different international river issues emerge on different scales, and that the solution of these issues was found on different scales. Causes and effects in an issue-area sometimes are restricted to a part of the basin, whilst the geographical scope of another issue-area may easily exceed the boundaries of a river basin. Furthermore, because most agreements on international Scheldt issues did involve issue-linkage, issues were addressed on scales exceeding the scales of the single issues at stake.

A third additional conclusion concerns the role of individual persons in the decision making process. Although the importance of this factor should not be overestimated, the case study showed ample evidence for the important role individual persons may play in a decision making process. Some persons did influence decision making in a positive way, whilst others had a rather negative influence on the course of the international decision making process.

Practical implications

The research may contribute to the solution of upstream-downstream problems in international river basins in three ways. First, it may stimulate those involved in decision making on these issues to look at decision making from a pluricentric perspective, and to develop strategies taking into account the (inter)dependencies between them. Secondly, the relevance of some lessons learned from the case study of decision making on international Scheldt issues may go beyond the case of the river Scheldt. Thirdly, the theory of decision making on international river issues made it possible to develop a typology of strategies that can be used to contribute to the solution of upstream-downstream problems in international river basins. This typology may be used as a checklist by those involved in decision making on these issues. Successively the strategies that can be used by individual basin states, the EU, and an independent intermediary are addressed.

In a final chapter the theories and research methods applied and the main research results are discussed and several recommendations for further research are formulated.

About the author

Sander V. Meijerink was born on 17 August 1970 in Lichtenvoorde, the Netherlands. After his secondary education from 1982 to 1988 in Aalten, he studied Organ at the Twente Academy of music in Enschede, where he passed the first-year examination. From 1989 to 1994 he studied Public Administration at Twente University in Enschede. He graduated in water and environmental management. During his studies he completed a traineeship at the Water Pollution Control Authority Eastern Gelderland. His M.Sc. thesis, which was prepared in cooperation with Delft Hydraulics in Emmeloord, dealt with the economic evaluation of the impacts of sea-level rise. From 1994 to 1998, he was employed as Ph.D. student at the Faculty of Civil Engineering and the Centre for Research on River Basin Administration, Analysis and Management (RBA Centre) of the Delft University of Technology. There, he participated in a multidisciplinary research team, and carried out research on the management of international river basins, in particular the river Scheldt basin. This research was funded by the Dutch Ministry of Transport, Public Works, and Water Management. During his Ph.D. studies he presented papers at international conferences and workshops in Budapest, Hamburg, and Montréal, and attended courses in Baja and Debrecen (Hungary), and Paris. He published in Dutch and in English on the financial aspects of integrated water management, river basin management, and intercultural management. Currently, he is employed at the Province of Overijssel as coordinator regional and urban water management and as project manager of the 'Waterpact van Twente', a pilot project for sustainable urban water management.

ENVIRONMENT & POLICY

KLUWER ACADEMIC PUBLISHERS – DORDRECHT / BOSTON / LONDON